NEURAL TUBE DEFECTS

The Ciba Foundation is an international scientific and educational charity (Registered Charity No. 313574). It was established in 1947 by the Swiss chemical and pharmaceutical company of CIBA Limited —now Ciba-Geigy Limited. The Foundation operates independently in London under English trust law.

The Ciba Foundation exists to promote international cooperation in biological, medical and chemical research. It organizes about eight international multidisciplinary symposia each year on topics that seem ready for discussion by a small group of research workers. The papers and discussions are published in the Ciba Foundation symposium series. The Foundation also holds many shorter meetings (not published), organized by the Foundation itself or by outside scientific organizations. The staff always welcome suggestions for future meetings.

The Foundation's house at 41 Portland Place, London W1N 4BN, provides facilities for meetings of all kinds. Its Media Resource Service supplies information to journalists on all scientific and technological topics. The library, open five days a week to any graduate in science or medicine, also provides information on scientific meetings throughout the world and answers general enquiries on biomedical and chemical subjects. Scientists from any part of the world may stay in the house during working visits to London.

Ciba Foundation Symposium 181

NEURAL TUBE DEFECTS

1994

JOHN WILEY & SONS

Chichester · New York · Brisbane · Toronto · Singapore

Published in 1994 by John Wiley & Sons Ltd
Baffins Lane, Chichester
West Sussex PO19 1UD, England
Telephone (+44) (243) 779777

Suggested series entry for library catalogues:
Ciba Foundation Symposia

Ciba Foundation Symposium 181
x + 299 pages, 32 figures, 29 tables

Library of Congress Cataloging-in-Publication Data

Neural tube defects.
p. cm.—(Ciba Foundation symposium; 181)
"Symposium on Neural Tube Defects held at the Ciba Foundation, London, 18–20 May 1993"—Contents p.
Editors: Gregory Bock and Joan Marsh.
Includes bibliographical references and index.
ISBN 0-471-94172-7
1. Neural tube—Abnormalities—Congresses. I. Bock, Gregory. II. Marsh, Joan. III. Symposium on Neural Tube Defects (1993: Ciba Foundation) IV. Series.
[DNLM: 1. Neural Tube Defects—congresses. W3 C161F v. 181 1994 / WL 101 N4916 1993]
QM695.N45N49 1994
612.6′4018—dc20
DNLM/DLC
for Library of Congress 93-21298
CIP

British Library Cataloguing in Publication Data

A catalogue record for this book is
available from the British Library

ISBN 0 471 94172 7

Phototypeset by Dobbie Typesetting Limited, Tavistock, Devon.
Printed and bound in Great Britain by Biddles Ltd, Guildford.

Contents

Symposium on Neural tube defects, held at the Ciba Foundation, London, 18–20 May 1993

This symposium is based on a proposal made by Dr Andrew Copp

Editors: Gregory Bock and Joan Marsh

Participants

D. E. Brenner Pharmacoepidemiology & Central Drug Monitoring, PH 3.54, CIBA-GEIGY Ltd, CH-4002 Basle, Switzerland

A. J. Copp Developmental Biology Unit, Division of Cell and Molecular Biology, Institute of Child Health, 30 Guilford Street, London WC1N 1EH, UK

H. S. Cuckle Institute of Epidemiology and Health Services Research, Department of Clinical Medicine, University of Leeds, 34 Hyde Terrace, Leeds LS2 9LN, UK

A. E. Czeizel WHO Collaborating Centre for the Community Control of Hereditary Diseases, Department of Human Genetics & Teratology, National Institute of Hygiene, Gyáli u.2-6, 1097 Budapest, Hungary

H. Dolk Environmental Epidemiology Unit, Department of Public Health and Policy, London School of Hygiene & Tropical Medicine, Keppel Street, London WC1E 7HT, UK

M. Goulding Molecular Neurobiology Laboratory, The Salk Institute for Biological Studies, 10010 North Torrey Pines Road, PO Box 85800, San Diego, CA 92186-5800, USA

J. G. Hall (*Chairwoman*) Department of Pediatrics, BC Children's Hospital, University of British Columbia, 4480 Oak Street, Room 2D15, Vancouver, BC, Canada V6H 3V4

L. B. Holmes Embryology–Teratology Unit, Harvard Medical School, Massachusetts General Hospital, Department of Pediatrics, 32 Fruit Street, Boston, MA 02114-2696, USA

A. G. Jacobson Center for Developmental Biology, Department of Zoology, The University of Texas at Austin, Austin, TX 78712-1064, USA

D. M. Juriloff Department of Medical Genetics, University of British Columbia, #226 Wesbrook Building, 6174 University Boulevard, Vancouver, BC, Canada V6T 1Z3

C. Kintner Molecular Neurobiology Laboratory, The Salk Institute for Biological Studies, 10010 North Torrey Pines Road, PO Box 85800, San Diego, CA 92186-5800, USA

D. Lindhout MGC-Institute of Clinical Genetics, Erasmus University Rotterdam, Academic Hospital Rotterdam Dijkzigt, PO Box 1738, 3000 DR Rotterdam, The Netherlands

J. L. Mills Prevention Research Program, Epidemiology Branch, National Institute of Child Health & Human Development, Room 7B03, 6100 Building, Bethesda, MD 20892, USA

G. M. Morriss-Kay Department of Human Anatomy, University of Oxford, South Parks Road, Oxford OX1 3QX, UK

H. Nau Institut für Toxikologie und Embryopharmakologie, Freie Universität Berlin, Universitätsklinikum Rudolf Virchow, Garystr 5, D-1000 Berlin 33, Germany

R. O'Rahilly Institut für Anatomie und Spezielle Embryologie, Universität Freiburg, Freiburg, Switzerland

G. P. Oakley Division of Birth Defects and Developmental Disabilities, National Center for Environmental Health, Centers for Disease Control and Prevention, Building 101, F34, 4770 Buford Highway, Atlanta GA 30341-3724, USA

J. M. Opitz Department of Medical Genetics, Shodair Children's Hospital, PO Box 5539, 840 Helena Avenue, Helena, MT 59604, USA

G. C. Schoenwolf Department of Neurobiology and Anatomy, School of Medicine, University of Utah, 50 North Medical Drive, Salt Lake City, Utah 84132, USA

J. M. Scott Department of Biochemistry, Trinity College, Dublin 2, Ireland

M. J. Seller Division of Medical & Molecular Genetics, United Medical & Dental Schools of Guy's & St Thomas's Hospitals, 7th Floor, Guy's Tower, Guy's Hospital, London SE1 9RT, UK

A. S. W. Shum (*Ciba Foundation Bursar*) Department of Anatomy, The Chinese University of Hong Kong, Shatin, NT, Hong Kong

D. B. Shurtleff Birth Defects Clinic, Division of Congenital Defects, Department of Pediatrics, University of Washington, PO Box 5371, CH-47, Seattle, WA 98195, USA

F. Stanley Department of Paediatrics, The University of Western Australia and Director, Western Australian Research Institute for Child Health, GPO Box D184, Perth, Western Australia 6001, Australia

D. G. Trasler Department of Biology, McGill University, 1205 Avenue Docteur Penfield, Montreal, Quebec, Canada H3A 1B1

H. van Straaten Department of Anatomy & Embryology, University of Limburg, PO Box 616, 6200 MD, Maastricht, The Netherlands

N. J. Wald Department of Environmental & Preventive Medicine, Wolfson Institute of Preventive Medicine, St Bartholomew's Hospital Medical College, Charterhouse Square, London EC1M 6BQ, UK

Introduction

Judith G. Hall

Department of Pediatrics, BC Children's Hospital, University of British Columbia, 4480 Oak Street, Room 2D15, Vancouver, BC, Canada, V6H 3V4

Each of us comes to this meeting with a particular interest and particular insights and each of us will therefore have the opportunity to catch up with the latest developments in a variety of areas. Each of us is aware of the impact that neural tube defects have had on humans. Some of us have more direct experience with humans, but we can all appreciate that part of the reason for this conference is the potential right now for relieving some of the effect neural tube defects have on humans. Many of us know families or even individuals who have been affected with neural tube defects. The suffering that occurs when such a person is not able to reach their full potential impacts each of us and our societies. So there is a very practical aspect of this meeting—the hope that something can be done to prevent these tragedies.

Neural tube defects are also a model for birth defects in general. Because there are very many unanswered scientific questions, there are various challenges posed by a meeting like this. One aim of this meeting is to provide a synthesis of recent work. We can also consider some of the implications of that work both for new studies and potentially even for formulating some recommendations.

I come as a representative of clinical genetics. We clinical geneticists have a particular view about the world. We believe that the patients with genetic disorders whom we see in the clinic provide an opportunity to learn about mechanisms of disease at the same time that care is being provided. We think of the clinic as a laboratory—as a place where we can see whether or not the things that are being found in basic science make sense in real families. Our job is, of course, to treat, to prevent complications and to help families and patients deal with their situation in the most effective way. So we end up synthesizing basic research with what is happening clinically.

One of the things we also try to do is to figure out whether or not a specific patient 'breaks the rules'—whether a particular family is an exception. So part of what happens in the clinic is that we identify those affected individuals who are unusual and view them as potentially helping to teach us about the normal situation and why things can go wrong.

One of the things therefore that comes out of the clinic is an incredible sense of heterogeneity—lots and lots of what we see does not fit with our current ideas.

We do have the traditional ways of thinking about genetic problems: as chromosomal problems, single gene problems or multifactorial ones. Clearly, these traditional explanations for genetic disease fit many cases of neural tube defects. We know there are chromosomal problems with neural tube defects; we know there are many single gene problems in humans with neural tube defects. But the vast majority of individuals with neural tube defects have what we call multifactorial genetic disorders. During this conference we will hear about known environmental influences; I suspect there are still a number of others that have not been recognized. The common category of human neural tube defects is multifactorial and we should be thinking about the multiple factors, trying to sort out the heterogeneity. That is part of what a conference like this can do; it can help us to identify the many factors suspected to be involved.

In the last five to ten years, clinical geneticists have become increasingly aware of what we call non-traditional types of genetics. Much of our understanding of mechanisms other than single gene defects stems from the work in developmental genetics. Studies on early development in animals are affecting the understanding of the human situation. Cytoplasmic inheritance, maternal effects, mosaicism, both programmed mosaicism and spontaneously arising mosaicism, parent-of-origin effects, growth factors—all these are becoming better understood because of the advances in molecular and developmental genetics. This work will have a large impact on our understanding of birth defects such as neural tube defects in humans.

I would like to take the prerogative of the chair to describe briefly some of our work in British Columbia. We have a very useful situation because all patients with neural tube defects come to one of two health care centres. We also have in British Columbia a registry of all birth defects and genetic diseases that allows us to recognize changing frequencies and, for practical purposes, have total ascertainment. We now have detailed information on over 700 families affected by neural tube defects. These have been well defined as to the level of lesion, other problems and recurrences within the family. British Columbia is a low incidence area. It is known that there are low and high incidence areas around the world; there are also ethnic differences with regard to occurrence and recurrence of neural tube defects. For the population in British Columbia we were able to separate those with high lesions from those with low lesions and to distinguish those families who have additional anomalies from those who do not (Hall et al 1988). We also have a Sikh population which came from India and they seem to have a unique type of anomaly. We suspect that there may be unique susceptibility to neural tube defects in the populations from many other areas of the world. Much of what we know now about neural tube defects involves a Northern European strain of humans. It is likely that there will be specific susceptibility of other strains of humans that we will come to understand. So those physicians who take care of humans need to be careful about identifying the background of a particular individual (which most researchers working with animals do, of course, very carefully).

I also want to tell you about our most recent venture using the results from studies on mice to understand the abnormalities seen in humans. For the mouse there is a multisite closure model that explains certain types of abnormalities that occur among neural tube defects in mice (Golden & Chernoff 1993, Juriloff et al 1993). We are finding that the same model seems to explain all of the neural tube defects among the children seen in our clinic, in terms of the specific site of the defect (Van Allen et al 1993). If we use the numbers of particular closure sites which were described by Golden & Chernoff (1993) to define the multiple sites of initiation of closure of the neural tube in the mouse, we can explain the defects observed in humans by either single-site closure anomalies or multiple-site anomalies. This model also explains the sites of encephaloceles. In humans, commonly encephaloceles are either in the nasal forehead area or in the occipital area; these appear to be areas of closure sites in humans. We think that each of these closure sites in humans has different genetic control and different environmental susceptibilities. This concept creates a framework for thinking about the results of animal studies and the effects of environmental factors on human development that we will hear about during this symposium. In this context closure 2 and closure 1 are the sites that seem to be susceptible to folic acid. We have proposed the hypothesis that these closure sites in humans can be identified, that the specific genes which regulate closure can be determined, and specific environmental components can be identified which influence each closure site.

Finally, this approach emphasizes the importance of integrating observations from many fields. Early developmental embryology appears to be very similar in many animal species, as do the genes that control the different processes. These observations support the importance of basic work being done in mice because of its applicability to the human situation. The hope, of course, of this symposium is to bring together workers from many different fields for cross fertilization, for consensus and status taking, and, if at all possible, to develop practical recommendations regarding neural tube defects in humans.

References

Golden JA, Chernoff GF 1993 Intermittent pattern of neural tube closure in two strains of mice. Teratology 47:73–80

Hall JG, Friedman JM, Kenna BA, Popkin J, Jawanda M, Arnold W 1988 Clinical, genetic and epidemiological factors in neural tube defects. Am J Hum Genet 43:827–837

Juriloff DM, Harris MJ, Harrod ML, Gunn TM, Miller JE 1993 Ataxia and a cerebellar defect in the exencephaly-prone SELH/Bc mouse stock. Teratology 47:333–340

Van Allen MI, Kalousek DK, Chernoff GF et al 1993 Evidence for multi-site closure of the neural tube in humans. Am J Med Genet 47:723–743

DISCUSSION

Opitz: Judy, in the Sikh population in British Columbia, you see an increased incidence of neural tube defects. What kind are these?

Hall: There is both an increased occurrence rate among the BC Sikh population and there is a subgroup which has a different kind of neural tube defect. They have high thoracic defects, with which the affected Sikh individuals cope extremely well. Somebody of Northern European ancestry who has a high thoracic neural tube defect is usually very incapacitated, whereas the BC Sikh affected individuals can walk, are of normal intelligence and generally do very well.

Opitz: Where are the areas of greatest variability in vertebral structure in Sikhs? This relates to the old hypothesis that neural tube defects occur in those sites in the neuraxis where there's the greatest normal developmental variability—the lumbosacral area and the occipital/cervical area. This is based on data from the 1920s and 1930s when people used to count these things and analyse them in great detail. Are the Sikhs, on the basis of their normal vertebral structure, different from other Indians or European populations?

Hall: Yes, they are. There is a very high incidence of spina bifida occulta among Sikhs. We went to Sikh families in which there had been recurrence of neural tube defects and X-rayed the parents. All of them had spina bifida occulta in the lumbar area. So we thought we would be able to screen and tell who is at increased risk among the Sikhs. We studied X-rays of this population that had been taken for some other reason and found an incidence of about 50% spina bifida occulta among the Sikhs. In Northern Europeans this incidence is more like 10%. So the background incidence among Sikhs of lumbar spina bifida occulta is high, but there does not appear to be an increased incidence of thoracic segmentation problems.

Opitz: The hypothesis that there is a direct correlation between the incidence of neural tube defects in humans and the normal variability of their spine should be tested. The hypothesis is that these malformations occur at sites of ongoing evolutionary modification of structure. According to the developmental field model, these sites would be less buffered or less well canalized. So one wonders if the lumbosacral spina bifida occulta is an indicator of an increased sensitivity due to some epigenetic dysregulation or susceptibility to spina bifida in the mid-thoracic area.

Hall: I don't think so. I think it is due to a specific recessive gene. In those specific families, there is a high incidence of stillbirth and also a high incidence of hydrocephalus.

Opitz: Is the hydrocephalus Arnold Chiari?

Hall: No, it's not and the siblings with hydrocephaly do not have spina bifida. When you add up all the recurrences, there is almost a one in four chance of some type of problem. So I think the predisposition to neural tube defects among the Sikhs in British Columbia is due to a different gene.

Opitz: It could be due to an autosomal recessive gene. On the other hand, it may be that inbreeding within this population has made them homozygous for many genes, thereby reducing buffering which normally keeps neural tube development proceeding correctly. One way to answer that question is to study the inbred and outbred populations. Given that the population in British Columbia is now outbreeding, is the incidence of neural tube defects decreasing?

Hall: We think that it's like the Irish when they moved to Massachussetts: the incidence of neural tube defects decreased, but it didn't go away altogether. Socially, it's unacceptable for East Indians to have any kind of congenital anomaly in their family, so these are 'family secrets' that are often not shared with the medical profession. It is very difficult to get good family histories from before they emigrated to North America. We know there is a high incidence of neural tube defects in the area of India from which the Sikhs in British Columbia come. Also, in British Columbia there is still a very strong influence to have arranged marriages, so the inbreeding may persist.

Mills: The Sikhs tend to be vegetarians. They also cook food very heavily, which raises the possibility of both folic acid and vitamin B_{12} deficiency in relation to their high incidence of neural tube defects. That would make them a potential risk group both in terms of genetics and in terms of diet. In the population of Sikhs in British Columbia, it would be interesting to look at those Sikhs who don't eat a traditional vegetarian diet.

Hall: We are interested in the diet and would like to look at this.

The buffering that John Opitz mentioned suggests that mammals have evolved with several points of closure of the neural tube, which provide back-up mechanisms in case one closure site fails. It does look as if there is more than one site of closure in humans. In particular individuals, there may be a predisposition to a closure not working properly. One thing in which we are interested is a possible interaction between diet and different closure sites. In amphibia and birds is there a similar back-up mechanism or is there really one closure process that 'zips' all the way up?

Jacobson: In the amphibia it appears as though it is one process. There is one site at which closure starts and the animal ends up with two neuropores.

Schoenwolf: In birds, macroscopically there are about three different sites of closure—anterior neuropore, posterior neuropore and something in between at the level of the auditory placodes. Henny van Straaten has looked very carefully at posterior neuropore closure with a high-definition microscope. Even within an area that macroscopically looks as if it's closing all at one time, there is pulsatile activity and there are localized sites of closure within that area. Closure is not as smooth as we initially thought.

Normal neurulation in amphibians

Antone G. Jacobson

Center for Developmental Biology, Department of Zoology, The University of Texas at Austin, Austin, TX 78712-1064, USA

Abstract. How does cell behaviour accomplish neurulation in amphibian embryos? During neurulation, the neural plate (while preserving the same volume) doubles its length, triples its thickness, narrows 10-fold, greatly decreases its surface and rolls into a tube. Cells that compose the neural plate produce these changes in three ways. They change shape, change neighbours and attempt to crawl beneath the contiguous epidermis. Plate width, length and area are decreased and the plate thickens when apical surfaces of plate cells contract radially, but plate length increases and width is further decreased when cells reposition themselves and collect along plate boundaries. Contraction of the apical surfaces of plate cells also helps roll the plate into a tube. Poisson buckling resulting from elongation of plate borders may contribute bending forces that help tube formation. The main folding force in tube formation is a rolling moment toward the midline produced by neural plate cells attempting to crawl beneath the contiguous epidermis. Experiments, observations and computer simulations support these assertions, reveal the organization of cell behaviour and implicate contraction of actin filaments as the main source of the necessary forces.

1994 Neural tube defects. Wiley, Chichester (Ciba Foundation Symposium 181) p 6–24

The study of neurulation requires analysis of the behaviour of the cells of the neural plate. Amphibian embryos offer several advantages for the study of the mechanics of neurulation. Their embryos are accessible and have large cells. Behaviour patterns of single cells are easy to observe. Variegation in egg pigmentation provides built-in markers for the tracking of individual cells. Perhaps the greatest advantage of most amphibian embryos is that there is no growth until well after neurulation has been completed. There is some cell division during neurulation, but the daughter cells do not enlarge, so there are no immediate morphogenetic consequences. Measurements of the volume of the neural plate of a newt (Jacobson & Gordon 1976) reveal that there is no increase in size during neurulation. In contrast, growth during neurulation in amniote embryos may be a confusing factor.

The neural plate is an epithelium that is one cell thick in most amphibians. Neural plate formation begins in newt embryos when a hemisphere of neural ectoderm converts to a disc of the same diameter. The cells of the plate double in height during this conversion (Jacobson & Gordon 1976). In the disc-shaped plate, the cells are taller than the cells of the contiguous epidermis. During a relatively long period, the neural disc reshapes itself into the form of a keyhole, remaining flat as it becomes longer, narrower and thicker. Neural folds begin forming at the plate edges. Then, in a relatively short period, the neural folds rise and roll toward the midline, closing the plate into a tube, while the plate abruptly elongates, narrows and thickens (Fig. 1). The rate of elongation of the midline of the neural plate is 10 times faster during the period when the plate is rolling into a tube than in the periods before or after tube formation (Jacobson & Gordon 1976). When the elastic neural plate is stretched along its midline or along its edge, as it is late in neurulation, the resulting Poisson buckling may help fold the plate toward the midline (Jacobson 1978).

Many studies of neurulation are based upon transverse sections and ignore other dimensions of the neural plate. There has been a general belief that the neural plate narrows and rolls into a tube because the apical surfaces of the cells of the plate contract radially. Such a contraction would also shorten the midline and fold the anterior and posterior ends of the neural plate upward. Instead, the midline elongates and the anterior and posterior ends of the neural plate bend downward over the spherical surface of the embryo. Different patterns of cell behaviour must be invoked to explain this and other aspects of neurulation. Several types of cell behaviour together are responsible for neurulation.

In a newt embryo, from the inception of the neural plate to completion of tube formation, the thickness of the plate almost triples, the length doubles, the width narrows at least 10-fold, the surface is reduced and the plate rolls into a tube.

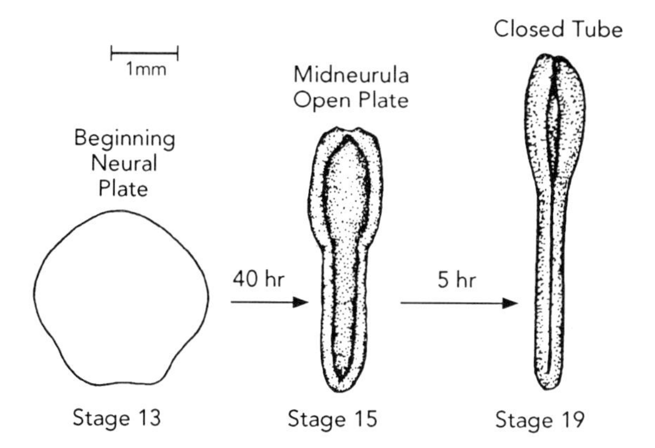

FIG. 1. Drawings, to the same scale, of dorsal views of the neural plate and tube of a newt, *Taricha torosa*. The plates were excised from the embryo and laid flat on an agar bed, then photographed immediately. This method avoids any optical foreshortening. The times between stages, at 17 °C, are from a time-lapse film of neurulation. Modified from Jacobson (1980).

Evidence given below suggests that cells of the amphibian neural plate themselves accomplish the deformations of neurulation. Some adjacent and underlying tissues are necessary to affect the behaviour of the plate cells, but they do not exert forces on the plate that are necessary for neurulation.

Cells have limited ways to apply the forces that accomplish the work involved in the tissue deformations of neurulation. In all cases, the forces provided by the constituent cells must be in the plane of the tissue.

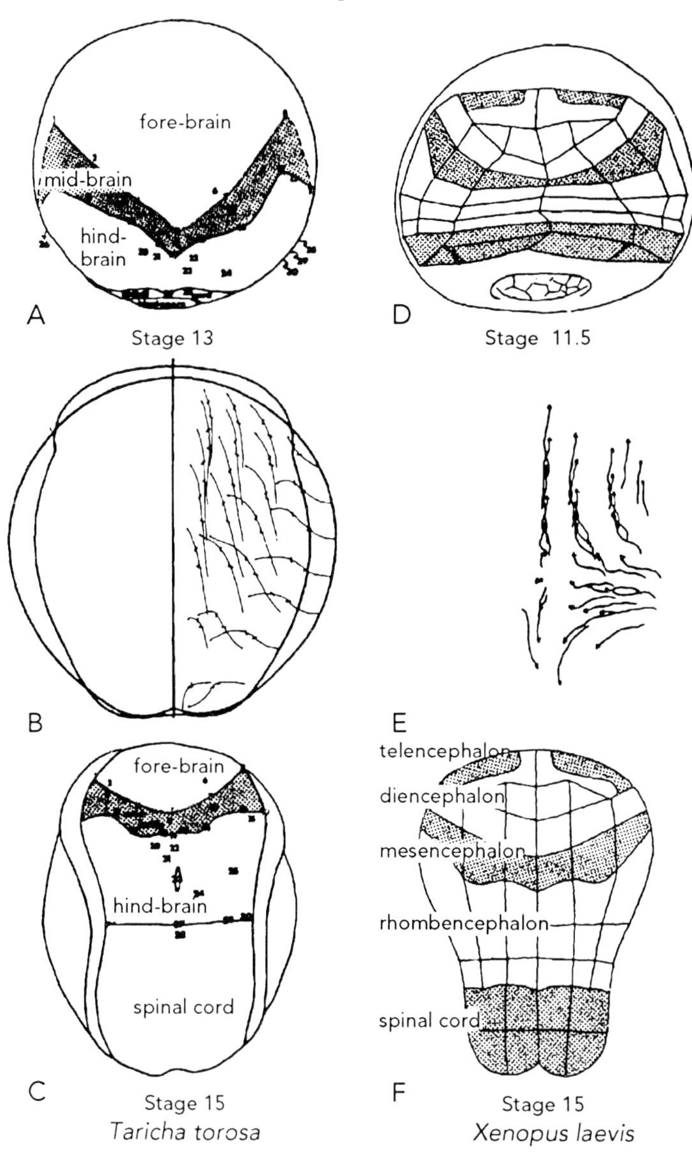

The necessity for maps of cell positions during neurulation

Any analysis of cell behaviour in a field of shifting cells requires that the cell positions be mapped through time. One cannot quantitatively describe changes of cell shape during neurulation without a map of the neural plate and a knowledge of trajectories of the individual cells. More than one investigator has described changes of cell shape by measuring sample cells in typical transverse sections at different times without regard for the changes in the locations of the cells. There is a general flow of cells during neurulation toward the midline and toward the anterior (C.-O. Jacobson 1962, Burnside & Jacobson 1968, Keller et al 1992). Keller has called this set of cell movements 'convergent extension'. During neurulation, cells flow through cross-sections, so it is necessary to measure cells in a section at an early stage, then deduce from cell trajectories the correct section in which to measure them at a later stage.

Maps of cell trajectories in the neural plate from gastrula to mid-neurula stages are available for a newt and a frog (Fig. 2B, E). A fate map of the mid-neurula axolotl neural plate was made by C.-O. Jacobson (1962). This map was projected onto a mid-neurula stage embryo of *Taricha torosa*, a newt, and re-mapped to late gastrula stage 13 by running backwards a time-lapse movie of neurulation and following many of the cells (Fig. 2C, A; Jacobson & Gordon 1976). Burnside & Jacobson (1968) mapped cell trajectories during early neurulation in the same newt (Fig. 2B). Eagleson & Harris (1990) made a fate map of the mid-neurula (stage 15) neural plate of the frog *Xenopus laevis* (Fig. 2F). Keller et al (1992) used a video tape of *Xenopus* neurulation to map this fate map back to mid-gastrula stage 11.5 by following cells at the intersections of the grid on the stage 15 embryo back to the earlier stage (Fig. 2 F, D). In maps of both species at gastrula stages, the brain occupies most of the prospective neural half of the animal hemisphere. The spinal cord and the hindbrain are narrow strips that lie perpendicular to the long axis near the blastopore. They occupy the region called by Keller the 'non-involuting marginal zone' (NIMZ). Convergent extension movements greatly lengthen the spinal cord and hindbrain along the axis during neurulation and the movements are largely confined to these regions.

FIG. 2. The fate map of the open neural plate stage of an axolotl (C.-O. Jacobson 1962) was projected onto the same stage in a time-lapse film of neurulation of the newt, *Taricha torosa* (C). The film was run backwards and cells at the numbered points were followed to a beginning neural plate stage (A) (from Jacobson & Gordon 1976). Cell trajectories between the two stages, as described by Burnside & Jacobson (1968), are shown in B. The beginning of each arrow is the cell position at the earlier stage; the arrow head marks the position at the later stage. D, E, F is a similar study from a video recording of neurulation of a frog, *Xenopus laevis*, by Keller et al (1992). Cells at the intersections of the grid lines were followed. The fate map of the stage 15 *Xenopus* embryo is from Eagleson & Harris (1990).

Cell behaviour during neurulation

In *Xenopus*, the movements of convergence and extension begin in the prospective posterior neural plate by stage 11.5 and continue through neurulation. Convergent extension is driven mostly by intercalations among cells in the mediolateral direction (Keller et al 1992).

Keller et al (1985) explanted large pieces of tissue from above the dorsal lip of the blastopore of *Xenopus* embryos. These explants contained both the prospective chordamesoderm and much of the area that later would be neural plate. Two explants were positioned with the basal ends of chordamesodermal cells in contact and with the basal ends of prospective neural plate cells together. These 'Keller sandwiches' were held flat in culture while convergence and extension movements of cells started in the chordamesoderm, followed by similar movements in the posterior regions of the prospective neural plate. The two regions extended themselves in opposite directions. If the sandwiches were cut along the line between chordamesoderm and prospective posterior neural plate before culturing, each separated part made its own movements of convergence and extension.

When grown in a culture medium adjusted to resemble blastocoel fluid, 'open face' sandwiches of chordamesoderm converge and extend, but the prospective neural plate parts will not extend unless some tissue is apposed to their basal surfaces (Keller et al 1985). In normal embryos the underlying tissue would be the involuted notochord. In *Xenopus* embryos lacking a notochord, Malacinski & Youn (1981) observed normal neurulation. These embryos had somites, rather than notochord, beneath the posterior neural plate. In Keller sandwiches, posterior neural plates face one another and the cells make normal movements. In all these cases with one or another of the various tissues beneath the neural plate, convergent extension movements occur and the movements are intrinsic to the cells of the prospective posterior neural plate.

Jacobson & Gordon (1976) analysed in a newt embryo the details of cell behaviour during the period from stage 13 when the plate is disc shaped to mid-neurula stage 15 when the plate is shaped like a keyhole but has not yet started to roll into a tube (Fig. 1). We called this the 'shaping period', it takes about 90% of the time of neurulation. The period of tube formation that follows takes only about 10% of the time.

Burnside (1973) measured the changes in shape of some cells between the disc and keyhole stages in the newt embryo. The lengths of cells were measured from sections and their apical areas were calculated as the ends of cylinders of the measured heights. As the cells elongate during this shaping period they preserve volume and remain cylindrical so the apical areas are inversely proportional to cell heights.

Because height measurements were made from sections, it was necessary to refer to the map of cell trajectories made by Burnside & Jacobson (1968) to find appropriate sections in which the same cells appeared at both times. Obviously, the measurements were made in different embryos, so some estimate of the consistency of cell trajectories from embryo to embryo was necessary. Burnside & Jacobson (1968) found that the cells at intersections of grids on three different embryos followed the same trajectories, so it is reasonable to make these measurements on different embryos.

Jacobson & Gordon (1976) similarly measured many cell groups over the entire neural plate at the beginning and at the end of the shaping period to map prospective shrinkage of apical areas of cells in the disc-shaped neural plate (Fig. 3B).

After viewing many time-lapse movies of neurulation of newt embryos, we noticed that the cells of the neural plate overlying the notochord behaved differently from the rest of the neural plate cells. At that time we called these cells the 'supra-notochordal cells', but later named them the 'notoplate' (Jacobson 1981). The notoplate eventually forms the initial floor plate of the neural tube.

The boundary between the notoplate and the rest of the neural plate is visible in time-lapse movies. We suspected that the notoplate has a role in elongating the midline of the neural plate during neurulation. At that time it was not yet clear that the convergent extension movements seen in and around the notoplate were intrinsic to it, rather than being driven by the underlying notochord. The observations from Keller sandwiches noted above make it clear that convergent extension movements may occur independently in the posterior neural plate (including the notoplate) as well as in the chordamesoderm.

Computer simulations and models of cell behaviour during neurulation

Jacobson & Gordon (1976) made a computer simulation of the shaping period of newt neurulation. We used the empirical data of the shrinkage map to simulate the shrinkage of the apical surfaces of the constituent cells during the shaping period. Observed changes in position of the border between the notoplate and the rest of the neural plate were used to simulate the convergence and extension that are seen in the posterior neural plate. These two sets of changes were simulated separately and together. Shrinkage alone (a computer experiment) did not produce a keyhole shape (Fig. 3F), whereas midline elongation alone (another computer experiment) did, but not one that matched the normal embryo (Fig. 3G). The computer simulations showed that apical shrinkage plus the midline elongation (which results from changes in the position of the notoplate border) were both necessary and together were sufficient to get the observed changes in shape of the neural plate from early neurula to mid-neurula stages (Fig. 3E).

These simulations raised questions about how cells accomplish the movements of convergence and extension seen in the posterior neural plate. Observations

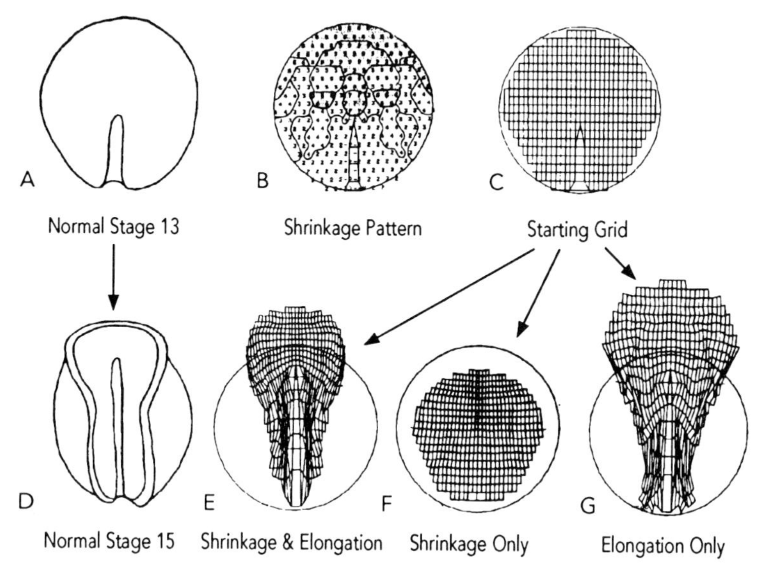

FIG. 3. Outline drawings from a time-lapse film of neurulation of a newt show the conversion of the plate from a disc shape (A) to a keyhole shape (D). The notoplate is outlined in the midline of the plate. (B) Map of a stage 13 neural plate outline indicating the amounts of shrinkage of cell surfaces (and concomitant elongation perpendicular to the surface of the cells) from stage 13 to 15. Cells whose surfaces shrink the most have higher numbers and are shaded. The rocket-shape in the lower midline represents the notoplate region. Photographs of a computer graphics screen: (C) A starting grid for computer simulations. (E) Distortion of the grid after the shrinkage pattern of B and the changes in position of the notoplate boundary that reflect midline elongation are both simulated. (F) A computer experiment that simulates shrinkage only. (G) A computer experiment simulating midline elongation only. Shrinkage and elongation are both necessary and sufficient to produce the shape change observed in the normal embryo. Figures modified from Jacobson (1980).

indicate that the neuroepithelial cells are changing neighbours in the plane of the tissue, despite being tied to one another by subapical junctional complexes. If cells changed neighbours in the plane of the tissue randomly, there would be no morphogenetic consequences. Some means of organizing cell rearrangements must exist. The neural plate elongates because more cells intercalate mediolaterally than anteroposteriorly. How then can cells *forcefully* intercalate to produce the necessary compression that extends the length of the posterior neural plate?

It was to answer these questions that Jacobson et al (1985, 1986) proposed the 'cortical tractor model' of epithelial morphogenesis. The model drew on what was known of mechanisms of cell motility in mesenchymal cells and applied this to epithelial cells. In this model, cell rearrangements in the plane of the epithelial tissue are due to lateral protrusions that spread apically, followed by displacement of the bulk of the cell to the site of the lateral protrusion (hence a *forceful* intercalation). During protrusion formation and cell rearrangement (cell movement), actin cortical microfilaments depolymerize and repolymerize at the protrusive tips; contraction waves in the actin network follow. Components of the inner cytosol flow to the site of the protrusion and become incorporated into the cortical cytoskeleton and the cell membrane. Cell–cell adhesion molecules are extended through the cell membrane and directly or indirectly attach to the cytoskeleton and to adhesion molecules on adjacent cells. Junctional molecules also insert into the membrane and are carried apically with the cytoskeleton. There is a time-averaged flow of cortical cytoplasm and membrane away from the tip of the protrusion and toward the body of the cell and also more generally toward the apical end of the epithelial cell. This 'cortical tractor' is propelled by waves of contraction in the actin filament network that start at the tips of protrusions. Such peristaltic contraction waves were long ago described in isolated neural plate cells as running from protrusions at the basal end of the isolated cell toward the apical end (Holtfreter 1946). Their effect is to move the cell toward the end of the protrusion.

At or near the apical ends of the cells, there must be a 'sink' or 'sinks' where the membrane, cytoskeleton and associated adhesion and junctional molecules are recycled into the interior cytoplasm. It is quite possible that this tractoring positions the subapical junctional complex. These junctions must be dynamic, constantly turning over, and thus allow cells to change neighbours without disrupting the apical seals. The subapical purse-string of actin microfilaments may also be put in position by the apically directed cortical tractoring; contraction of this bundle may then help reduce the apical surfaces of neural plate cells.

It is clear from direct observation of time-lapse movies that the notoplate border elongates along the axis during neurulation. For a border to elongate, cells must be added next to the border and stay there. We proposed that the boundary between the notoplate and the rest of the neural plate organized the otherwise random intercalations of cells by trapping cells that encounter the border. Cells from either the notoplate side or the neural plate side of the boundary would randomly intercalate among themselves and those that encountered the border would stay there. Cells would not cross the boundary. This might be due to adhesive differences in the two domains of cells or to contact inhibition between the two cell types.

We made observations and computer simulations that confirm important parts of the model (Jacobson et al 1986). We found, as predicted, that there are lateral protrusions in the notoplate that extend two or more cell diameters and that the direction of their extension is random except at the border. We re-examined time-lapse movies of newt neurulation and followed individual cells in both the neural plate and in the notoplate. Cells of each kind move about randomly until they reach the boundary, then they stay at the boundary, as predicted.

We made computer simulations testing the physical properties we proposed. These simulations showed that a rolling moment from the plate edge toward the midline, imposed either by basal tractoring or by apical constriction, can roll the plate into a tube. This was especially realistic if the midline was anchored, which, in fact, it is by the tight adhesion of the notoplate cells to the underlying notochord.

The chordamesoderm that underlies the neural plate also makes convergence and extension movements. Keller & Tibbetts (1989) found that small clumps of labelled cells inserted into prospective notochord at the beginning of gastrulation respect the border between notochord and somite as they rearrange during ensuing convergence and extension. They behave as the cells do at the boundary between the notoplate and the neural plate in newt and axolotl embryos.

Keller et al (1992) followed cells during gastrulation and early neurulation of *Xenopus* embryos and did not see any notoplate boundary. Nevertheless, their maps of cell trajectories suggest that no cells cross the lines where one would expect the notoplate boundaries to be.

Cuts were made the length of the notoplate boundary in the mid-neurula newt embryo to see if a boundary with the notoplate must be present for elongation of the neural plate to occur. In the newt, the side of the neural plate lacking a notoplate boundary elongated little or not at all; the side of the neural plate having a notoplate boundary elongated. If both notoplate boundaries were severed, then three pieces were created—right and left neural plate pieces and a central notoplate piece—none with a boundary. None of these separated pieces elongated. In control experiments, the notoplate was severed down the midline so that right and left neural plate pieces each had some boundary; each side then elongated (Jacobson 1991). Similar experiments with axolotl embryos have given more ambiguous results that suggest that this embryo relies more on the neural plate/epidermal boundary for elongation (A. G. Jacobson & J. D. Moury, unpublished work 1993).

Events at the epidermal border with the neural plate

The observations discussed above called attention to the border of the neural plate with the epidermis. We have measured during neurulation the change in length of the border between the neural plate and the epidermis in a newt and in axolotl embryos. We find that the brain border does not elongate, but the border of the prospective spinal cord does (A. G. Jacobson & J. D. Moury, unpublished work 1993). As Keller et al (1992) point out, convergence and extension movements are restricted in *Xenopus* to the spinal cord and hindbrain regions of the neural plate. The measurements we have made on urodeles suggest the same for them.

Elongation of the spinal cord border must involve intercalation of cells into the border from both sides, as seen at the notoplate/neural plate border. The brain

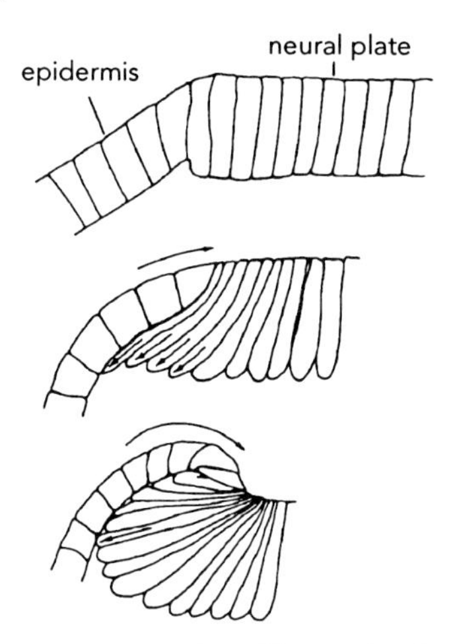

FIG. 4. Drawings from transverse sections illustrating how neural plate cells at the epidermal border crawl beneath the epidermis, elongate themselves, raise neural folds and create a rolling moment (arrows) toward the neural plate midline. Figures modified from Jacobson et al (1986).

border, which shortens, could be driven to do so entirely by the large decreases in apical surfaces of the brain plate cells along it. The epidermal cells on the other side of the border would have to make many adjustments, including associating with increasing numbers of brain plate cells; some cells might leave the border. We are trying to quantify the numbers of brain plate cells associated with each epidermal cell at early and at late neurulation.

Jacobson et al (1986) observed that neural plate cells at the plate edge try to crawl beneath the epidermis and thus produce a rolling moment toward the midline that could be a large force in rolling the plate into a tube (Fig. 4). Several complex things thus occur at the neural plate border simultaneously. The brain plate border shortens, the spinal cord border elongates and a neural fold lifts as plate cells attempt to crawl beneath the epidermis. In addition, as the neural folds form, some cells next to the border ingress beneath the epithelia from both the epidermis and the neural plate; these are later part of the neural crest.

Some have proposed that epidermis 'pushes' the neural plate edge, lifting a fold. This idea has been discounted because slits made in the adjacent epidermis gape instantaneously, demonstrating that the epidermis is under tension, not compression (Jacobson & Gordon 1976). To show that association with some contiguous epidermis is necessary for the plate cells at the borders to undergo the complicated behaviours noted, we have isolated neural plates with underlying mesoderm so they will have the proper associations of their basal surfaces, either with or without a strip of contiguous epidermis. These isolated plates never form a tube if there is no epidermis; they also fail to elongate as much as normal. If there is a strip of epidermis included with the isolate, then the explanted neural plates form neural tubes and elongate normally (A. G. Jacobson & J. D. Moury, unpublished work 1993).

Neural plate cells nearest the border with epidermis greatly increase their heights and reduce their apical surfaces to mere points. Continued basally directed crawling concentrates tugging forces at these apical points until the subapical junctions break, which allows the cells to pull themselves out of the epithelium (Jacobson et al 1986). The ingressed cells most likely become neural crest cells.

Associations with cells outside the neural plate may help guide the behaviours of cells within the neural plate

When two domains of different sorts of cells form a common boundary, the cells that abut the boundary on each side behave differently from the rest of the cells in the two domains. As noted above, cells from each side intercalate along the border between the neural plate and the notoplate: they do not cross the border and they remain at the border. Cell behaviour is thus organized along the border so that the border elongates. The epidermal border with prospective spinal cord also elongates, but the border between brain plate and epidermis does not in the amphibians examined. Jacobson & Tam (1982) found that the brain plate of the mouse embryo does elongate along its border with the epidermis. Convergence and extension movements among the cells of the notoplate and the posterior neural plate occur only when some other cells are beneath their basal surfaces. Cells that will satisfy this requirement of basal association include those from notochord (as in a normal embryo), somitic mesoderm (as in notochord-free embryos) or a duplicate of the posterior neural plate (as in Keller sandwiches).

Tube formation depends on lateral contact between neural plate cells and epidermal cells. Neural plate cells attempt to crawl beneath epidermal cells in normal embryos; in the process they elongate as they produce a rolling moment toward the midline that is a principal driving force of tube formation. An explant of neural plate with only a rim of epidermis attached will close into a tube and the sequence of shape changes in neural plate cells that normally starts at the epidermal border occurs in these explants (A. G. Jacobson & J. D. Moury, unpublished work 1993).

Because the plate edge is parallel to the midline, the rolling moment produced where neural plate cells contact epidermal cells bends the neural plate upward and toward the midline. This bend increases during neurulation until the plate becomes a tube. The bend produces a fold line on the surface of the neural plate comparable to the 'lateral hinge points' described by Schoenwolf (1982, 1991) in chick embryos. This fold line begins early in neurulation near the edge of the neural plate, then advances toward the midline. The position of the fold line of the plate coincides exactly with the apical end of the most ventral neural plate cell whose basal surface contacts the epidermis (Fig. 5). We have quantified

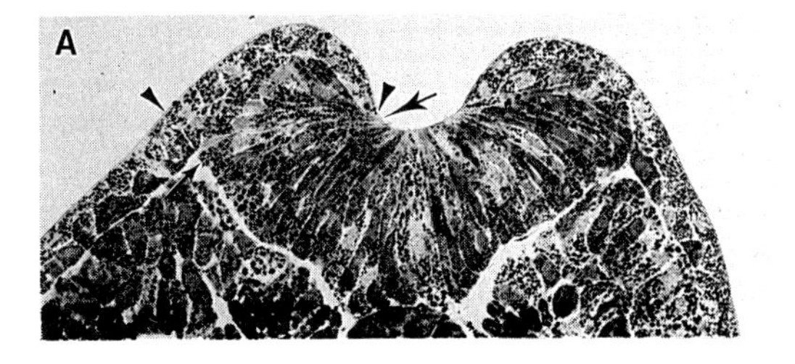

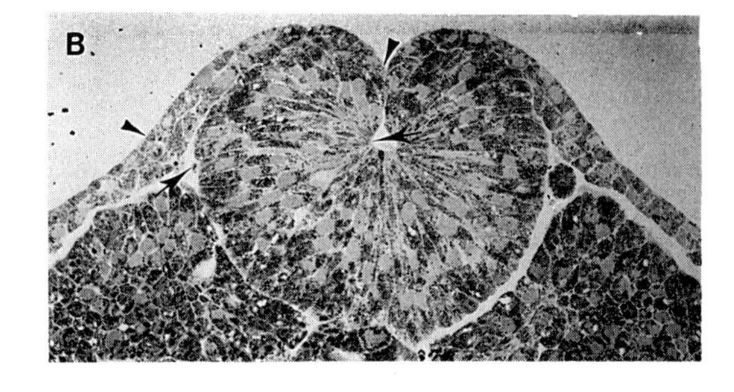

FIG. 5. Plastic transverse sections through the neural plates of newt embryos at stage 17 (A) and stage 20 (B). The cell whose apex is at the line of folding in the neural plate (right arrows) is the most ventral cell to have basal contact with the epidermis (left arrows). The amount of epidermis–neural plate contact (between arrow heads) changes with the stage, as do the positions of the fold lines.

the amount of contact between neural plate cells and epidermis along the anteroposterior axis through the course of neurulation in axolotl and newt embryos (A. G. Jacobson, J. D. Moury & Y. Lu, unpublished work 1993). The amount of contact varies with time in all regions. The regions that develop the greatest amount of contact are where the plate is widest. Our findings are consistent with a change in behaviour of neural plate cells that advances as a wave from the plate edge toward the midline. This results in elongation of the plate cells and plate folding as the basal ends of the plate cells sequentially encounter the epidermis.

Because the border between the neural plate and the epidermis seems to organize plate folding, we reasoned that any created border between neural plate and epidermis should raise a fold. Using transplants between different genetic strains (pigmented and albino) of the axolotl embryo, Moury & Jacobson (1989, 1990) demonstrated that artificial confrontation of any portion of the neural plate and epidermis will result in formation of new neural folds and neural crest cells at the created boundary. These results overturn old ideas of neural crest induction and fold formation. Both these processes result from local interactions between neural plate cells and epidermal cells. We also found that neural crest arises from both neural plate and epidermis.

Because the properties of the border between neural plate and epidermis greatly affect neurulation, it is important to know how the initial position of this boundary is established. Zhang & Jacobson (1993) hypothesized that the anterior border of the neural plate is positioned by interactions of the same signals that induce ventral and dorsal mesoderm in *Xenopus laevis*. We suggested that these signals continue through the prospective ectoderm of the animal hemisphere and interact to establish the anterior border of the neural plate near the animal pole.

To increase ventral signals at the animal pole, we transplanted ventral vegetal cells into the animal poles of early blastula (stage 7) embryos. The neural boundary was driven back toward the blastopore, producing embryos with short axes. A mark placed in prospective anterior neural plate ends up in the anterior neural plate of a normal embryo, but in the anterior ventral epidermis in the experimental embryos, suggesting that prospective neural plate has been converted to epidermis. We did the same transplantation at a later stage—just before gastrulation commences (stage 9)—and the embryos developed normally, indicating that the anterior neural plate border is established prior to gastrulation.

To reduce ventral signals that would reach the animal pole, we removed a few ventral vegetal cells at the 64-cell stage. Our prediction that these embryos would have axes that were longer than normal was confirmed. A mark placed in anterior prospective ventral epidermis ends up in the epidermis near the anterior end of the neural plate in a normal embryo, but in the anterior neural plate in these experimental embryos, suggesting that the experiment converts prospective ventral epidermis into anterior neural plate. If dorsal rather than ventral signals are reduced by removing dorsal vegetal cells, then the axes that form are shortened, as predicted.

The experiments above indicate that the same signals that induce ventral mesoderm and dorsal mesoderm during cleavage and blastula stages interact at the animal pole to position the border of the neural plate before gastrulation begins. Once the neural plate boundary is established, and epidermal and neural plate differentiation begin, planar inductions that specify neural crest occur in both directions across the boundary. Other planar interactions between the two cell types raise neural folds. In *Xenopus*, prior to gastrulation, planar induction from contiguous chordamesoderm is responsible for the induction of convergence and extension movements in the posterior neural plate (Keller et al 1992). Planar induction thus plays a major role in setting up the boundaries of the *Xenopus* neural plate and inducing the cellular movements that elongate the neural plate.

Conclusions

The changes in the shape of the neural plate during neurulation may be explained by the behaviour of the neural plate cells themselves, with some important patterns of behaviour being influenced by cells contiguous with or lying beneath the neural plate.

The neural plate becomes taller when its constituent cells elongate; the elongation of cells occurs in three phases and in a patterned way. When the neural plate begins, its thickness doubles as the plate converts from a hemisphere to a disc of the same diameter. During the shaping period of neurulation, as the disc converts to a keyhole shape, various regions of the plate become still taller in a complex way (Fig. 3B). During tube formation, additional elongation of the cells begins at the edge of the plate in contact with epidermis and this becomes a wave of cell elongation that sweeps toward the midline (Fig. 4).

Elongation of the constituent cells accounts for all of the increase in height of the neural plate. Cell elongation and concomitant decrease in apical surface area may occur in two ways. Constriction of subapical purse-strings of microfilaments may drive the changes, or cells may extend their basal ends, elongating themselves and reducing their apical surfaces since cell volume is preserved. Microfilaments are largely responsible for both methods. Microtubules may have a role in sustaining the new lengths. During the shaping phase of neurulation, the neural plate cells remain cylindrical and cell height stays inversely proportional to apical surface area. Microtubules may have a role in maintaining this relationship (Burnside 1973). As the plate becomes a tube, cells lose their cylindrical shape and of necessity become wedge shaped. The cells may assume a wedge shape passively, but if so, there must be changes in the relative lengths of microtubules to accommodate the change.

The width of the neural plate decreases at least 10-fold during neurulation. As recounted above, much of this decrease is accounted for by the decrease in diameter and apical surfaces of the cells. Changes in cell positions within the plate are another important cause of narrowing of the plate. The cell rearrangements described as convergence and extension convert plate width to length as the cells intercalate permanently, mostly in the mediolateral direction. The borders between the neural plate and the notoplate and between the spinal cord region of the neural plate and the epidermis have a role in organizing these cell intercalations.

The neural plate roughly doubles in length during neurulation. Cell rearrangements along the notoplate and epidermal borders with the neural plate lengthen the borders and both elongate and narrow the neural plate.

The neural plate rolls into a tube partly because it is prepared to do so by the activities that narrow and elongate it. The borders of neural plate with epidermis and with notoplate are aligned along the long axis during plate shaping. The borders are then in positions such that elongation of the notoplate borders and of the spinal cord portions of the plate edges produces Poisson buckling forces that may assist tube formation. The main driving force in the rolling of the plate into a tube appears to be the rolling moment toward the midline produced by the cells that contact the epidermis at the plate edge.

Several types of cell behaviour, mainly driven by microfilament activities that change the shape and the positions of the cells, together shape the neural plate and roll it into a neural tube. While the plate cells themselves appear to be responsible for neurulation, contacts with contiguous and underlying tissue are necessary to produce the appropriate behaviour patterns of the plate cells.

Acknowledgements

The work of the author was supported by grant HD 25902 from the National Institutes of Health. Much of the work cited was done with my colleagues David Moury, Jon Nuelle, Jing Zhang, Ying Lu, Richard Gordon, George Oster, Garrett Odell and Louis Cheng.

References

Burnside B 1973 Microtubules and microfilaments in amphibian neurulation. Am Zool 13:989–1006

Burnside B, Jacobson AG 1968 Analysis of morphogenetic movements in the neural plate of the newt *Taricha torosa*. Dev Biol 18:537–552

Eagleson GW, Harris WA 1990 Mapping of the presumptive brain regions in the neural plate of *Xenopus laevis*. J Neurobiol 21:427–440

Holtfreter J 1946 Structure, motility and locomotion in isolated embryonic amphibian cells. J Morphol 79:27–62

Jacobson C-O 1962 Cell migration in the neural plate and the process of neurulation in the axolotl larva. Zool Bidr Upps 35:433–449

Jacobson AG 1978 Some forces that shape the nervous system. Zoon 6:13–21

Jacobson AG 1980 Computer modeling of morphogenesis. Am Zool 20:669–677

Jacobson AG 1981 Morphogenesis of the neural plate and tube. In: Connelly TG, Brinkley LL, Carlson BM (eds) Morphogenesis and pattern formation. Raven Press, New York, p 233–263

Jacobson AG 1991 Experimental analysis of the shaping of the neural plate and tube. Am Zool 31:628–643

Jacobson AG, Gordon R 1976 Changes in the shape of the developing vertebrate nervous system analysed experimentally, mathematically, and by computer simulation. J Exp Zool 197:191–246

Jacobson AG, Tam PPL 1982 Cephalic neurulation in the mouse embryo analysed by SEM and morphometry. Anat Rec 203:375–396

Jacobson AG, Odell GM, Oster GF 1985 The cortical tractor model for epithelial folding: application to the neural plate. In: Edelman GM (ed) Molecular determinants of animal form. Alan R Liss, New York, p 143–166

Jacobson AG, Oster GF, Odell GM, Cheng LY 1986 Neurulation and the cortical tractor model for epithelial folding. J Embryol Exp Morphol 96:19–49

Keller RE, Tibbetts P 1989 Mediolateral cell intercalation is a property of the dorsal, axial mesoderm of *Xenopus laevis*. Dev Biol 131:539–549

Keller RE, Danilchik M, Gimlich R, Shih J 1985 The function of convergent extension during gastrulation of *Xenopus laevis*. J Embryol Exp Morphol (suppl) 89:185–209

Keller RE, Shih J, Sater A 1992 The cellular basis of the convergence and extension of the *Xenopus* neural plate. Dev Dyn 193:199–217

Malacinski GM, Youn BW 1981 Neural plate morphogenesis and axial stretching in 'notochord-defective' *Xenopus laevis* embryos. Dev Biol 88:352–357

Moury JD, Jacobson AG 1989 Neural fold formation at newly created boundaries between neural plate and epidermis in the axolotl. Dev Biol 133:44–57
Moury JD, Jacobson AG 1990 The origins of neural crest cells in the axolotl. Dev Biol 141:243–253
Schoenwolf GC 1982 On the morphogenesis of the early rudiments of the developing central nervous system. Scanning Electron Microsc 1:289–308
Schoenwolf GC 1991 Neurepithelial cell behavior during avian neurulation. In: Gerhart J (ed) Cell–cell interactions in early development. Wiley, New York, p 63–78
Zhang J, Jacobson AG 1993 Evidence that the border of the neural plate may be positioned by the interaction between signals that induce ventral and dorsal mesoderm. Dev Dyn 196:79–90

DISCUSSION

Oakley: Did you mention the rate of cell division as a part of this process?

Jacobson: There is no growth. There is some cell division during neurulation, approximately one and a half rounds, but because the cells don't enlarge, that has no immediate morphogenetic consequences. There is a stored morphogenetic potential because if one area makes a lot of cells, when those cells do begin to enlarge, that area will enlarge faster.

Oakley: So slowing down the rate of cell division should not have any impact on neurulation?

Jacobson: Not in the amphibian. In the mammal and the bird, there is growth because cells do enlarge after division. I think that makes it much harder for those species to neurulate. In those cases, the rate of growth can greatly affect neurulation.

Kintner: In *Xenopus*, Harris & Hartenstein (1991) blocked cell division using hydroxyurea and aphidicoline at the beginning of gastrulation. The embryos, even in the absence of cell division, formed what seemed to be a perfectly normal neural tube. Have similar experiments been tried in any other species?

Schoenwolf: If you treat chick embryos with colchicine (a microtubule inhibitor), they don't neurulate. But that's a pretty crude treatment.

The similarity of neurulation among different species is a very important concept to keep in mind. There is a tremendous amount of similarity, not just among mammals, but among all vertebrates. One needs to exploit the advantages and disadvantages of those organisms for different types of study. For example, mice have good genetics, so one can use them for genetic experiments. Amphibians and chicks are useful for microsurgical procedures.

But we should also keep in mind some of the tremendous differences amongst species, as well as the similarities. I was struck by how similar neurulation in the amphibian is to the process in the chick. But there are some major differences, even between different amphibians and between amphibians and birds. One thing that particularly struck me was the movement of the neural folds during closure. In the frog, the folds arise, then there is a tremendous amount of contraction and cell rearrangement that causes the folds to move together such that the intervening portion of the neural plate is shrunk. It's like two hills on either side of a valley and the valley floor then disappears, bringing the two hills closer together. In the chick or in the mammal, the tips of the neural folds move together, so a considerable amount of expansion must occur.

Jacobson: I think that the neurulation mechanisms used by amphibians are also used by birds and mammals, but in birds and mammals growth interferes and makes the process look different.

Hall: How much difference is there among amphibians?

Jacobson: Typically, one only talks about the laboratory species. If one looked at all amphibians, I'm sure you would find species that neurulate like reptiles do and some that neurulate in other ways. In most Amphibia, the neuroepithelium is just one cell thick as it is in birds and mammals. *Xenopus* is an exception, the neuroepithelium is two cells thick.

Copp: You suggested that two signals may be responsible for defining the anterior border of the neural plate. There has been a great deal of work on mesoderm-inducing factors in the amphibian (see Dawid 1992); the concept is emerging that different factors may be responsible for induction of dorsal and ventral mesoderm—activin and fibroblast growth factors, respectively. Have you tried explant cultures in which you mix different quantities or ratios of activin and fibroblast growth factors to see whether you can move the boundary?

Jacobson: Jing Zhang in my laboratory intends to do this, but we haven't so far.

van Straaten: In the mouse embryo, Martins-Green (1988) found that the neural fold arises not by folding at the ectoderm/neurectoderm junction, but by progressive formation of an intermittent basal lamina in the incipient neural fold. This lamina would separate cells more and more in the dorsal direction, which would result in the ectodermal and neurectodermal parts of the fold. You said that neurectodermal cells crawl underneath the ectoderm: could they be using the same mechanism? Could the crawling cells build up basement membrane along which they crawl?

Jacobson: I have looked at lots of scanning electron micrographs, but I haven't looked at many transmission electron micrographs. I don't notice much basement membrane at these stages. In order to crawl, the neurectodermal cells have to have something that they can pull against, either the epidermal cells themselves or something on the epidermis.

From the scanning electron micrographs, it appears that the cells are crawling directly against the cell surface underlying the epidermis.

Stanley: How similar is this type of migration to neural crest migration, where the basement membrane is obviously very important? In both cases there is a sheet of cells migrating.

Jacobson: The neural crest doesn't migrate through that space where the epidermis and neural plate are in contact until the space opens up and basement membranes form.

Schoenwolf: Do you think that cortical tractoring also plays a role in brain neural fold elevation in mouse, where neural crest cells are dropping out at the same time as the folds are going up? If those dorsolateral cells are leaving the epithelial sheet, how can they generate forces for elevation?

Jacobson: I think the ingressing and dropping out of those cells are a result of the tractoring.

Schoenwolf: Once they become neural crest, can they still exert forces on the epidermis?

Jacobson: Once a neural crest cell detaches from the epithelium, any force exerted as it crawls will likely move the crest cell rather than distort the epithelium it crawls against.

Hall: Are neural crest cells in mammals slightly different from neural crest in amphibians in how they originate?

Schoenwolf: Yes. In mammals, the crest cells leave the neural folds in the head and tail regions very early; in the trunk region they come out only after the neural folds have fused.

Jacobson: We've looked at the contact area very carefully. Evolutionary biologists say that the epidermal placode is closely related to the neural crest. Placodes are being induced during the time the contact is taking place.

Even more interesting is that at the ventral site where the last cell of the neural plate was in contact with the epidermis, the epidermal cells are trying to crawl under the neural plate. They look as if they are coming out; furthermore, they become double layered in that region. That whole layer can be thought of either as a linear placode running down the length of the spinal region or as a separate source of neural crest. We are studying that, but it's very hard to study.

Copp: In the tadpole, is the spinal cord in the tail formed by a process analogous to secondary neurulation in higher vertebrates?

Schoenwolf: I haven't studied the amphibian, but from the literature it seems that most of the neural tube in the tail is formed by extension of the primary tube. There is no formation of a cavity. Amphibians are also unusual in that the caudal neural tube ultimately breaks down and forms muscle.

Copp: So even in the amphibian, where neurulation appears superficially to be a simple process of neural folding along the whole body axis, all at one time, things are actually not that simple. There is a secondary process of neurulation that follows the primary neural folding event.

References

Dawid IB 1992 Mesoderm induction and axis determination in *Xenopus laevis*. BioEssays 14:687–691

Harris WA, Hartenstein V 1991 Neuronal determination without cell division in Xenopus embryos. Neuron 6:499–515

Martins-Green M 1988 Origin of the dorsal surface of the neural tube by progressive delamination of epidermal ectoderm and neuroepithelium: implications for neurulation and neural tube defects. Development 103:687–706

Formation and patterning of the avian neuraxis: one dozen hypotheses

Gary C. Schoenwolf

Department of Neurobiology and Anatomy, University of Utah, School of Medicine, 50 North Medical Drive, Salt Lake City, UT 84132, USA

Abstract. Formation of the neuraxis is dependent on cell–cell interactions and cell movements beginning during stages of gastrulation. Cell movements bring together new combinations of cells, allowing sequential inductive interactions to occur and leading to the specification of the neural plate and to its ultimate mediolateral (subsequently dorsoventral) and rostrocaudal patterning. Formation of the neural plate involves changes in the shape of its constituent cells and the first appearance of neural-specific cell markers. Shortly after the neural plate forms it undergoes 'shaping', in which the pseudostratified columnar epithelium constituting the neural plate thickens apicobasally, narrows transversely and extends longitudinally. Shaping is driven by three principal intrinsic types of cell behaviour: changes in cell shape, position and number. The next stage of neurulation begins while shaping is underway—bending of the neural plate. Bending involves two main processes, furrowing and folding. Furrowing of the neural plate is associated with the formation of the hinge points; these are localized, longitudinal areas where the neuroepithelium is attached to adjacent tissues and where wedging of neuroepithelial cells occurs. Cell wedging in the median hinge point occurs as a result of inductive interactions with the notochord; such wedging drives *furrowing*, thereby facilitating subsequent folding. Folding of the neural plate requires extrinsic forces generated largely by the surface ectoderm. Types of cell behaviour that could provide such forces include changes in cell shape, position and number. As a result of shaping and bending of the neural plate, the neural folds are brought into apposition in the dorsal midline. Final closure of the neural groove is mediated by cell surface glycoconjugates coating the apical surfaces of the neural folds. Patterning of the neuraxis begins during shaping of the neural plate and continues throughout stages of neurulation and into early postneurula stages. Patterning probably involves inductive interactions with adjacent tissues and the expression of putative positional identity genes such as homeobox-containing genes.

1994 Neural tube defects. Wiley, Chichester (Ciba Foundation Symposium 181) p 25–50

Neurulation—formation of the neural tube—occurs in two phases in avian embryos (Schoenwolf 1983). The first phase, primary neurulation, results in formation of the entire brain and much of the spinal cord (i.e. the cervical, thoracic and cranial lumbosacral levels). During this phase, the neural plate first appears and shortly thereafter it 'rolls up into a neural tube'. The second phase, secondary neurulation, leads to formation of the most caudal levels of the spinal cord. This phase begins near the time (and rostrocaudal level) when the caudal neuropore closes; it involves the formation and subsequent canalization of a solid aggregate of neuroepithelial cells (i.e. the medullary cord) associated with the tail bud.

This paper focuses on primary neurulation, a process that occurs in four different stages, which partially overlap both temporally and spatially: formation of the neural plate from ectoderm; shaping of the neural plate, ultimately resulting in an elongated central nervous system extending rostrocaudally along the back of the organism; bending of the neural plate, with the establishment of the neural groove; and closure of the neural groove, with formation of the roof plate of the neural tube, neural crest and overlying surface ectoderm. Additionally, as neurulation is occurring, the neuraxis becomes patterned in both the mediolateral (ultimately dorsoventral) and rostrocaudal planes. Coordinated with neurulation is the process of gastrulation—the formation of the three primary germ layers, the ectoderm, mesoderm and endoderm. The morphogenetic movements of gastrulation play an essential role in the formation of the neural plate and in neuraxial patterning by reshuffling cell layers and thereby bringing different tissues together, allowing new cell–cell inductive interactions to occur.

My approach will be to list twelve current hypotheses of avian neurulation and to provide a brief account of some of the evidence on which they are based. Owing to constraints imposed by publication guidelines, I will cite only selected references (typically, review articles and articles on avian neurulation from which additional references can be obtained) and my discussion will be concise. I have chosen the term 'hypotheses' to categorize the following twelve statements because, as defined by Webster's dictionary (McKechnie 1979), most of them, although already based on substantial experimental data, remain in the ultimate sense 'something not proved, but assumed for the purpose of argument'.

One dozen hypotheses

The focus of earlier studies of neurulation by many investigators was largely restricted to a single region of the embryo, the neural plate, to a single cell behaviour, change in neuroepithelial cell shape (i.e. 'cell wedging') and to a single underlying mechanism, the constriction of circumferential, apical bands of microfilaments within neuroepithelial cells (for reviews see Schoenwolf 1982, Schoenwolf & Smith 1990a). Although the neural plate, neuroepithelial cell wedging and apical microfilaments certainly all play important roles in neurulation, our view of how neurulation occurs has been greatly broadened by experiments conducted during recent years.

The central hypothesis

Hypothesis 1: Neurulation is a complex multifactorial process driven by multiple cell behaviours occurring both within the neural plate (i.e. intrinsic *to the neural plate) and outside this structure (i.e.* extrinsic *to the neural plate, within adjacent tissues).* This is the central hypothesis—hypotheses 2–12 are encompassed by it (for avian neurulation see reviews by Schoenwolf 1991, Schoenwolf & Smith 1990a,b, Schoenwolf & Alvarez 1992; for amphibian and mammalian neurulation see Schroeder 1970, Jacobson & Gordon 1976, Jacobson & Tam 1982).

Formation of the neural plate

Hypothesis 2: Formation of the neural plate requires cell–cell inductive interactions. Neurulation begins with the formation of the neural plate. Little is known about this process, especially in birds. On the basis of presumed parallels with amphibian embryos, it is generally believed that the avian neural plate is induced to form from the epiblast owing to inductive interactions with the underlying mesoderm (and perhaps endoderm). Experimental observations certainly support this belief, e.g. the formation of ectopic embryos after transplantation of Hensen's node to the germ cell crescent (Gallera 1971, Dias & Schoenwolf 1990) and the change in cell fate after plugs of prospective neural plate are grafted to prospective surface ectoderm and vice versa (Schoenwolf & Alvarez 1991). However, the precise stage of neural plate formation is unknown: a rudiment morphologically and immunologically recognizable as neural plate is already present by the linear primitive streak stage. The location of its precursor cells within the blastoderm is unknown; whether its precursor cells undergo extensive displacement, as do prospective mesodermal and endodermal cells during formation of the primitive streak, is unknown; and whether its precursor cells are initially clustered or widely dispersed is unknown. Also unknown are the exact spatial relationships among cells of the prospective ectoderm, mesoderm and endoderm during neural induction. Thus, what tissue interactions occur during this process, when they occur and the mode of inductive signal transmission (i.e. whether it is vertical or planar) are virtually complete mysteries.

Shaping and bending of the neural plate

Soon after the neural plate forms it undergoes shaping and bending (Schoenwolf 1985). These two processes collectively convert the neural plate from a flat epithelial sheet into an elongated trough (and, eventually, in conjunction with fusion of the neural folds flanking the trough, into an elongated tube). Shaping and bending of the avian neural plate are subjects covered under hypotheses 3 and 4.

Hypothesis 3: Shaping of the flat neural plate requires changes in cell behaviour, chiefly within the neural plate. Shaping involves three principal events: apicobasal thickening, transverse narrowing and longitudinal extension. At least three *intrinsic* behaviours drive shaping (Table 1) (Schoenwolf & Powers 1987, Smith & Schoenwolf 1987, Schoenwolf 1988, Schoenwolf & Alvarez 1989). There are changes in neuroepithelial cell shape (an increase in neuroepithelial cell height, which assists in neural plate thickening and narrowing) and changes in neuroepithelial cell number (neuroepithelial cells divide throughout neurulation at 8–12 h intervals, which at least partially accounts for the increase in the volume and surface area of the neural plate that occurs during neurulation; if cell division is oriented so that daughter cells are positioned longitudinally rather than transversely within the neural plate, this would assist in neural plate extension). There are also changes in neuroepithelial cell position (neuroepithelial cells undergo cell–cell intercalation in the transverse plane, which assists in neural plate narrowing and concomitant extension). Experimental results provide direct evidence that extensive cell–cell intercalation occurs in epithelial sheets during avian neurulation and that neurulation involves large-scale movements of cells, typical of convergent extension.

Hypothesis 4: Bending of the neural plate requires changes in cell behaviour both within and outside the neural plate. Bending involves two principal events—localized furrowing of the neural plate and subsequent folding, i.e. the rotation of the neural plate around loci termed hinge points (see below). Furrowing and folding collectively result in elevation and convergence of the neural folds toward the dorsal midline of the embryo. Furrowing is generated by neuroepithelial cell wedging (an *intrinsic* cell behaviour), whereas folding is generated principally by cell behaviours within the adjacent (lateral) surface ectoderm. Neuroepithelial cells within three restricted mediolateral areas of the neural plate change their shape from column-like to wedge-like (Schoenwolf & Franks 1984). This change results in the formation of longitudinal furrows, around which subsequent folding of the neural plate occurs.

At least three *extrinsic* cell behaviours drive *folding* (Table 1) (Schoenwolf 1988, Schoenwolf & Alvarez 1991, Smith & Schoenwolf 1991, Alvarez & Schoenwolf 1992). Again, there are changes in cell shape (surface ectodermal cells undergo spreading, thereby flattening apicobasally) and changes in cell number (surface ectodermal cells divide during neurulation at 10 h intervals, which accounts at least partially for the increase in the volume and surface area of the surface ectoderm during neurulation). Changes in cell position also occur (surface ectodermal cells intercalate in the transverse plane in concert with similar intercalations occurring within the neural plate). Experimental results on the surface ectoderm provide direct evidence that extensive cell–cell intercalation is a common feature within the entire ectodermal sheet, throughout all stages of gastrulation

TABLE 1 Summary of the morphogenetic behaviours of extraembryonic surface ectodermal, intraembryonic surface ectodermal and neuroepithelial cells during shaping and bending of the avian neural plate and closure of the neural groove

	Cell division			*Changes in cell shape*	
Cell type	*Rounds*	*Inferred direction*[a]	*Amount of convergent extension*[b]	*Height*	*Configuration*
Extraembryonic surface ectoderm	2	Radial	0	Decreased	Squamous
Intraembryonic surface ectoderm	2	Longitudinal	+ +	Unchanged	Cuboidal or low columnar
Lateral neuroepithelium	2–3	Longitudinal	+ + +	Increased	High columnar, spindle-shaped
Median hinge-point neuroepithelium	2–3	Longitudinal	+ + +	Decreased	Low columnar, wedge-shaped

This table is based on the studies of Schoenwolf & Alvarez (1989, 1991). It emphasizes several important conclusions. Plugs of extraembryonic surface ectoderm expand radially in surface view because of at least two processes: cell division, presumably oriented radially and parallel to the plane of the epithelium, and change in cell height such that cells transform from cuboidal to squamous. Plugs of intraembryonic surface ectoderm narrow transversely and lengthen longitudinally owing to at least two processes: cell division, presumably oriented longitudinally, and cell–cell intercalation. Cell heights are essentially unchanged (on average; their exact height depends on their particular position in the epithelial sheet) during these processes, with these cells remaining cuboidal (or low columnar). Plugs of neuroepithelium (contributing to levels of the neuraxis caudal to the forebrain level) also narrow transversely and lengthen longitudinally and, again, the same two processes that act in plugs of intraembryonic surface ectoderm are involved. Additionally, median hinge-point cells change their height (to a low columnar configuration) and become wedge shaped, whereas lateral neuroepithelial cells increase their height (to a high columnar configuration) and remain spindle shaped. As a consequence of the coordinated movements of these three types of cells, convergent extension occurs throughout the entire area pellucida, concurrently with gastrulation and neurulation, and the blastoderm (especially its extraembryonic portions including the area opaca) expands radially over the yolk.

[a]Radial and longitudinal indicate, respectively, that daughter cells are positioned radially (increasing graft diameter uniformly) or principally within the longitudinal plane (thereby increasing graft length but not graft width).

[b]0, + + and + + + indicate, respectively, little or no, moderate and substantial (about two rounds) convergent extension (i.e. transverse narrowing and longitudinal lengthening).

and neurulation, and they suggest that such cell movements play a major role in ectodermal expansion.

Changes in the extracellular matrix probably contribute to the extrinsic forces required for folding. Depletion of the matrix results in a severe inhibition of bending of the neural plate (Schoenwolf & Fisher 1983). Changes in the matrix could act in a solely mechanical way to cause bending, e.g. through hydrostatic pressure resulting from hydration, or such changes could regulate cell behaviour, thereby acting more indirectly through a cascade of events.

Closure of the neural groove

Hypothesis 5: Final closure of the neural groove is mediated by cell surface glycoconjugates. Final closure of the neural groove has received comparatively little attention. Shaping and bending of the neural plate bring the paired neural folds into apposition along the dorsal midline, but complete closure requires their subsequent adherence and fusion. These processes involve the synthesis of cell surface (apparently adhesive) coats (Takahashi & Howes 1986, Takahashi 1988, 1992) and localized protrusive activity of cells in the neural fold transition zone (Schoenwolf 1979), but little else is known about the process of closure.

Patterning of the neuraxis

Patterning of the avian neuraxis begins during shaping of the neural plate and continues throughout stages of neurulation and into early postneurula stages. Patterning of the neuraxis is covered under hypotheses 6–8.

Hypothesis 6: The neural plate initially becomes patterned in the transverse plane and this process begins during neural plate shaping and continues during its bending. During shaping, two populations of cells first become distinct in the early neural plate (Schoenwolf & Franks 1984), indicating that patterning has already begun in the transverse plane of the neural plate. Median hinge-point (MHP) cells are neuroepithelial cells that become anchored to the underlying notochord (or prechordal plate mesoderm, more rostrally); most (about 70%) of them become wedge shaped. The second population consists of lateral neuroepithelial cells, about 70% of which remain spindle shaped. Later, an additional population of cells appears, at certain rostrocaudal levels of the neuraxis, near the lateral edges of the neural plate. These dorsolateral hinge-point cells are neuroepithelial cells that become anchored to the adjacent surface ectoderm of the forming neural folds; about 55% of them become wedge shaped.

Rostrocaudal patterning of the neuraxis begins at later stages of neurulation than does transverse patterning. Near the time of neural groove closure, the neuraxis first exhibits overt *rostrocaudal* patterning. The expression of positional identity genes, such as the homeobox-containing gene *Engrailed-2*, first occurs in the avian neuraxis at a rostrocaudally restricted level, temporally and spatially coincident with closure of the neural groove (Gardner et al 1988). Additionally, fate-mapping studies reveal that MHP and lateral cells become distinct populations during shaping and that rostrocaudal subdivisions of the neuraxis, i.e. the forebrain, midbrain, hindbrain and spinal cord, do not become morphologically distinct until near the time of neural groove closure (Schoenwolf & Alvarez 1989, 1991).

Hypothesis 7: The wedging of MHP cells, one of the first overt signs of neuraxial patterning, is induced by the notochord, but it occurs independently of morphogenetic forces generated in other tissues. The formation of MHP cells is dependent on inductive interactions with the notochord. In the absence of the notochord, MHP cells do not differentiate; in the presence of extra notochords, supernumerary strips of MHP cells form overlying the extra notochords (van Straaten et al 1988, Smith & Schoenwolf 1989, Placzek et al 1990, Yamada et al 1991). Moreover, MHP cell wedging and neural plate furrowing occur autonomously when the prenodal midline strip of neural plate is isolated from all other lateral tissues, provided that the notochord comes to underlie the midline plate isolate. This demonstrates that neuroepithelial cell wedging (and concomitant neural plate furrowing) is not merely the passive consequence of pressure generated on the midline of the neural plate as the neural folds are rising (Schoenwolf 1988). MHP cells have a complex origin at the late gastrula/early neurula stage (Schoenwolf & Alvarez 1989, Schoenwolf & Sheard 1990, Schoenwolf et al 1992, Garcia-Martinez et al 1993): from the area just rostral to Hensen's node, from Hensen's node itself, and from areas bilateral to the approximate rostral quarter of the primitive streak (i.e. approximately its rostral 375 μm). MHP cells, in addition to their characteristic short statures and wedge shapes, also have longer (about 65%) cell cycles than do lateral cells, owing to prolongation of both the S and other, M and especially G2, phases (Smith & Schoenwolf 1987). Moreover, MHP cells, many or perhaps all of which eventually contribute to the floor plate, display a number of unique antigens, against which monoclonal antibodies have been generated (Yamada et al 1991). Finally, MHP cell wedging involves both apical constriction, presumably assisted in part by the contraction of apical bands of microfilaments (Lee & Nagele 1985), and basal expansion, presumably owing to the repositioning of each cell's nucleus to the base of the cell through cell cycle regulated, interkinetic nuclear migration. The latter process would also contribute to passive apical constriction of the cell as its nucleus migrates away from the apical pole of the cell after division (Smith & Schoenwolf 1988).

Hypothesis 8: The spatial distribution of the expression in the dorsoventral axis of the neural tube of Engrailed-2 in the chick is regulated by the cranial notochord, but the distribution of its rostrocaudal expression is not. One of the most difficult problems of early development concerns the mechanisms by which changes in a number of diverse cell behaviours within multiple tissues are precisely coordinated in both space and time, leading to the highly orchestrated morphogenetic movements of gastrulation and neurulation. One intriguing possibility is that such orchestration is achieved through regional gene expression. An exciting finding of recent years is that vertebrate embryos contain homologues of *Drosophila* genes, such as the homeobox-containing genes, known to be directly involved in early morphogenesis and patterning of the fly (see review by Lawrence 1992) and that these homologues are expressed in temporally and spatially restricted domains of vertebrate embryos during early development (see review by McGinnis & Krumlauf 1992). These genes will be referred to as positional identity genes.

An example of a positional identity gene is the homeobox-containing gene, *Engrailed-2*. It is expressed in the early embryonic neuraxis of several vertebrates (Patel et al 1989, Gardner et al 1988). In chick embryos lacking the cranial portion of the notochord, *Engrailed-2* is still expressed in a rostrocaudal pattern closely resembling that of control embryos (Darnell et al 1992). However, the midline region of the most ventral part of the neural tube, which in control embryos does not express *Engrailed-2* at any time or at any rostrocaudal level of the neuraxis, expresses *Engrailed-2* (at the normal rostrocaudal level) in the absence of the normally underlying notochord. This suggests that the notochord normally suppresses *Engrailed-2* expression through induction of MHP cells.

The origin, pathway of migration and state of commitment of ectodermal and mesodermal cells of the early embryo

Hypothesis 9: Although MHP cells acquire many of their specific cell-type characteristics as a result of induction by the notochord, prospective MHP cells already possess the ability to undergo cell–cell intercalation in a predefined pattern prior to inductive interactions with the notochord. When prospective MHP cells are grafted heterotopically into prospective lateral cell territory, they can move back to their normal site of origin and then migrate appropriately, intercalating with other MHP cells, as if they had never been disturbed (Alvarez & Schoenwolf 1991). Similarly, prospective lateral cells transplanted heterotopically into prospective MHP cell territory also can move back toward their normal site of origin where they mix with other lateral cells.

Hypothesis 10: Cells in the avian epiblast at late gastrula and early neurula stages are clustered into topographically distinct zones, each with a unique prospective fate (and therefore, it has been possible to construct prospective fate maps). Fate-mapping studies have revealed the locations of several topographically distinct zones within the early avian blastoderm. However, precise boundaries among adjacent areas cannot be delineated (Schoenwolf &

Alvarez 1989, 1991, Schoenwolf & Sheard 1990, Garcia-Martinez et al 1993). This suggests that cell fate is ultimately fine tuned at the boundaries by tissue interactions, which depend on the precise location a particular cell eventually occupies. This hypothesis is supported by heterotopic grafting studies in which, for most cell types, cell fate is still labile at the neurula stage (Alvarez & Schoenwolf 1991, Schoenwolf & Alvarez 1991).

Hypothesis 11: The precise pathway of migration during ingression of prospective mesodermal cells into the interior of the embryo is determined by the rostrocaudal level of departure of cells from the primitive streak and mesodermal cell fate is mostly labile. Mesodermal mediolateral subdivisions, i.e. notochord, somite, intermediate mesoderm, lateral plate mesoderm and extraembryonic mesoderm, arise from the primitive streak, where their precursor cells are arrayed in rostrocaudal order (Schoenwolf et al 1992). Mesodermal cell fate is still labile at late gastrula and early neurula stages, except for the notochord, whose fate becomes determined very early in development (Garcia-Martinez & Schoenwolf 1992).

Neural tube defects

Hypothesis 12: Studies on normal neurulation provide insight into possible mechanisms involved in the generation of dysraphic (open) neural tube defects. Abnormal neurulation, resulting ultimately in anencephaly and spina bifida, could result from defects in a number of cell behaviours, including cell–cell intercalation, cell division and/or change in cell shape within either the neural plate or surface ectoderm. The type of defect formed would depend on precisely when during development the abnormal cell behaviour occurred or the exact rostrocaudal level of the embryo at which neurulation went awry. A detailed account of the range of neural tube defects is given by Copp et al (1990).

The hinge-point model for bending of the neural plate

The high probability that both intrinsic and extrinsic forces are involved in bending of the neural plate has led a colleague and me to formulate what we have termed the hinge-point model for bending of the neural plate (Schoenwolf & Smith 1990b). This model is based on two main premises (Figs. 1, 2): 1) the neural plate is firmly anchored to adjacent tissues at the hinge points, i.e. MHP cells to the prechordal plate mesoderm and notochord, and dorsolateral hinge-point cells to the surface ectoderm of the neural folds; 2) forces for folding are generated outside of, i.e. lateral to, the hinge points. Thus, the model predicts that each hinge point is an area that directs and facilitates bending, much like creasing a sheet of paper facilitates its subsequent folding, and that each 'hinge' acts like a standard door hinge rather than like a self-closing, spring hinge (i.e. forces for 'closing' arise outside of the hinge point).

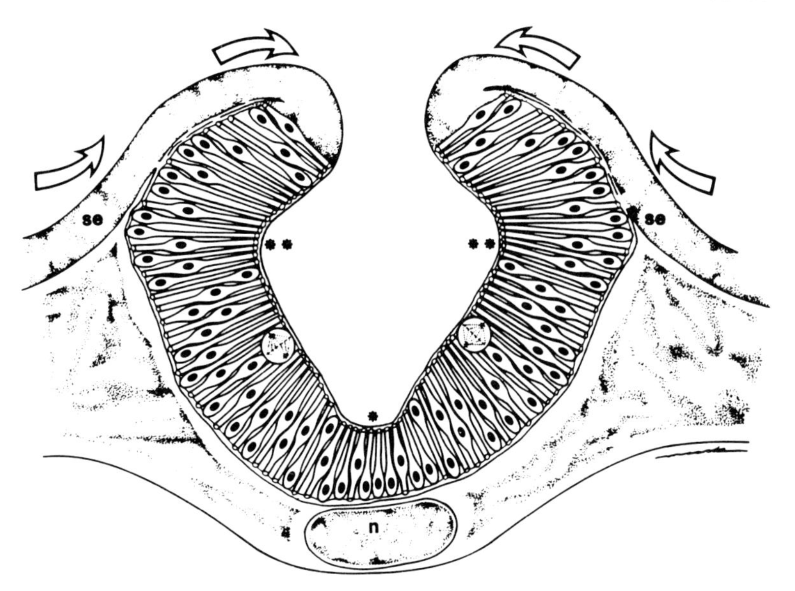

FIG. 1. Schematic representation based on a transverse scanning electron micrograph section through the future hindbrain level of a chick embryo during bending of the neural plate. Note the median hinge-point region (asterisk), notochord (n), dorsolateral hinge-point regions (double asterisks) and expanding surface ectoderm (se, arrows). Most hinge-point neuroepithelial cells tend to be wedge shaped (each with its nucleus at the base of the cell), whereas most lateral neuroepithelial cells (*between* the hinge points) tend to be spindle shaped (each with its nucleus at the bulging 'waist' of the cell). Adapted from Schoenwolf & Smith (1990a), with permission of the Company of Biologists Ltd.

The above discussion on the complexity of neurulation leads inescapably to the conclusion that the traditional viewpoint of neurulation, namely, that bending of the neural plate is driven principally by microfilament-mediated cell wedging exclusively within the neural plate, is no longer tenable. The traditional viewpoint seems to be partially correct for one aspect of neural plate bending—*furrowing*. The MHP region of the neural plate furrows because its cells become wedge shaped. However, it is hypothesized that 1) such wedging does not drive subsequent folding, it only facilitates this process, and 2) the mechanism of cell wedging within furrowing regions of the neural plate involves both apical constriction and basal expansion.

Conclusions

The process of neurulation has been the subject of intensive study over the last two decades. It has become increasingly clear that neurulation is a multifactorial process driven by a number of diverse, fundamental cell behaviours. Such

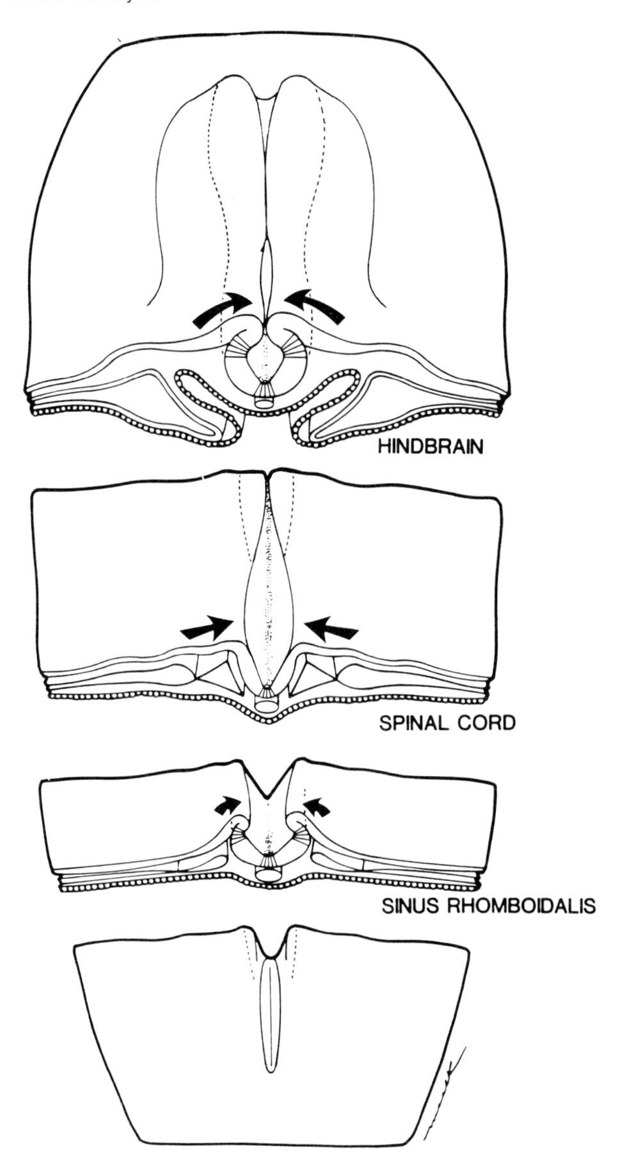

FIG. 2. Schematic three-dimensional representation of a chick embryo during bending of the neural plate showing three cross-sectional views. Arrows indicate directions of expansion of the surface ectoderm. The median hinge point is present at all three rostrocaudal levels illustrated, but the dorsolateral hinge points are absent throughout the length of the spinal cord, except at its extreme caudal end (the level of the sinus rhomboidalis). Adapted from Schoenwolf & Alvarez (1992), with permission of Academic Press.

morphogenetic behaviours occur in both the neural plate and surrounding tissues and include changes in cell shape, size, position and number, as well as changes in cell–cell and cell–extracellular matrix interactions. Previous studies have characterized many of the relevant behaviours underlying shaping and bending of the avian neural plate, and several studies have examined pattern formation within the developing neuraxis, where patterning has been assessed according to morphological, behavioural and immunochemical criteria. The hinge-point model of neurulation provides a mechanical basis for visualizing the coordination and diversity of cell behaviours involved in normal neurulation and provides insight into what behaviours might go awry leading to dysraphic neural tube defects. Such anomalies are among the most common congenital malformations present in new-born humans. Developing methods for preventing these extremely serious birth defects is the ultimate goal of studies of neurulation. To achieve this goal, we will need to understand fully the processes underlying normal neurulation and how such processes are altered during abnormal neural tube development. Despite a considerable advance in our knowledge of neurulation in recent years, it is clear that much remains to be learned about this intricate process.

Acknowledgments

I wish to thank my colleagues whose diligent work and stimulating discussions provided many of the results and ideas discussed above. I wish to acknowledge especially Drs Ignacio S. Alvarez, Virginio Garcia-Martinez and Jodi L. Smith. The research of my laboratory is supported by grant nos. NS 11812 and HD 28845 from the National Institutes of Health.

References

Alvarez IS, Schoenwolf GC 1991 Patterns of neurepithelial cell rearrangement during avian neurulation are established independently of notochordal inductive interactions. Dev Biol 143:78–92

Alvarez IS, Schoenwolf GC 1992 Expansion of surface epithelium provides the major extrinsic force for bending of the neural plate. J Exp Zool 261:340–348

Copp AJ, Brook RA, Estibeiro JP, Shum ASW, Cockroft DL 1990 The embryonic development of mammalian neural tube defects. Prog Neurobiol 35:363–403

Darnell DK, Schoenwolf GC, Ordahl CP 1992 Changes in dorsoventral but not rostrocaudal regionalization of the chick neural tube in the absence of cranial notochord, as revealed by expression of *Engrailed-2*. Dev Dyn 193:389–396

Dias MS, Schoenwolf GC 1990 Formation of ectopic neurepithelium in chick blastoderms: age-related capacities for induction and self-differentiation following transplantation of quail Hensen's nodes. Anat Rec 229:437–448

Gallera J 1971 Primary induction in birds. Adv Morphog 9:149–180

Garcia-Martinez V, Schoenwolf GC 1992 Positional control of mesoderm movement and fate during avian gastrulation and neurulation. Dev Dyn 193:249–256

Garcia-Martinez V, Alvarez IS, Schoenwolf GC 1993 Locations of the ectodermal and non-ectodermal subdivisions of the epiblast stages 3 and 4 of avian gastrulation and neurulation. J Exp Zool, in press
Gardner CA, Darnell DK, Poole SJ, Ordahl CP, Barald KF 1988 Expression of an *engrailed*-like gene during development of the early embryonic chick nervous system. J Neurosci Res 21:421–437
Jacobson AG, Gordon R 1976 Changes in the shape of the developing vertebrate nervous system analyzed experimentally, mathematically and by computer simulation. J Exp Zool 197:191–246
Jacobson AJ , Tam PPL 1982 Cephalic neurulation in the mouse embryo analyzed by SEM and morphometry. Anat Rec 203:375–396
Lawrence PA 1992 The making of a fly: the genetics of animal design. Blackwell Scientific Publications, Oxford
Lee H-Y, Nagele RG 1985 Studies on the mechanisms of neurulation in the chick: interrelationship of contractile proteins, microfilaments, and the shape of neuroepithelial cells. J Exp Zool 235:205–215
McGinnis W, Krumlauf R 1992 Homeobox genes and axial patterning. Cell 68:283–302
McKechnie JL (ed) 1979 Webster's new twentieth century dictionary of the English language, unabridged, 2nd edn. William Collins, New York
Patel NH, Martin-Blanco E, Coleman KG et al 1989 Expression of *engrailed* proteins in arthropods, annelids, and chordates. Cell 58:955–968
Placzek M, Tessier-Lavigne M, Yamada T, Jessell T, Dodd J 1990 Mesodermal control of neural cell identity: floor plate induction by the notochord. Science 250:985–988
Schoenwolf GC 1979 Observations on closure of the neuropores in the chick embryo. Am J Anat 155:445–466
Schoenwolf GC 1982 On the morphogenesis of the early rudiments of the developing central nervous system. Scanning Electron Microsc 1:289–308
Schoenwolf GC 1983 The chick epiblast: a model for examining epithelial morphogenesis. Scanning Electron Microsc 3:1371–1385
Schoenwolf GC 1985 Shaping and bending of the avian neuroepithelium: morphometric analyses. Dev Biol 109:127–139
Schoenwolf GC 1988 Microsurgical analyses of avian neurulation: separation of medial and lateral tissues. J Comp Neurol 276:498–507
Schoenwolf GC 1991 Neurepithelial cell behavior during avian neurulation. In: Gerhart J (ed) Cell–cell interactions in early development. Alan R Liss, New York, p 63–78
Schoenwolf GC, Alvarez IS 1989 Roles of neuroepithelial cell rearrangement and division in shaping of the avian neural plate. Development 106:427–439
Schoenwolf GC, Alvarez IS 1991 Specification of neurepithelium and surface epithelium in avian transplantation chimeras. Development 112:713–722
Schoenwolf GC, Alvarez IS 1992 Role of cell rearrangement in axial morphogenesis. In: Pedersen RA (ed) Current Topics in Developmental Biology. Academic Press, Orlando, FL, vol 27:129–173
Schoenwolf GC, Fisher M 1983 Analysis of the effects of *Streptomyces* hyaluronidase on formation of the neural tube. J Embryol Exp Morphol 73:1–15
Schoenwolf GC, Franks MV 1984 Quantitative analyses of changes in cell shapes during bending of the avian neural plate. Dev Biol 105:257–272
Schoenwolf GC, Powers ML 1987 Shaping of the chick neuroepithelium during primary and secondary neurulation: role of cell elongation. Anat Rec 218:182–195
Schoenwolf GC, Sheard P 1990 Fate mapping the avian epiblast with focal injections of a fluorescent-histochemical marker: ectodermal derivatives. J Exp Zool 255:323–339
Schoenwolf GC, Smith JL 1990a Mechanisms of neurulation: traditional viewpoint and recent advances. Development 109:243–270

Schoenwolf GC, Smith JL 1990b Epithelial cell wedging: a fundamental cell behavior contributing to hinge point formation during epithelial morphogenesis. In: Keller RE, Fristrom D (eds) Control of morphogenesis by specific cell behaviors. Saunders, London (Semin Dev Biol 1) p 325–334
Schoenwolf GC, Garcia-Martinez V, Dias MS 1992 Mesoderm movement and fate during avian gastrulation and neurulation. Dev Dyn 193:235–248
Schroeder TE 1970 Neurulation in *Xenopus laevis*. An analysis and model based upon light and electron microscopy. J Embryol Exp Morphol 23:427–462
Smith JL, Schoenwolf GC 1987 Cell cycle and neuroepithelial cell shape during bending of the chick neural plate. Anat Rec 218:196–206
Smith JL, Schoenwolf GC 1988 Role of cell-cycle in regulating neuroepithelial cell shape during bending of the chick neural plate. Cell Tissue Res 252:491–500
Smith JL, Schoenwolf GC 1989 Notochordal induction of cell wedging in the chick neural plate and its role in neural tube formation. J Exp Zool 250:49–62
Smith JL, Schoenwolf GC 1991 Further evidence of extrinsic forces in bending of the neural plate. J Comp Neurol 307:225–236
Takahashi H 1988 Changes in peanut lectin binding sites on the neuroectoderm during neural tube formation in the bantam chick embryo. Anat Embryol 178:353–358
Takahashi H 1992 The masking effect of sialic acid on Con A, PNA and SBA ectoderm binding sites during neurulation in the bantam chick embryo. Anat Embryol 185:389–400
Takahashi H, Howes RI 1986 Binding pattern of ferritin-labeled lectins (RCA_1 and WGA) during neural tube closure in the bantam embryo. Anat Embryol 174:283–288
van Straaten HWM, Hekking JWM, Wiertz-Hoessels EJLM, Thors F, Drukker J 1988 Effect of the notochord on the differentiation of a floor plate area in the neural tube of the chick embryo. Anat Embryol 177:317–324
Yamada T, Placzek M, Tanaka H, Dodd J, Jessell TM 1991 Control of cell pattern in the developing nervous system: polarizing activity of the floor plate and notochord. Cell 64:635–647

DISCUSSION

Shum: Many grafting experiments use quail/chick chimeras. How normal is the behaviour of the quail cells in the chick environment? What is the difference in the cell cycle time of quail and chick cells? Also, are the cells of different sizes and would this affect cell movement?

Schoenwolf: Clearly, there are differences between quail and chick cells; the nucleolar marker is one. At early stages, the quail blastoderm is smaller than the chick blastoderm, so there's a size difference. They also develop at different rates: the quail has a total incubation period to hatching of about 16 days, whereas for chick it is 21, so the quail develops faster.

We have not looked carefully at the cell cycle for possible differences between quail and chick. We are concerned about differences and we have done several controls. For example, one can do unilateral grafts of cells from quail to chick. If there are differences in cell behaviour, such as the rate of cell division, and if cell division plays a role in morphogenesis, then a unilateral graft should cause an asymmetry to develop. We don't see that. Probably, the quail cells adapt very quickly to the chick environment and start acting like chick cells in terms of things like cell division.

Another way to address this is by transplanting plugs of chick tissue to quail hosts and seeing if you get the same result as when you graft quail tissue to chick hosts. It is fairly easy to do this with neural plate because the cells stay together. It's more difficult with mesoderm, where the cells disperse; that is, it is difficult to discern the nucleolar marker in isolated quail cells in a chick field. But using an anti-quail antibody, we can look at the chimeras and get a negative image of the quail cells at single cell resolution. Their behaviour appears to be the same as that of the chick cells.

The third thing we can do is to label chick plugs with a fluorescent dye and transplant them into chick hosts. We graft chick tissue to chick tissue, and again we get the same result. If you looked at every cell's behaviour throughout the whole period of culture, you might see some variation, but you can't do that, you can only look at the final outcome. When we look at the final outcome, we can't detect any differences.

The final thing we can do is microinject cellular markers such as DiI (dioctadecyl-3,3,3′,3′-tetramethylindocarbocyanine perchlorate). Again, we get the same results as with chimeras.

Shum: You have shown that in embryos lacking a notochord, where the median hinge point was absent (Smith & Schoenwolf 1989), the neural folds still elevated, converged and fused to form a closed neural tube. Also, the dorsolateral hinge points do not form at all levels of the body axis, only at certain regions. So what do you think is the role of hinge points in neurulation?

Schoenwolf: My belief is that hinge points facilitate the process of neurulation, as opposed to driving it. If you cut out hinge points, or if you take notochords out so hinge points don't form, neurulation still occurs. However, the efficacy is reduced: instead of about 90% of embryos in culture forming closed neural tubes, only about 60% do. I think there is not a single switch for neurulation—there are many things occurring and if you alter the cells' behaviour or eliminate things like hinge points, you reduce the efficacy of neurulation.

More directly, I think the hinge points provide an anchoring site that the plate can rotate around. When you remove the hinge points, other things take their place. If you cut out the floor plate, then at the medial cut sites the neural plate will heal, with endoderm acting as part of the hinge point. If you prevent a notochord from forming, the neuroepithelial cells remain tall, which probably adversely affects neurulation; shorter cells would facilitate bending. But, you still have an anchorage point, because the endoderm is still there. So I think that hinge points facilitate neurulation, but they are not essential.

But hinge points do a lot more than just provide anchorage sites. The cells of the median hinge point give rise to the floor plate, and that's a major signalling area for dorsoventral polarization of the neural tube and organization of the neural elements. I think the embryo uses hinge points to make it slightly easier to neurulate, then it uses them for other purposes later in development.

I don't think you could argue that dorsolateral hinge points aren't necessary because they are not found at all levels of the body axis. If you look at neurulation in the brain, where there are hinge points, and neurulation in the spinal cord, where there aren't, there are major differences in the neural plate and the shape of the neural tube in those regions. The neural plate in the brain region of the chick is very broad compared with the spinal cord. In the brain, the neural folds have to come together from quite far apart and you get a broad lumen which is particularly wide in the forebrain and midbrain. But in the spinal cord region, the neural folds clap shut like a book. There is occlusion in the spinal canal. Mary Desmond and Antone Jacobson (1977) looked at enlargement and inflation of the brain due to pressure of the cerebrospinal fluid. It is thought that the brain starts to enlarge before the posterior neuropore closes. How does the embryo confine fluid to the brain when the posterior neuropore is open? It occludes the lumen. So by not having dorsolateral hinge points at the spinal cord level, the embryo seals off its neural canal prematurely, allowing the brain to start to expand. One can test that by injecting dyes into the brain and one sees that occlusion is an effective seal (Schoenwolf & Desmond 1984).

Shum: In the spinal cord region of chick embryos, the median hinge point is present and the relatively flat neural folds elevate and fuse at the dorsal midline without forming dorsolateral hinge points, resulting in a slit-like neural lumen. However, in the mouse there are regional differences along the spinal cord. Rostrally, the neural tube closes in a fashion similar to that seen in the chick, but more caudally dorsolateral hinge points are formed, resulting in a diamond-shaped neural tube. Further down the spinal cord, neither median nor dorsolateral hinge points are formed; instead, the neural folds exhibit general curvature to form a round or oval-shaped neural tube. I have looked at three mouse strains and it seems that in different strains, the transition from one type of closure to the other occurs at different levels. Why do dorsolateral hinge points form in certain regions and not others to create differently shaped neural tubes?

Schoenwolf: I think the embryo needs to have a brain that is shaped differently from the spinal cord and it uses hinge points which facilitate neurulation to achieve that effect. In the chick, there are also hinge points at the caudal end of the spinal cord, in the area called the sinus rhomboidalis (Fig. 2). The caudal portion of the chick neural plate looks very similar to the frog hyla. There is this precocious what we would call convergence, the dorsolateral hinge points forming essentially prior to elevation.

It is important to remember there are differences among species and there are also differences at different rostrocaudal levels. With the chick in particular, we tend to think that because development proceeds in a rostrocaudal sequence we can look at one rostrocaudal level of the axis at different times and understand all of neurulation. This is not true.

van Straaten: There is both flexibility and the possibility for adaptation during neurulation. In the chick sinus rhomboidalis area, when the neural folds were prevented from fusing by widening the neuropore, the folds converged much further than they would have done normally. This means that the lateral hinge points bent more than they did normally—that's a kind of flexibility. I found that even in the somitic region, where only apposition occurs during closure, there was some convergence and formation of lateral hinge points after experimental widening of the neural gutter (unpublished results). This illustrates the flexibility of one mechanism used for neurulation, but it is one of several mechanisms.

Morriss-Kay: Alisa Shum mentioned the V-shaped or diamond-shaped neural folds that are seen at more cranial levels of the developing spinal cord and the more gentle curvature seen more caudally. I've noticed this too in both rat and mouse embryos. The only correlation I could suggest is that there are more cells at the more cranial levels, so the neural folds are larger. When the upper trunk neural tube closes, it is a squared-off oval shape, whereas more caudally, where the neural groove and neural folds are U-shaped, the closed neural tube is more circular and consists of fewer cells. This is not a suggestion about cause and effect, it is just a simple correlation; but there may be some causal element to it.

Schoenwolf: I have seen similar things in the caudal region. At that point the notochord has not formed definitively; there is some overlap with the primitive streak and eventually with the tail bud. This relates to whether hinge points are necessary. When there is no hinge point at that site, the neural tube still closes.

Morriss-Kay: You have to be careful comparing the avian and mammalian notochord. The mammalian one is a very flimsy structure and very small compared to the avian notochord.

Shum: I believe neurulation is driven by a combination of different mechanisms which vary along the body axis. Therefore, when we study the mechanisms of neurulation, we should state which level we are looking at. There are many seemingly contradictory results, for example, concerning the effects of drugs that disrupt microfilaments or microtubules on neurulation. Some of the discrepancy may be due to the fact that people are looking at different levels of the body axis, where the extent of involvement of microfilaments or microtubules in neurulation may vary.

Jacobson: The notoplate or floor plate is attached intimately to the notochord. We have found that computer simulations are useful for dealing with these complex issues. In the simulation we did looking at the physical parameters for the rolling toward the midline, the midline was anchored in some cases, but not in others (Jacobson 1994, this volume). We got neurulation in both ways but it was a much better mimic when the midline was anchored.

Hall: Anchored to the notochord?

Jacobson: With the computer you just anchor it; normally it's anchored to the notochord.

Hall: If the notochord is more flimsy in mammals, does that mean that in mammals you are less likely to need anchoring?

Morriss-Kay: There is anchoring at early stages of neurulation in mammals, because the notochord is a very flat structure. Where there is an obvious early floor plate with hinge points, the basal surface of the midline neural plate cells is attached to this broad, flat epithelial notochord; only later does it become a more rod-shaped, thin structure.

O'Rahilly: Do you think that the mesenchyme in the neural fold has an important role in neurulation?

Schoenwolf: Until one does a series of rigorous experiments testing virtually everything in the embryo that's doing something at the time of neurulation, one cannot say that something does not play a role. My approach has been to start with a broad outlook and try to narrow it down by eliminating things. However, I do think that, particularly in the head where there are relatively massive neural folds, at the very least mesenchyme provides a scaffold. The neural fold is lifted up, by whatever means, then this scaffold is put in place to hold the fold in that position. It's a ratchet-type of process: the fold is raised a little, then held, raised and held.

There is the old issue of whether extracellular matrix is also involved in this process. A very simplistic view of extracellular matrix regards it as jelly; if this were present in the neural folds, it would work as a scaffold too. The matrix is not necessarily driving the process, it might again be a facilitator, or it could be driving. It is a difficult problem to study because you have to use inhibitors and you don't necessarily know what your inhibitors are doing.

Morriss-Kay: I used to think that the extracellular matrix, especially the jelly component of it—the hydrated hyaluronate—was very important in providing a scaffold. We then used a specific enzyme, *Streptomyces* hyaluronidase, to remove the hyaluronate. This takes the water away with it so the cells are very condensed. The neural tube still closes and adopts a reasonably normal shape (Morriss-Kay et al 1986).

Schoenwolf: 100% of the time?

Morriss-Kay: Yes; so I no longer believe that the jelly component of the extracellular matrix is important as scaffolding. The embryos form a normal looking neural tube. It is reduced in size because the cell cycle time is lengthened, but the shape is really quite good.

Shum: How about the rate of neurulation?

Morriss-Kay: That's delayed, but only very slightly.

Shum: I am concerned about the rate of neurulation. The neural tube is surrounded by different tissues, so maybe when you remove one tissue, other tissues or other factors can compensate. Just because you see 'normal' neurulation after removing a tissue, it does not mean that during normal development that tissue is not important or does not play a role. This is why one should look at the rate of neurulation rather than just at whether neurulation is completed or not.

Jacobson: Professor O'Rahilly asked about a possible role in neurulation of the mesenchyme in the neural fold. There is an old literature on this. Nieuwkoop, Johnen & Albers (1985) pointed out that neural crest mesenchyme contacting the prospective brain may suppress mitosis, so the thinning of the roof plate of the brain occurs where the neural crest first contacts the brain tissue. The continued patterned apposition of the neural crest mesenchyme and of the mesodermal inducers of the nervous system, the notochord and somites, has a role in the shaping of the neural tube, which Holtfreter & Hamburger (1955) called 'formative effects'.

Hall: Certainly, in some humans with neural tube defects there are mesenchymal abnormalities, in heart, kidney and other tissues. This suggests there may be a mesenchymal defect. But these are usually sporadic.

O'Rahilly: The skull is involved too. We have no quarrel with the occlusion that has been mentioned for the chick embryo, but we see absolutely no evidence that this mechanism works in the human embryo, contrary to published reports.

Hall: So in human embryos there is no build up of pressure?

O'Rahilly: We don't believe there is.

Schoenwolf: Do you say that because the anterior neuropore is open at the time of occlusion?

O'Rahilly: Yes; there is only partial occlusion, never total.

Hall: Then the concept of rupture is probably not relevant to humans: is that what you are suggesting?

O'Rahilly: It can't always be excluded, but we doubt it is a common factor.

van Straaten: Many different tissues are involved in neurulation; if one tissue does not act properly, neurulation may fail or it may not. We partially isolated the caudal neural tube: the remainder of the open neuropore still closed, in spite of the fact that mesoderm wasn't present (van Straaten et al 1993)! This doesn't mean the mesoderm is not important in neurulation, but it indicates that mechanisms which were left *in situ* were able to accomplish neurulation, possibly by compensating for the mesodermal action.

Opitz: Judith, you mentioned correlated anomalies. We have to ask whether in a specific case these are due to pleiotropy, where all of the malformations present are manifestations of a single underlying cause, e.g. a gene mutation, or whether they represent defects of earlier gastrulation. I always find it helpful to look at neurulation basically as the end result of gastrulation. Many events occur before gastrulation that affect both neural tube development and mesodermal development surrounding the neural tube. Hence the correlated anomalies you see in associations, e.g. the high incidence of neural tube defects in infants of diabetic mothers or the schisis associations that Andrew Czeizel (1981) has defined.

I think there are probably determinative inductive events before neural tube formation which affect both mesodermal differentiation and neural tube formation.

Schoenwolf: This is why, when studying neurulation, one must also consider gastrulation—it is because of the cellular interactions that occur then. As we map back to earlier and earlier stages, we find that the prospective neural plate at pre-streak stages is all located at the caudal end of the embryo, intimately associated with the mesoderm. This fits with the current concept of planar inductive events. A lot of the patterning is probably being established very early, even before overt gastrulation events.

Seller: With regard to the related defects associated with neural tube defects, I am sure there is a definite association with the site of the lesion in the neural tube. Dagmar Kalousek and I made a study of this. It's very unusual to have a cleft palate with a low-level lumbar spina bifida. You get a cleft palate with an anencephaly or with a cervical spina bifida.

Hall: I am struck that a fetus can have anencephaly, that is, basically not have neurulated properly anteriorly, and yet the rest of its body can be fairly normal. It's amazing that you can have virtually no brain during development and have everything else in pretty good shape.

Mills: The adrenal glands don't function.

Seller: Most of the multifactorial type of neural tube defects are isolated lesions. In only about 20% are there multiple abnormalities. I do feel this must be saying something to us.

Hall: From what John Opitz was saying, this implies that gastrulation has proceeded normally in those 80% of embryos.

Opitz: Yes; the apparently non-complex varieties are defects of organogenesis; the earlier blastogenesis went reasonably well.

Jacobson: Setting up the dorsal and ventral mesoderm coincides with setting up the neural plate boundary, at least in *Xenopus*.

Schoenwolf: Does the surface epidermis itself receive an inductive stimulatory signal or does an inhibitory signal diffuse through it? How do you visualize those signals interacting?

Jacobson: I visualize those signals interacting in the same way that signals interact in setting up the axes of *Drosophila* (Nüsslein-Volhard 1991): there are some things that turn off genes and some things that turn them on. When you have both positive and negative kinds of interaction, they set up a line very discretely. This is a role of the ventral signals that has not really been appreciated before.

Schoenwolf: But in *Drosophila* those signals occur in a syncytial blastoderm: in a vertebrate embryo there would be more of a problem communicating the signals.

Hall: It's quite clear that the same signals get used.

Copp: Antone Jacobson (this volume) described experiments with amphibian embryos in which he cut out the neural plate with or without some surface ectoderm still attached and got very distinct effects. The neural plate that had surface ectoderm attached formed a closed neural tube, whereas the totally isolated neural plate did not. Gary, could you comment on the role of the surface ectoderm in the closure of the folds in the chick embryo?

Schoenwolf: In *Xenopus*, I was struck by how the ectoderm rolls over the dorsal lip. For neural folds to get up to the dorsal midline, there has to be expansion of the surface ectoderm.

Jacobson: As long as it's attached.

Schoenwolf: Yes. This is true for the chick and even more so in the mouse, particularly in the head folds. The neural folds start very low and neural crest cells are emigrating from the folds as the folds elevate. The lateral ectoderm is expanding. The issue is whether tissues outside the neural plate have a role in bending or whether this process is purely an intrinsic neuroepithelial event. I believe it requires tissues outside of the neural plate but there is some scepticism in the audience about that.

First, we have to accept the concept that for neurulation to occur, the surface ectoderm has to expand. If it doesn't expand, the neural folds are going to be held down.

Morriss-Kay: In the mouse and rat head region, the large neural folds are closing very rapidly towards the end of neurulation. The surface ectoderm close to the neuroepithelium appears crumpled, as if it's expanded in anticipation of that very rapid closure moment (Morriss & New 1979). This precludes the need for stretching of the surface ectoderm by the forces set up by the neuroepithelium as the neural folds move together in the midline.

Schoenwolf: Therefore, if you are compiling a catalogue of what's important in neurulation, you have to include expansion of the surface ectoderm. One can argue about whether this expansion occurs in advance of fold elevation and whether it actually contributes to that process. Our experiments suggest that it does contribute (e.g. Smith & Schoenwolf 1991). So do the experiments of Rudy Brun, which showed that in amphibians cell wedging in the hinge points is blocked by colchicine (Brun & Garson 1983).

Shum: Lewis (1947) made a unilateral incision between the surface ectoderm and the neural fold of the amphibian embryo and found that the wound gaped open. The neural folds moved more rapidly than normal towards the midline and formed a morphologically intact neural tube. He suggested that the surface ectoderm, rather than assisting, actually exerts contractile tension and opposes bending of the neural plate.

Schoenwolf: There are many interpretations of those experiments. You can cut out a piece of *endoderm* from the chick and it will very quickly roll up and form a tube. Epithelial sheets in isolation roll up. I don't know whether that is a normal neurulation event. I can't say that it isn't, but I don't know that it is. The fact that isolated ectodermal sheets roll up inside out is also puzzling if this behaviour represents a normal neurulation event.

van Straaten: We repeated Lewis' experiment in the mouse embryo and observed immediate apposition of the neural folds after a lateral incision. This suggests there is tension. As Andrew Copp (1994, this volume) will show, tensions caused by the underlying axial curvature might play an important role in caudal neurulation. Also, in cranial neurulation in the mouse, Jacobson & Tam (1982) found a correlation between decrease of mesencephalic angle and closure of the mesencephalic neural folds. The main tension restraining the neural folds from medial displacement seemed to be generated by the marked axial curvature and not by the ectoderm.

Schoenwolf: But if you make a hole in the belly further down in the surface ectoderm, the cut ends then roll. That is not neurulation. So why are you saying that when a fold is bending over closer to the neuropore, that is neurulation?

Jacobson: This question is really: is the epidermis pushing? If you say that it is, then you are obliged to explain how cells can produce a pushing force. If the cells were pushing, you would expect them to be under compression. If you make a slit in the epidermis close to the neural plate, it gapes instantaneously, indicating that it's under tension not compression. Therefore I would argue that the epidermis cannot be pushing.

Hall: Is this true in all species?

Schoenwolf: It's true in the frog, but when you make that gape, you are not only cutting the epidermis, you are also going through the basement membrane and you are probably going through the lateral plate mesoderm. There is yolk present that leaks out, so you are breaching the surface. Remember that in normal development there's a vitelline membrane. The ectoderm is plastered, more or less, against the vitelline membrane and its subvitelline fluid. So the internal hydrostatic pressure of the embryo is balanced by the external pressure.

Jacobson: In the amphibian, the ectoderm is not in contact with the vitelline membrane. The fertilization membrane is raised.

Trasler: In the *splotch* mouse, where many things appear to go wrong, when you look at the histology, the mesoderm is reduced in cell number, the neuroepithelium is abnormal and the notochord is reduced in cell number. In this case, there is a single defective gene, but it's affecting all of those three tissues during neural tube closure, as well as its effects on neural crest.

Copp: Are you asking us to accept that mesoderm expansion may be a common requirement for neurulation? I have trouble with that concept. The craniocaudal variation in the morphology of neurulation in the mouse is incompatible with this idea. Initiation of neurulation, which happens in the cervical region, occurs at a level where the mesoderm is segmented—there are somites present. It is difficult for me to see how mesodermal expansion is occurring at the time of neural fold elevation and fusion at this level, because somites are epithelial structures that appear to contract as they segment. On the other hand, in the cranial region there is massive mesodermal expansion and in the lower spine the neural plate is flanked by unsegmented mesoderm. So the mesoderm is doing three completely different things during neurulation at different levels of the body axis.

Schoenwolf: I didn't say that the mesoderm or mesenchyme was important for neurulation. I said that the mesoderm is involved in the patterning of the neuraxis. There are mesodermal–ectodermal interactions that are initiated very early in gastrulation, so it's a patterning phenomenon.

Hall: In mouse there are also strain differences. Are there strain differences in birds?

Schoenwolf: In chickens there are, definitely. Ann Gibbins in Ontario in the Poultry Science Department looks at spontaneous neural tube defects in chickens. She finds that the strain of the bird makes a tremendous difference.

Kintner: You describe a motor at the dorsal midline that drives elongation of the embryo. It seems to be a universal motor found in vertebrates. To what extent do you think the motor is the same in different species? Was it invented once and is basically the same in all vertebrate species or do you think there are major differences?

Schoenwolf: I don't think that in the chick the notoplate is the sole motor; the lateral regions of the neural plate are also involved. Convergent extension is the main driver of shaping of the neural plate. Convergent extension, though, can come about through cell rearrangement, changes in cell shape and changes in cell number. All of these events are occurring during shaping of the chick neural plate and they lead to a convergent extension movement.

Recently, we have taken out what should be floor plate of the neural tube. We punched out Hensen's node, which will give rise to some floor plate cells, and we replaced it with a plug of cells from the lateral neural plate region. This produces lateral cells in the midline as well as at the sides of the midline: all lateral cells still converge and extend together and form a neuraxis. So it's not necessary in the chick to have the boundary between notoplate and lateral regions. However, normally there is a boundary between those two regions and the rate of extension is somewhat different in the chick between the lateral and the medial region. If you remove midline cells in front of Hensen's node, cells that normally contribute along with Hensen's node to floor plate, then replace them with cells from the lateral neural plate region, these cells can't extend down the length of the axis because they encounter midline cells from Hensen's node. So the grafted cells merge laterally with lateral neural plate cells and extend down the length of the axis, lateral to the floor plate. Thus, there are differences in these two populations of neuroepithelial cells, as Antone Jacobson has been saying for years, that prevent them from intermixing. But in the chick, those differences are not required for rostrocaudal extension of the neural plate. Therefore, in the chick I can't see that such differences are driving the extension. It is the cell behaviour underlying convergent extension overall that is driving extension.

Goulding: How can you be sure that the cells that you have placed into the area of Hensen's node have all the characteristics of laterally positioned cells? They may have taken on some of the characteristics of midline cells.

Schoenwolf: Midline cells only become midline cells if they are induced by notochord. When we take out Hensen's node which gives rise to midline cells—midline neural plate cells and notochord—at least some of the embryos have no notochord, yet the extension still occurs.

Opitz: These midline cells have inducing properties found in the zone of polarizing activity (ZPA). Do they retain these properties as they stream rapidly down the midline?

Schoenwolf: Floor plate and Hensen's node both have ZPA activity (Wagner et al 1990, Hornbruch & Wolpert 1986).

Lindhout: Do you think these cells need a lot more energy than other cells?

Schoenwolf: I don't have the slightest idea. Neurulation is driven by a series of fundamental cell behaviours, such as cell division and cell migration, in addition to things like changes in cell shape. There are many ways in which cells can change shape, but one way is by apical constriction. Another is by basal expansion. Many very simple kinds of things that cells do make sense when one thinks about the variety of teratogens that can cause defects. For example, every teratogen is going to interfere with cell division.

Jacobson: When you say the words convergence and extension, you imply that there's much more mediolateral than anteroposterior intercalation. Cells aren't just going to do that by themselves, there has to be something that directs them to do it. I suggested that cell movements are organized when cells collect at the boundaries of the neural plate with the notoplate and with the epidermis. The thing that struck me was that the boundaries are formed extraordinarily early. In *Xenopus* the notoplate is induced probably at the same time as the caudal mesoderm by the signals from the dorsal vegetal cells. That boundary is there; the cells do not cross the boundary and they do collect at it. There is probably contact inhibition between those two domains of cells and/or there are adhesion differences between the two domains of cells. This provides an organizing principle which people must come up with for any other model. I think the notochord/somite boundary is organizing, at least part of the time during convergent extension of the caudal mesoderm. Keller et al (1992) stated that they couldn't visualize the notoplate/neural plate boundary, but they did visualize it with their trajectory map because none of those cells crossed the area where the boundary should be.

Holmes: It has been reported that infants with anencephaly consistently have abnormalities of the bones in the base of the cranium. Presumably, these cranial defects are caused by the same developmental process which produces the brain malformation.

Opitz: Müller & O'Rahilly (1991) have written a beautiful paper on two early anencephalic embryos. One had a nearly normal brain and a severely defective skull; the other was the opposite. They made a very important point about the skull in anencephaly, namely that a fetus may have a well preserved brain with holoacranial anencephalic skull or virtually no brain with a nearly normal skull.

O'Rahilly: We were surprised that the anomalies of the brain and skull don't seem to go hand in glove, as we would have expected. The brain and skull seem to be rather independent, at least to a certain degree and during a presumably limited time.

Opitz: I have done dissections/autopsies on anencephalic infants/fetuses in which I have found surprisingly well developed parietal and occipital bones that had collapsed in a flat stack onto the base of the skull, apparently *after* disappearance of the brain. Thus, it seems that after initial failure of rostral brain/neural tube development, a degenerative process is responsible for the disappearance of brain which occurs later than initiation of formation of the chondrocranium.

O'Rahilly: It is my impression that even in cases with anencephaly, the brain develops quite well during the embryonic period proper (the first eight weeks after fertilization). Of course, by the time of birth it is defective.

Hall: It appears that several things may be happening while the embryo or fetus with a neural tube defect is growing *in utero*. If there is an open defect, the nerve cells migrate out. In addition, the amniotic fluid surrounding the open defect is not a particularly friendly environment for neural cells in developing brain tissue. The surrounding skull and skin also don't develop normally without the brain pushing against them. Nevertheless, it would appear that no matter what the embryo was like at the start, they all end up looking the same by the time they abort or are born.

References

Brun RB, Garson JA 1983 Neurulation in the Mexican salamander (*Ambystoma mexicanum*): a drug study and cell shape analysis of the epidermis and neural plate. J Embryol Exp Morphol 74:275–295

Copp AJ 1994 Genetic models of mammalian neural tube defects. In: Neural tube defects. Wiley, Chichester (Ciba Found Symp 181) p 118–143

Czeizel AE 1981 Schisis-association. Am J Med Genet 10:25–35

Desmond ME, Jacobson AG 1977 Embryonic brain enlargement requires cerebrospinal fluid pressure. Dev Biol 57:188–198

Holtfreter J, Hamburger V 1955 Amphibians. In: Willier BH, Weiss PA, Hamburger V (eds) Analysis of development. WB Saunders, Philadelphia, PA, p 230–296

Hornbruch A, Wolpert L 1986 Positional signalling by Hensen's node when grafted to the chick limb bud. J Embryol Exp Morphol 94:257–265

Jacobson AG 1993 Somitomeres: mesodermal segments of the head and trunk. In: Hanken J, Hall B (eds) The vertebrate skull. University of Chicago Press, Chicago, IL, vol 1:42–96

Jacobson AG 1994 Normal neurulation in amphibia. In: Neural tube defects. Wiley, Chichester (Ciba Found Symp 181) p 6–24

Jacobson AG, Tam PPL 1982 Cephalic neurulation in the mouse embryo analyzed by SEM and morphometry. Anat Rec 203:375–396

Keller R, Shih J, Sater A 1992 The cellular basis of the convergence and extension of the *Xenopus* neural plate. Dev Dyn 193:199–217

Lewis WH 1947 Mechanics of invagination. Anat Rec 97:139–156

Morriss GM, New DAT 1979 Effect of oxygen concentration on morphogenesis of cranial neural folds and neural crest in cultured rat embryos. J Embryol Exp Morphol 54:17–35
Morriss-Kay GM, Tuckett F, Solursh M 1986 The effects of *Streptomyces* hyaluronidase on tissue organization and cell cycle time in rat embryos. J Embryol Exp Morphol 98:57–70
Müller F, O'Rahilly R 1991 Development of anencephaly and its variants. Am J Anat 190:193–218
Nieuwkoop PD, Johnen AG, Albers B 1985 The epigenetic nature of early chordate development. Cambridge University Press, Cambridge, p 190
Nüsslein-Volhard C 1991 Determination of the embryonic axes of Drosophila. Dev Suppl 1:1–10
Schoenwolf GC, Desmond ME 1984 Neural tube occlusion precedes rapid brain enlargement. J Exp Zool 230:405–407
Smith JL, Schoenwolf GC 1989 Notochordal induction of cell wedging in the chick neural plate and its role in neural tube formation. J Exp Zool 250:49–62
Smith JL, Schoenwolf GC 1991 Further evidence of extrinsic forces in bending of the neural plate. J Comp Neurol 307:225–236
van Straaten HWM, Hekking JWM, Consten C, Copp AJ 1993 Intrinsic and extrinsic factors in the mechanism of neurulation: effect of curvature of the body axis on closure of the posterior neuropore. Development 117:1163–1172
Wagner M, Thaller C, Jessell T, Eichele G 1990 Polarizing activity and retinoid synthesis in the floor plate of the neural tube. Nature 345:819–822

Normal neurulation in mammals

Gillian Morriss-Kay, Heather Wood and Wei-Hwa Chen

Department of Human Anatomy, University of Oxford, South Parks Road, Oxford OX1 3QX, UK

Abstract. During mammalian neurulation regional differences are evident between the cranial region, in which neurulation is most complex, the trunk as far as the caudal neuropore and the secondary neurulation region of the caudal trunk plus tail. Differences among these three regions are characterized by specific patterns of morphogenesis and by specific patterns of gene expression. During cranial neurulation distinct regions develop in the brain and the presomitic hindbrain forms seven rhombomeric divisions. The first clear morphological boundary is the preotic sulcus (later transformed into the gyrus between rhombomeres 2 and 3), which may limit cell movement as neuroepithelial cells rostral to it flow towards and into the rapidly expanding forebrain region. The formation of rhombomeres as morphological entities and the development of a normal rhombomere-specific pattern of homeobox and other gene expression domains depend on relatively low levels of retinoic acid. Retinoic acid receptors, which are retinoic acid-activated transcription factors, and retinoid binding proteins, which control the availability of retinoic acid to the receptors, show regional patterns of expression in the cranial, trunk and caudal regions of the neuroepithelium during neurulation. These patterns suggest a possible mechanism for region-specific gene expression during neurulation.

1994 Neural tube defects. Wiley, Chichester (Ciba Foundation Symposium 181) p 51–69

Mammalian neurulation can be considered as a two-part process: primary neurulation, in which the tube forms as a result of morphogenetic movements of the neuroepithelium, and secondary neurulation, in which a cord of cells within the tail bud mesenchyme condenses to form an epithelial tube continuous with that of the primary neurulation region. The transition zone between the two processes (the level of closure of the caudal neuropore) is in the sacral region. Its precise somitic level shows some variation among species: it is at the level of somites 33–35 (mid-sacral region) in the mouse (Copp et al 1990) and at the level of somites 30–31 (second sacral vertebra region) in human embryos (Müller & O'Rahilly 1987). This review will consider only primary neurulation, i.e. the processes involved in formation of the neural tube between the forebrain and the caudal neuropore. The central nervous system becomes regionalized during neurulation, with respect to shape, gene expression domains and potential structure and function. The pattern of neurulation is itself region specific, showing differences in forebrain, midbrain/upper hindbrain, mid-hindbrain, lower (occipital) hindbrain, spinal cord and caudal neuropore regions. Some of these regional differences can be seen soon after the neural tube has begun to close (Fig. 1).

The account that follows is based mainly on studies in rodent embryos.

Formation of the neural plate

Before and during the stage of neural plate formation, the embryo undergoes gastrulation, which converts it from an apparently simple two-layered epithelial structure, consisting of epiblast and hypoblast, to a three-layered embryo consisting of both epithelial and mesenchymal tissues (ectoderm, endoderm and mesoderm). The whole three-layered embryo forms from the epiblast: movement of epiblast cells through the primitive streak leads to the formation of primary mesenchyme (mesoderm) and to the replacement of the hypoblast by definitive endoderm (Lawson & Pedersen 1992 and references therein). The notochord is at first a midline epithelial structure within the endoderm layer; its separation to become a rod-like structure occurs during neurulation and its changes in shape prior to separation are closely linked with those of the early stages of neurulation. It is important to note that the mammalian notochord is a very thin rod compared with that of amphibians and birds and is unlikely to have any skeletal function.

The process of neural induction (determination of neural plate within the epiblast) is not understood in mammals, but the anatomical relationships of the tissues discussed above, and the extracellular matrix and cell surface molecules known to be synthesized by them, are characteristic of interacting tissues. Neural induction is only the first of a series of tissue interactions involved in differentiation of the neuroepithelium; others are concerned with the morphogenetic movements of neurulation and with the determination of regional patterns of gene expression and differentiation. At the late presomite stage, the rostral neural plate differentiates from epiblast as a shield-shaped area of pseudostratified epithelium, clearly demarcated from the thinner epithelium lateral and rostral to it, which will become surface ectoderm (Morriss-Kay 1981). It has a midline neural groove overlying its midline attachment to the notochord. As the embryo elongates, new neural plate continues to differentiate from epiblast until the primitive streak is converted into tail bud mesenchyme, after which new neural tube forms by secondary neurulation.

Early stages of neurulation

The earliest neural plate consists only of tissue that will develop into forebrain, midbrain and the most rostral part of the hindbrain. As the first few pairs of

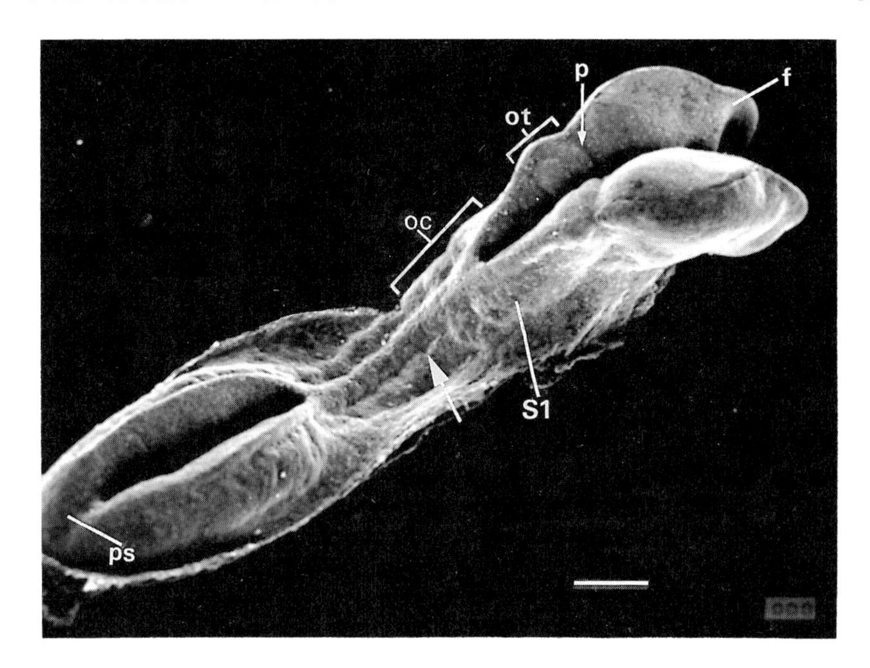

FIG. 1. Scanning electron micrograph of an 8-somite-stage rat embryo, illustrating some of the regional differences in neurulation. The forebrain region (f) is growing rapidly; it forms two hemispherical structures that will meet in the midline and fuse at the 14-somite stage, closing the rostral neuropore. The midbrain and rostral hindbrain, as far as the preotic sulcus (p) (future position of the rhombomere 2/3 boundary), form a pair of broad neural folds from which neural crest cells are at this stage migrating into the first arch and over the basal surface of the forebrain. The otic hindbrain (ot) forms a second broad region (pro-rhombomere B); neural crest cells are about to emigrate at this stage. The occipital hindbrain (oc) lies at the level of the first four pairs of somites (S1 indicates the position of the first somite); the neural tube closes in a caudorostral sequence here and in a rostrocaudal sequence in the trunk, having initiated closure at the level of the hindbrain/spinal cord boundary (arrow). The caudal neuropore is large, with V-shaped neural folds. New neuroepithelium is differentiated from epiblast just cranial to the primitive streak (ps) until the caudal neuropore closes at around the 30-somite-stage. Bar = 100 μm.

somites form, the flat neural plate is converted into a pair of convex neural folds which become very broad, as if growing rapidly in the transverse plane. In fact, the transverse expansion is due to a change in epithelial structure from pseudostratified to cuboidal, the cell number remaining constant. A constant cell number in the transverse plane is maintained throughout neurulation, even during emigration of neural crest cells (Morriss-Kay & Tuckett 1985) (Fig. 2).

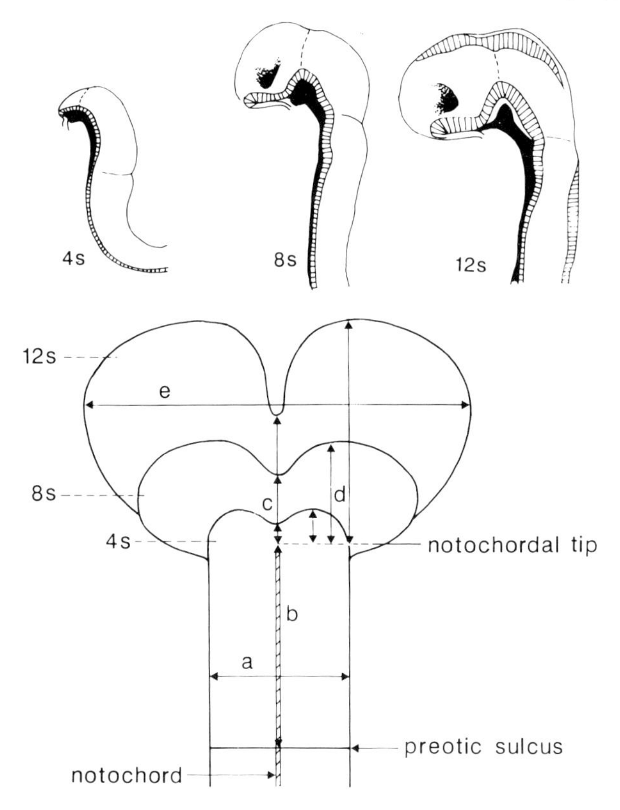

FIG. 2. Growth of the forebrain region during neurulation. Top: tracings from scanning electron micrographs of sagittally halved rat embryos at the 4-somite, 8-somite and 12-somite stages (see Morriss-Kay 1981 for original micrographs). The main diagram was compiled from counts of cell number at the same three stages (mean values from five embryos per stage), in planes a, b, c and e; the dimensions of plane d were calculated from size comparisons with c at each stage. Cell number in plane a (125) and plane b (175) was unchanged throughout neurulation. Expansion of the forebrain is illustrated by increase of cell number in plane c (16, 57 and 98 cells at the three successive stages) and in plane e (125, 260 and 330 cells).

As the cranial neural folds become convex, the amount of primary mesenchyme beneath them increases. The volume of extracellular matrix (ECM) in the mesenchyme beneath the broad cranial neural fold region is greater than that in more caudal regions; when the preotic sulcus forms (coincidentally with formation of the first two pairs of somites), it overlies the division between ECM-rich mesenchyme rostrally and more condensed mesenchyme caudally. The early morphogenetic changes of the cranial neural folds should be seen as part of the morphogenesis of the cranial region as a whole, including formation of the foregut together with caudal movement of the heart. The presumptive surface ectoderm, which is initially in the same plane as the neural ectoderm, changes its orientation to become perpendicular to it, forming the lateral walls of the developing head (Figs 1 and 5).

Formation of the forebrain region

The forebrain region consists initially of a small amount of tissue at the rostral edge of the neural plate. The notochord extends into this region, though not quite to the tip: the midline neuroepithelium rostral to the notochord is the chiasmatic plate (Müller & O'Rahilly 1985). Except for this midline area, the forebrain region expands dramatically during neurulation, increasing its cell number much more rapidly than can be accounted for by cell division alone. Between the 4-somite stage and the 8-somite stage, the surface area of the forebrain neuroepithelium shows a 20-fold increase; during the same period, surface area and cell number between the preotic sulcus and the tip of the notochord remain constant in both longitudinal and transverse planes (Morriss-Kay 1981) (Fig. 2). The neuroepithelial cell cycle time is six hours in all regions; cell labelling studies have shown that the neuroepithelium is a fluid sheet of cells and that the restriction of growth in the forebrain region is due to the rostral flow of cells within the midbrain/rostral hindbrain region (Morriss-Kay & Tuckett 1987). The preotic sulcus appears to be the caudal boundary of the region of forward cell flow. By the time the cranial neuropore closes, the forebrain region is very large, having grown in both transverse and longitudinal planes.

Midbrain and hindbrain neurulation

Neural tube closure begins close to the hindbrain/spinal cord junction. It extends caudally down the trunk as the embryo elongates and also extends rostrally within the occipital hindbrain. At the same time, the neural folds at the level of the forebrain/midbrain junction approach each other and form an independent fusion point, leaving a spindle-shaped opening which closes independently of the rostral neuropore. Conversion of the convex neural folds to a V-shape involves conversion of the cuboidal neuroepithelium into a pseudostratified form, which increases in thickness until closure is complete (Morriss-Kay 1981). Development and maintenance of the concave curvature that precedes closure depend on the development and contraction of adherens junction-related microfilament bundles (Fig. 3). If microfilament bundle integrity is disrupted by cytochalasin D, concave curvature is lost; regeneration is possible only if the forebrain/midbrain apposition has already been made (Morriss-Kay & Tuckett 1985). Emigration of neural crest cells is an important part of the mechanism of closure of the spindle-shaped midbrain/rostral hindbrain neuropore (see below).

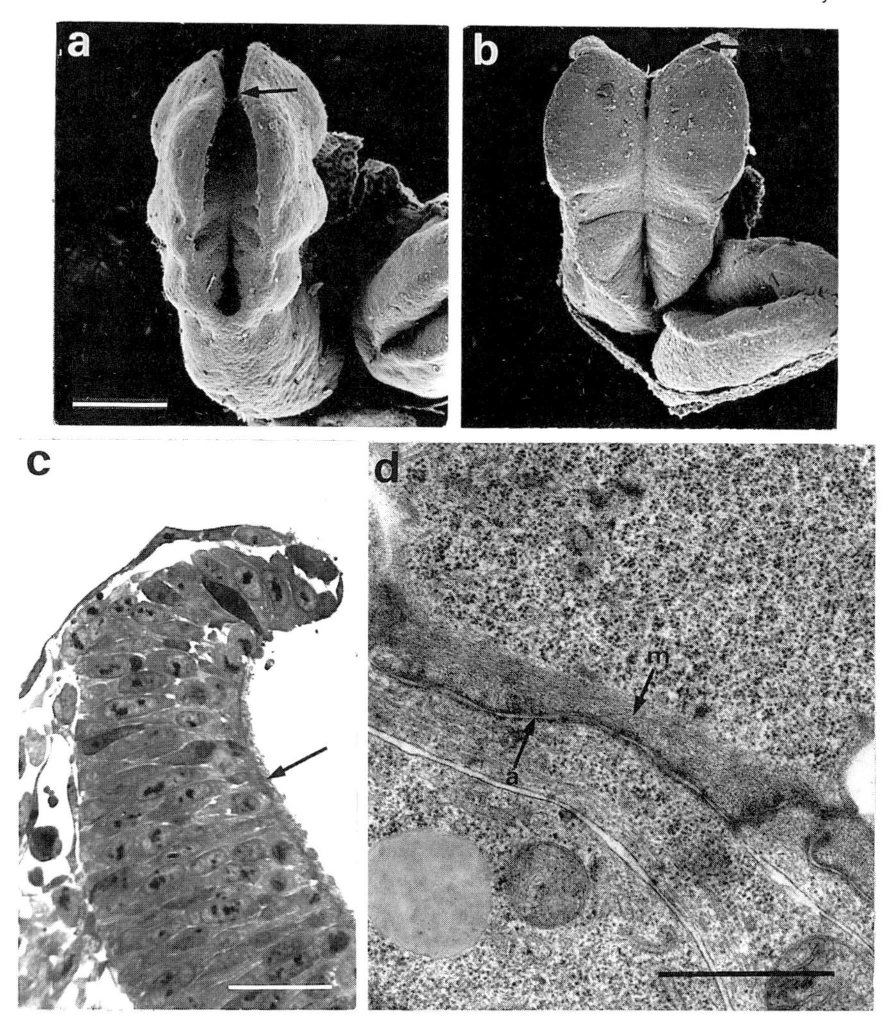

FIG. 3. Contraction of adherens junction-related microfilament bundles is responsible for generation and maintenance of concave curvature of the neuroepithelium prior to neural tube closure. (a) Normal 10-somite-stage rat embryo just before apposition at the forebrain/midbrain junction (arrowed here and in (b)). The internal surface of the midbrain/upper hindbrain region is concave and shows rhombomeric sulci. (b) After a one hour exposure to 0.15 μg ml^{-1} cytochalasin D, microfilament bundles are lost, resulting in loss of the concave curvature and conversion of the rhombomeric sulci to bulges. (c) A continuous line of adherens junctions and microfilament bundles (arrow) along the contracting apical surface except at the lateral edge (presumptive roof plate) can be seen by light microscopy and (d) is shown in detail by transmission electron microscopy (the apical surface is upwards). The micrograph shown in (c) is a toluidine blue-stained 1 μm section from an embryo embedded in Spurr resin. a, adherens junction; m, microfilament bundle. (a, b) Bar = 100 μm; (c) bar = 10 μm; (d) bar = 1 μm.

During the phase in which the biconvex neural folds are converted to a V-shape and then become concave prior to closure (i.e. after the 6-somite stage), the preotic sulcus flattens out and a series of seven sulci and gyri form in the preoccipital hindbrain. In the older (and some recent) literature these are referred to as neuromeres, but they are now more commonly called rhombomeres. The concave shape of the sulci depends on microfilaments; gyrus shape is maintained by microtubules (Tuckett & Morriss-Kay 1985). Rhombomeres are not only morphological segments, but also boundaries of cell movement (Fraser et al 1990), of gap junctional permeability (Martinez et al 1992) and of gene expression (Wilkinson et al 1989). Their development is inhibited if levels of retinoic acid are raised at the presomite, early neural plate stage (Morriss-Kay et al 1991).

The relationship between neural crest cell migration and neurulation

Neural crest cell emigration is part of the normal process of neurulation in the developing mammalian head. Broadly speaking, cranial neurulation takes place in a caudorostral direction; neural crest cell emigration, however, occurs as a series of events beginning with the population rostral to the preotic sulcus, which migrates ventrally from wide open midbrain/upper hindbrain neural folds to form the first pharyngeal arch mesectoderm and rostrally over thc basal surface of the forebrain to form the frontonasal mesenchyme (Tan & Morriss-Kay 1985, 1986). In mouse embryos, the second group of neural crest cells to emigrate is the population just caudal to the preotic sulcus, which moves down into the second pharyngeal arch, but in rat embryos these cells do not emigrate until after emigration of the cells just caudal to them. This timing difference between two closely related species emphasizes the distinct nature of the three cranial neural crest cell populations. The third (caudal hindbrain) group emigrates from neural folds that are approaching closure in the immediately post-otic region and are already closed in the occipital region of the hindbrain. This group emigrates in a rostrocaudal sequence continuous with that of the trunk, where no neural crest cell migration occurs before neurulation is complete.

Clearly, neural crest cell emigration from the closed occipital hindbrain and trunk regions of the neural tube cannot contribute to the process of neurulation. There is good evidence, however, that neural crest cell migration from midbrain and preoccipital hindbrain levels is essential for normal neurulation. Closure in these regions fails, or is delayed, when crest cell emigration is inhibited by exposure of the embryos to raised oxygen levels or when the ECM component chondroitin sulphate proteoglycan is degraded (Morriss & New 1979, Morriss-Kay & Tuckett 1989). Neural crest cell emigration from the midbrain/upper hindbrain neural folds appears to be essential for flexibility of the lateral edges, allowing them to curve medially and fuse in the dorsal midline.

Neurulation in the trunk and closure of the caudal neuropore

In the trunk, neural tube formation in mammals is essentially the same as in birds. The midline neuroepithelium is anchored to the notochord and in most sections appears to bend upwards from it at hinge points. As more caudal levels of the trunk neural plate form, cell number in the transverse plane decreases and the neural groove is U-shaped. Throughout the trunk, the generation of concave curvature in the neuroepithelium depends on microfilaments, resembling the V-shaped and concave stages of the cranial region.

The caudal (posterior) neuropore is present as soon as neural tube closure begins in the cervical region (Fig. 1). As the embryo elongates and the closed area of the neural tube extends caudally, the caudal neuropore is always at a level caudal to the most recently formed pair of somites. The primitive streak lies at the caudal edge of the caudally moving neuropore. When the neuropore nears the level at which it closes, the epiblast changes orientation to form a vertical epithelial wall between the neuropore and the tail bud mesenchyme. The apical layer of this epithelium shows stronger reactivity for actin than does the remainder of the epithelium or the tail bud mesenchyme (our unpublished data), suggesting that contraction of microfilament bundles around the neuropore plays an active role in its closure. The final stage of closure of the caudal neuropore resembles closure of the midbrain/hindbrain neuropore, being bidirectional (Sakai 1989). Caudal neuropore closure is inhibited when the basement membrane component heparan sulphate proteoglycan is degraded using heparitinase (Tuckett & Morriss-Kay 1989).

The role of extracellular matrix in mammalian neurulation

The ECM associated with the neuroepithelium during neurulation lies within the basement membrane, stretching down from this as a meshwork around the underlying mesenchyme cells. It is rich in heparan and chondroitin sulphate proteoglycans (HSPG and CSPG) (Fig. 4) and also contains hyaluronan, fibronectin and laminin (Solursh & Morriss 1977, Morriss & Solursh 1978, Tuckett & Morriss-Kay 1986). Studies on abnormal cranial neurulation show that the presence of HSPG in and subjacent to the neuroepithelial basement membrane is essential for conversion of the convex neural folds to the V-shaped and then concave form that precedes closure, and that, as described above, CSPG is essential for neural crest cell emigration and therefore for the normal closure events of the midbrain and upper hindbrain regions. This evidence comes from studies on the effects of degradative enzymes (heparitinase and chondroitinase ABC) specific for these two ECM components (Morriss-Kay & Tuckett 1989, Tuckett & Morriss-Kay 1989). Cranial neurulation is delayed in embryos exposed to retinoic acid; immunochemical staining shows a loss of HSPG and CSPG, but no effect on fibronectin or laminin (Morriss-Kay &

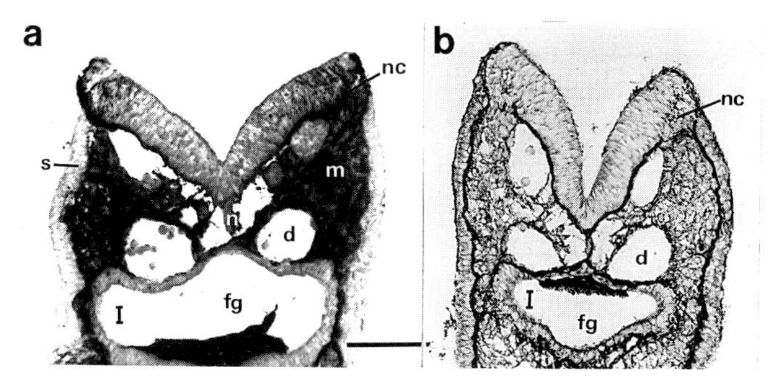

FIG. 4. Hindbrain sections (just caudal to the preotic sulcus) of 8-somite-stage embryos stained immunohistochemically with antibodies to (a) chondroitin sulphate proteoglycan (CSPG) (CS-56, ICN Biomedicals) and (b) heparan sulphate proteoglycan (HSPG) (gift of John Hassel). Both epitopes are present in the neuroepithelial (and other) basement membranes and form part of the mesenchymal extracellular matrix and notochordal sheath. CSPG is also associated with neural crest cell emigration from the lateral region of the neuroepithelium, both before (left side of (a)) and during (right side) emigration. The HSPG-containing neuroepithelial basement membrane is lost as the neural crest cells emigrate (b). d, dorsal aorta; fg, foregut; m, cranial mesenchyme; n, notochord; nc, neural crest; s, surface ectoderm; I, first pharyngeal pouch. Bar = 100 μm.

Mahmood 1992). Degradation of hyaluronate, with concomitant loss of the hydrated extracellular spaces, does not inhibit either neurulation or neural crest cell migration, even though it results in a lengthening of the cell cycle and therefore a much smaller embryo (Morriss-Kay et al 1986).

The histological, histochemical and immunohistochemical appearance of the neural folds and associated ECM during neurulation is typical of tissues engaged in epitheliomesenchymal interactions. The studies referred to above show that these interactions are important for normal neurulation. Frohman et al (1990) have suggested that inductive influences from mesenchyme may also be involved in genetic patterning of the neuroepithelium at the time of its differentiation from epiblast. The *Wnt-1* gene product is a good example of a short-range signalling molecule that is essential for formation of a normal neural tube. It is a secreted glycoprotein that binds to the ECM and plasma membrane (Bradley & Brown 1990). Mutant mouse embryos in which this gene is deleted develop a neural tube that lacks most of the midbrain and part of the rostral hindbrain (McMahon & Bradley 1990).

Genetic patterning of the neuroepithelium during neurulation

By the end of neurulation, there is a distinct pattern of gene expression in the neural tube that gives every segmental level a different genetic code

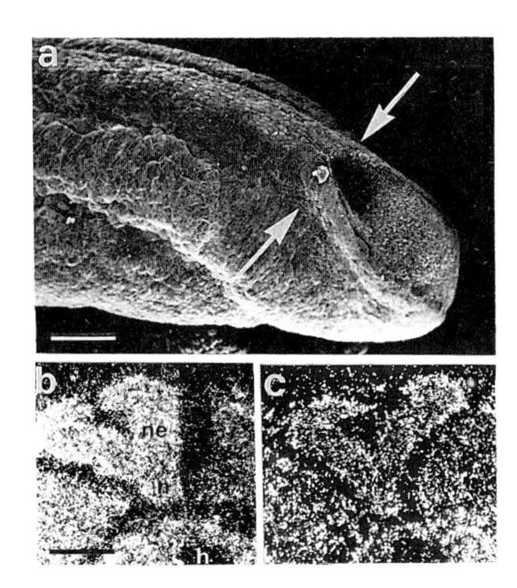

FIG. 5. (a) Scanning electron micrograph showing the caudal neuropore of a 28-somite-stage C57BL/6 mouse embryo (Day 10). The arrows indicate the plane of section shown in (b) and (c). (b) Section through the caudal region of the closing caudal neuropore of a *curly tail* mouse mutant embryo, hybridized with ^{35}S-labelled riboprobe to show RAR-γ transcripts. (c) Equivalent section from a *curly tail* mouse embryo with a severely enlarged caudal neuropore, showing lower intensity of labelling for RAR-γ transcripts. (a) Bar = 50 μm; (b, c) bar = 65 μm.

(Kessel & Gruss 1990). This pattern is reflected in the paraxial mesoderm (Kessel & Gruss 1991) and, in the cranial region at least, in the neural crest (Hunt et al 1991). Most of these genes have homeodomains. They include members of the *Hox*, *Pax*, *Msx*, *engrailed*, *even-skipped*, *POU*, *Dlx*, *Emx* and *Otx* families (Kessel & Gruss 1990, MacKenzie et al 1991, Simeone et al 1992, 1993). All these genes have a DNA-binding motif in the homeodomain and so probably all encode transcription factors. *Krox-20*, which is expressed only in rhombomeres 3 and 5 of the hindbrain, also encodes a transcription factor, but of the zinc finger type (Wilkinson et al 1989). It has recently been shown to be involved in regulating the segmental hindbrain expression of *Hoxb-2* (Sham et al 1993). The role of segmental gene expression in control of morphological development of the neural tube is unknown.

Cellular retinoic acid-binding protein type I (CRABP I) also shows region-specific mRNA expression domains in the neuroepithelium during neurulation (Ruberte et al 1991). CRABP I binds retinoic acid with high affinity and also mediates its degradation by catabolic enzymes (Napoli et al 1993), thereby decreasing the availability of retinoic acid for binding to the nuclear receptors (Boylan & Gudas 1991). Nuclear (active) levels of retinoic acid are therefore thought to be low where CRABP I is expressed, i.e. hindbrain and midbrain. Both morphological aspects of neurulation and neuroepithelial gene expression patterns are specifically affected in these regions when retinoic acid levels are raised (Morriss-Kay et al 1991, Marshall et al 1992), presumably because retinoic acid is able to gain access to and activate the nuclear receptors where it does not normally do so. Much higher levels of exogenous retinoic acid are required to alter morphological segment identity in the trunk (Kessel & Gruss 1991).

Retinoic acid receptors (RARs) are retinoic acid-inducible transcription factors that bind to retinoic acid response elements of target genes (see Ruberte et al 1991 for references, and for details of the following results). RAR-α and RAR-β show specific patterns of expression in the hindbrain that may be important in relation to expression of retinoic acid-inducible homeobox genes there. Of even more direct interest to understanding neurulation are the reciprocal expression domains of RAR-β and RAR-γ in the trunk: as the embryo elongates, RAR-γ is always restricted to the open neuropore region and RAR-β to the closed neural tube. In the *curly tail* mouse mutant, embryos show reduced expression of RAR-γ in the open caudal neuropore region, especially in the enlarged caudal neuropore of embryos with the prospective spina bifida phenotype (W.-H. Chen, A. J. Copp, G. Morris-Kay, unpublished data, 1993) (Fig. 5).

Correlations between normal and altered gene expression patterns and normal and abnormal patterns of neurulation are clearly beginning to provide evidence that will lead to significant insights into the roles of specific genes in the mechanisms underlying mammalian neurulation.

Acknowledgements

Heather Wood is supported by the Human Frontier Science Program Organisation, Wei-Hwa Chen by the government of Taiwan, Republic of China. Previously published work described in this article was funded by the Medical Research Council and Hoffman-La Roche. We thank Martin Barker for excellent technical assistance.

References

Boylan JF, Gudas L 1991 Overexpression of the cellular retinoic acid binding protein I (CRABP-I) results in a reduction in differentiation-specific gene expression in F9 teratocarcinoma cells. J Biol Chem 112:965–979

Bradley RM, Brown AS 1990 The proto-oncogene *int-1* encodes a secreted protein associated with the extracellular matrix. EMBO (Eur Mol Biol Organ) J 9:1569–1575

Copp AJ, Brook FA, Estibeiro P, Shum ASW, Cockroft DL 1990 The embryonic development of mammalian neural tube defects. Prog Neurobiol 35:363–403

Fraser SE, Keynes R, Lumsden A 1990 Segmentation in the chick embryo hindbrain is defined by cell lineage restrictions. Nature 344:431–435

Frohman MA, Boyle M, Martin GR 1990 Isolation of the mouse *Hox-2.9* gene: analysis of embryonic expression suggests that positional information along the anterior–posterior axis is specified by mesoderm. Development 110:589–607

Hunt P, Whiting J, Muchamore I, Marshall H, Krumlauf R 1991 Homeobox genes and models for patterning the hindbrain and branchial arches. Dev Suppl 1:187–196

Kessel M, Gruss P 1990 Murine developmental control genes. Science 249:374–379

Kessel M, Gruss P 1991 Homeotic transformations of murine vertebrae and concomitant alterations of *Hox* codes induced by retinoic acid. Cell 67:89–104

Lawson KA, Pedersen RA 1992 Clonal analysis of cell fate during gastrulation and early neurulation in the mouse. In: Postimplantation development in the mouse. Wiley, Chichester (Ciba Found Symp 165) p 3–26

MacKenzie A, Ferguson MWJ, Sharpe P 1991 *Hox-7* expression during murine craniofacial development. Development 113:601–611

Marshall H, Nonchev S, Sham MH, Muchamore I, Lumsden A, Krumlauf R 1992 Retinoic acid alters the hindbrain *Hox* code and induces the transformation of rhombomeres 2/3 into a rhombomere 4/5 identity. Nature 360:737–741

Martinez S, Geijo E, Sánchez-Vives MV, Puelles L, Gallego R 1992 Reduced junctional permeability at interrhombomeric boundaries. Development 116:1069–1076

McMahon AP, Bradley A 1990 The *Wnt-1* (*int-1*) proto-oncogene is required for development of a large region of the mouse brain. Cell 62:1073–1085

Morriss GM, Solursh M 1978 Regional differences in mesenchymal cell morphology and glycosaminoglycans in early neural-fold stage rat embryos. J Embryol Exp Morphol 46:37–52

Morriss GM, New DAT 1979 Effect of oxygen concentration on morphogenesis of cranial neural folds and neural crest in cultured rat embryos. J Embryol Exp Morphol 54:17–35

Morriss-Kay GM 1981 Growth and development of pattern in the cranial neural epithelium of rat embryos during neurulation. J Embryol Exp Morphol (suppl) 65:225–241

Morriss-Kay GM, Mahmood R 1992 Morphogenesis-related changes in extracellular matrix induced by retinoic acid. In: Morriss-Kay GM (ed) Retinoids in normal development and teratogenesis. Oxford University Press, Oxford, p 165–180

Morriss-Kay GM, Tuckett F 1985 The role of microfilaments in cranial neurulation in rat embryos: effects of short-term exposure to cytochalasin D. J Embryol Exp Morphol 88:333–348

Morriss-Kay GM, Tuckett F 1987 Fluidity of the neural epithelium during forebrain formation in rat embryos. J Cell Sci Suppl 8:433–449

Morriss-Kay GM, Tuckett F 1989 Immunohistochemical localization of chondroitin sulphate proteoglycans and the effects of chondroitinase ABC in 9- to 11-day rat embryos. Development 106:787–798

Morriss-Kay GM, Tuckett F, Solursh M 1986 The effects of *Streptomyces* hyaluronidase on tissue organization and cell cycle time in rat embryos. J Embryol Exp Morphol 98:57–70

Morriss-Kay GM, Murphy P, Hill RE, Davidson D 1991 Effects of retinoic acid excess on expression of *Hox-2.9* and *Krox-20* and on morphological segmentation in the hindbrain of mouse embryos. EMBO (Eur Mol Biol Organ) J 10:2985–2995

Müller F, O'Rahilly R 1985 The first appearance of the neural tube and optic primordium in the human embryo at stage 10. Anat Embryol 172:157–169

Müller F, O'Rahilly R 1987 The development of the human brain, the closure of the caudal neuropore, and the beginning of secondary neurulation at stage 12. Anat Embryol 176:413–430

Napoli JL, Posch KC, Fiorella PD, Boerman MHEM, Salerno GJ, Burns RD 1993 Roles of cellular retinol binding protein and cellular retinoic acid binding protein in the metabolic channeling of retinoids. In: Livrea MA, Packer L (eds) Retinoids: progress in research and clinical applications. Marcel Dekker, New York, p 29–48

Ruberte E, Dollé P, Chambon P, Morriss-Kay G 1991 Retinoic acid receptors and binding proteins. II. Their differential pattern of transcription during early morphogenesis in mouse embryos. Development 111:45–60
Sakai Y 1989 Neurulation in the mouse: manner and timing of neural tube closure. Anat Rec 223:194–203
Sham MH, Vesque C, Nonchev S et al 1993 The zinc finger gene *Krox-20* regulates *Hox-B2* (*Hox-2.8*) during hindbrain segmentation. Cell 72:183–196
Simeone A, Gulisano M, Acampora D et al 1992 Two vertebrate homeobox genes related to the *Drosophila empty spiracles* gene are expressed in the embryonic cerebral cortex. EMBO (Eur Mol Biol Organ) J 11:2541–2550
Simeone A, Acampora D, Mallamaci A et al 1993 A vertebrate gene related to *orthodenticle* contains a homeodomain of the *bicoid* class and demarcates anterior neuroectoderm in the gastrulating mouse embryo. EMBO (Eur Mol Biol Organ) J 12:2735–2747
Solursh M, Morriss GM 1977 Glycosaminoglycan synthesis in rat embryos during formation of the primary mesenchyme and neural folds. Dev Biol 57:75–86
Tan SS, Morriss-Kay GM 1985 The development and distribution of the cranial neural crest in the rat embryo. Cell Tissue Res 240:403–416
Tan SS, Morriss-Kay GM 1986 Analysis of cranial neural crest cell migration and early fates in postimplantation rat chimaeras. J Embryol Exp Morphol 98:21–58
Tuckett F, Morriss-Kay GM 1985 The ontogenesis of cranial neuromeres in the rat embryo. II. A transmission electron microscope study. J Embryol Exp Morphol 88:231–247
Tuckett F, Morriss-Kay GM 1986 The distribution of fibronectin, laminin and entactin in the neurulating rat embryo studied by indirect immunofluorescence. J Embryol Exp Morphol 94:95–112
Tuckett F, Morriss-Kay GM 1989 A role for heparan sulphate proteoglycan in the rat embryo: effects of heparitinase treatment during early organogenesis. Anat Embryol 180:393–400
Wilkinson DG, Bhatt S, Cook M, Boncinelli E, Krumlauf R 1989 Segmental expression of Hox-2 homeobox-containing genes in the developing mouse hindbrain. Nature 341:405–409

DISCUSSION

Lindhout: You said that CRABP is meant to trap retinoic acid and offer it for catabolism to prevent it from binding to the retinoic acid receptors in the nucleus. Is there good evidence for that? Or could the opposite be true, that CRABP acts as a reservoir or as a mechanism just to present or to transport retinoic acid to the nucleus?

Morriss-Kay: That was the first hypothesis about the role of CRABP. We have to call it CRABP I now because there are two forms. Little is known about how CRABP II works; it has a much lower binding affinity for retinoic acid than has CRABP I (Bailey & Siu 1988). Although they are both expressed in the hindbrain, they are expressed in very different patterns (Ruberte et al 1992).

The evidence for my statement is all from *in vitro* work, but it is very good and it fits the pattern of what we actually see in the retinoic acid-treated embryos.

Retinoic acid binds to CRABP I with high affinity; holo-CRABP I binds by protein–protein interaction to the catabolic enzymes that break down retinoic acid, so that CRABP I both sequesters retinoic acid and mediates its inactivation (Napoli et al 1993). As long as there is free CRABP I present, retinoic acid will be sequestered as it enters the cell and will not therefore be available to the nucleus. The other evidence comes from a paper by Boylan & Gudas (1991). They showed that overexpression of CRABP I in F9 cells inhibits the normal pattern of retinoic acid-induced differentiation. In order to induce differentiation, they had to increase the concentration of retinoic acid in the medium.

The reason these observations fit well with those from embryos is that there is a very good correlation between where CRABP I is expressed and the tissues that are vulnerable to abnormality of cell behaviour, morphogenesis and gene expression patterns when retinoic acid is given in excess. My interpretation is that when embryos are exposed to high concentrations of retinoic acid, the binding capacity of CRABP I is saturated; some retinoic acid is thereby enabled to reach the nucleus in cells in which it would not normally do so, altering the pattern of gene expression in these locations and initiating abnormal morphogenesis. High CRABP levels (hence low retinoic acid levels) in the midbrain and hindbrain during neurulation also correlate well with the observation that in vitamin A-deficient embryos development of the brain is normal at these early stages (our unpublished data). The first abnormality that we can see in deficient embryos is of the choroid plexus and of the structural layers in the developing cortex: that's at a much later stage.

The other important protein is cellular retinol-binding protein (CRBP I). It is involved in conversion of retinol to retinoic acid (Napoli et al 1993), so in embryos it is likely that retinoic acid is synthesized from maternally derived retinol in the cells that require it at the highest concentrations.

Oakley: What sort of abnormal morphogenesis do you get by exposure to retinoic acid at an early stage? Could you get holoprosencephaly?

Morriss-Kay: What happens depends on when you give the retinoic acid. This is interesting in the context of thinking about neural tube defects. There are clearly different stages during neurulation which are vulnerable to going wrong in different ways. If the embryo is exposed to retinoic acid at the early neural plate stage, before there is any embryonic segmentation, i.e. before the first somites begin to segment, rhombomeres fail to form in the hindbrain. The hindbrain retains a smooth surface instead of developing its normal beautiful series of seven rhombomeric sulci. The gene expression patterns are altered: in particular, *Hoxb-1*, which is normally expressed only in rhombomere 4 at the later stages, is expressed in a much larger area and is rostrally displaced to the midbrain/hindbrain junction (Morriss-Kay et al 1991). The absence of morphological segmentation (rhombomeres) correlates with the absence of clearly defined boundaries between domains of gene expression. Normally, *Krox-20* is expressed in rhombomere 3, *Hoxb-1* in rhombomere 4 and *Krox-20* in rhombomere 5. In embryos exposed to retinoic acid before the onset of segmentation, there are intermingled patches of cells expressing the two genes. No cells express both genes, but there is a mosaic pattern where there ought to be a clear boundary. Also, there is a loss of gene expression that ought to be in the first three rhombomeres.

If the embryos are exposed to retinoic acid only about two hours later, once segmentation has begun, then the morphological pattern of segmentation is not altered, but there is an abnormal pattern of gene expression. The developing cranial neural tube looks quite normal, and so does the face; *Hoxb-1* is expressed in both rhombomeres 2 and 4, instead of just in rhombomere 4 (our unpublished observations; see also Marshall et al 1992).

Nau: Could you say a little more about the distribution of the α and β retinoic acid receptors and the possible significance for neural tube development?

Morriss-Kay: RAR-α1 was at one time thought to be a housekeeping gene: it is almost ubiquitously expressed and has some of the characteristics of housekeeping genes in its promoter region (Leroy et al 1992). However, in mouse embryos its transcripts cannot be detected rostral to rhombomere 4 at the end of cranial neurulation (Ruberte et al 1991) and it is no longer considered to be a housekeeping gene.

Curiously, we cannot detect any of the receptors rostral to rhombomere 4. So if retinoic acid is having any effect on rostral hindbrain, midbrain and forebrain development, it must be acting through quite low levels of receptor protein or using the retinoid 'X' receptor (RXR) system. *RAR-β* is expressed only up into the lower part of rhombomere 7, i.e. in the brain it is confined to the unsegmented occipital region (this is somitically segmented, but is not segmented in the neuro-epithelium itself), and then all the way down the trunk neural tube. *RAR-β* is itself induced by retinoic acid. Therefore, it is no coincidence that there is hardly any overlap between the expression of the gene for CRABP I and *RAR-β*.

We have suggested that the lack of overlap between expression domains of *RAR-β* and *RAR-γ*, and the juxtaposition of the two of them at the rostral end of the caudal neuropore, is to do with some function that RAR-β may have in maturation or differentiation of cells within the closed neural tube—but that is merely conjecture. RAR-γ we thought might have something to do with neurulation processes in the caudal neuropore. That was before Wei-Hwa Chen had done the *RAR-γ in situ* hybridizations in the *curly tail* mouse embryo (Fig. 5), so it was very nice to get that result.

Hall: Gillian, what's the relationship between retinoic acid and expression of the *Hox* genes?

Morriss-Kay: The most 3′ *Hox* genes, e.g. *Hoxa-1* and *Hoxb-1*, are induced directly by retinoic acid. The more 5′ genes are induced too, but probably indirectly, since they require higher concentrations of retinoic acid for activation and activation is slower (Simeone et al 1990, Arcioni et al 1992). These observations were made on embryonal carcinoma cells: they correlate well with the colinearity of expression in the neural tube (expression of 3′ *Hox* genes extends to the most rostral levels). It has been suggested that differential periods of exposure to retinoic acid are responsible for the colinearity phenomenon in the embryo (Hogan et al 1992).

Juriloff: I would like to comment on the site of initiation of closure of the cranial neural tube at the midbrain/forebrain boundary. We studied the location of this site of neural fold contact in three mouse strains (Juriloff et al 1991). These are normal strains, they don't have neural tube closure defects. We found that the site, although in the general area of the midbrain/forebrain boundary, differs among strains. So this important event in neural tube closure is not uniform in normal genotypes.

Hall: Do you know anything about the expression of retinoic acid-binding proteins in that area?

Juriloff: It has not been studied in those strains. They do respond differently to some teratogenic stimuli. One strain is more sensitive to heat shock-induced exencephaly.

Copp: If you make F1 hybrids between the different strains, does the characteristic site of closure transmit as a single genetic trait?

Juriloff: We don't know.

Lindhout: I presume that this is reproducible within the strain?

Juriloff: That's correct: each strain has its own way of doing things and is entirely consistent about it within the strain.

Jacobson: In this discussion, it has been noted several times that there are important differences in neurulation and in the expression of neurulation defects along the anteroposterior axis. Also, much recent attention has been given to the expression of homeobox-containing genes (*Hox-2* series) and zinc finger genes such as K*rox*-20 in relation to the rhombomeres of amniote embryos (Wilkinson et al 1989a,b). The exit points of cranial motor nerve roots and the sites of formation of cranial ganglia have also been mapped in relation to the rhombomeres (Lumsden & Keynes 1989).

The question is, what causes these differences to arise along the axis? There is a general consensus that the underlying mesoderm induces forebrain at one end and more posterior nervous system at the other and that the rest of the differences in the brain and spinal cord arise later. However, the paraxial mesoderm is differentiating while gastrulation occurs and a series of segments, the somitomeres, is formed in order from the tip of the head to the tip of the tail (reviewed in Jacobson 1994). These somitomeric segments are likely candidates to impress their segmentation on the neural plate that forms later, especially since the initial rhombomeres form in register with the adjacent somitomeres.

Somitomeres in the occipital region of the head and in the trunk condense into somites, but in the head region anterior to the ears, the somitomeres remain dispersed. The positions of the cranial nerves illustrated by Lumsden & Keynes are adjacent to or within the anterior compartments of the associated somitomeres, just as spinal motor roots and dorsal ganglia form only in anterior parts of the sclerotomes of trunk somites (Keynes & Stern 1984, Teillet et al 1987).

I am suggesting that the somitomeres, which appear during gastrulation, are most likely the source of inductive events that organize the neural plate along the anteroposterior axis.

Copp: How does the somitomeric division of the mesoderm in the hindbrain region relate to the neural crest cells that will migrate from this region to provide much of the mesoderm of the head? The neural crest cells are patterned in the same way by *Hox* gene expression (Hunt et al 1991).

Jacobson: Neural crest begins its migration at the neural tube, then it encounters the somitomeric mesoderm. Somitomeres have been shown in six classes of vertebrates in quite a number of animals. In each of those animals, there is a correlation between where the neural crest comes out and which somitomeres it encounters.

Another interesting thing about somitomeres is that there are seven preotic somitomeres or segments in the amniotes and in the teleosts, but in the amphibians and in the shark there are just four. The rule is that number one is the same. The second and third in the amniote correspond to only one somitomere in amphibians. That follows down so that the limbs, which are attached to cervical vertebrae 3,4 and 5 in the salamander, are at exactly the same level in humans.

Hall: Is there any relation between segmentation and neural tube defects? Are neural tube defects more likely to arise in any particular segment?

Shurtleff: I know of no data to indicate one way or the other. There is a problem with talking about the level of neural tube defects in that one set of designations represents the nervous system level or neuromotor level and another represents the anatomical level, which relates more to the somites or vertebrae. There is much confusion in the literature about that 'level of lesion'.

Hall: There is often separation of sensory and motor function in particular patients.

Shurtleff: You can see that also between right and left sides. Many lesions are asymmetric for both motor and sensory levels.

Opitz: You have to be very careful with the concept of a structural homology here. One can get an identical limb, with the same segments, muscles, nerves and bones, in variations which may range over 12 vertebrae. Structurally, they are homologous, so the level of innervation makes no difference to how the limb develops.

Jacobson: It does within a given species, but there are many differences among species.

Shurtleff: Even within the human, there is variation in the innervation of muscles, particularly in the lower extremities. It is not as standard as John

Sharrard's original description suggests (McDonald et al 1991, Sharrard 1964, Asher & Olson 1983, Huff & Ramsey 1978).

Hall: Does this have anything to do with primary and secondary neurulation? Is secondary neurulation so low that when it is defective, it doesn't really manifest a problem in humans?

Shurtleff: The assessment of the neuromotor level for sacral level lesions is extremely difficult because of the involvement of the small muscles around the calf and foot. Below L5, S1, it becomes very vague. We've found (McDonald et al 1986) that there is also variability in muscle strength patterns recorded for the same child over time due, probably, to the extent to which children will cooperate.

Opitz: In those cases of extreme 'caudal regression', a person may have no sacral vertebrae and no lumbar vertebrae, yet seem to have reasonably well functioning legs. Is the innervation of those lower limbs standard or are there some motor nerves missing? Is the neural tube short in those cases?

Shurtleff: I cannot agree with your statement. I have seen caudal regressions as high as L1, L2, but none of the children had normally innervated lower extremities.

Opitz: But the motor nerves are normally differentiated.

References

Arcioni L, Simeone A, Guazzii S, Zappavigna V, Boncinelli E, Mavilio F 1992 The upstream region of the human homeobox gene HOX3D is a target for regulation by retinoic acid and HOX homeoproteins. EMBO (Eur Mol Biol Organ) J 11:265–277

Asher M, Olson J 1983 Factors affecting the ambulatory status of patients with spina bifida cystica. J Bone Jt Surg Am Vol 65:350–356

Bailey JS, Siu CH 1988 Purification and partial characterization of a novel binding protein for retinoic acid from neonatal rat. J Biol Chem 263:9326–9332

Boylan JF, Gudas L 1991 Overexpression of the cellular retinoic acid binding protein I (CRABP-I) results in a reduction in differentiation-specific gene expression in F9 teratocarcinoma cells. J Biol Chem 112:965–979

Hogan BLM, Thaller C, Eichele G 1992 Evidence that Hensen's node is a site of retinoic acid synthesis. Nature 359:237–241

Huff CW, Ramsey PL 1978 Myelodysplasia: the influence of the quadriceps and hip abductor muscles on ambulatory function and stability of the hip. J Bone Jt Surg Am Vol 60:432–443

Hunt P, Wilkinson D, Krumlauf R 1991 Patterning the vertebrate head: murine Hox 2 genes mark distinct subpopulations of premigratory and migrating cranial neural crest. Development 112:43–50

Jacobson AG 1994 Normal neurulation in amphibians. In: Neural tube defects. Wiley, Chichester (Ciba Found Symp 181) p 3–24

Juriloff DM, Harris MJ, Tom C, Macdonald KB 1991 Normal mouse strains differ in the site of initiation of closure of the cranial neural tube. Teratology 44:225–233

Keynes RJ, Stern CD 1984 Segmentation of the vertebrate nervous system. Nature 310:786–789

Leroy P, Krust A, Kastner P, Mendelsohn C, Zelent A, Chambon P 1992 Retinoic acid receptors. In: Morriss-Kay GM (ed) Retinoids in normal development and teratogenesis. Oxford University Press, Oxford, p 7–25

Lumsden A, Keynes R 1989 Segmental patterns of neuronal development in the chick hindbrain. Nature 337:424–428

Marshall H, Nonchev S, Sham MH, Muchamore I, Lumsden A, Krumlauf R 1992 Retinoic acid alters hindbrain *Hox* code and induces transformation of rhombomeres 2/3 into a 4/5 identity. Nature 360:737–741

McDonald CM, Jaffe KM, Shurtleff DB 1986 Assessment of muscle strength in children with meningomyelocele: accuracy and stability of measurements over time. Arch Phys Med Rehabil 67:855–861

McDonald CM, Jaffe KM, Shurtleff DB, Menelaus MB 1991 Modifications to the traditional description of neurosegmental innervation in myelomeningocele. Dev Med Child Neurol 33:473–481

Morriss-Kay GM, Murphy P, Hill RE, Davidson D 1991 Effects of retinoic acid excess on expression of *Hox-2.9* and *Krox-20* and on morphological segmentation of the hindbrain of mouse embryos. EMBO (Eur Mol Biol Organ) J 10:2985–2995

Napoli JL, Posch KC, Fiorella PD, Boerman MHEM, Salerno GH, Burns RD 1993 Roles of cellular retinol binding protein and cellular retinoic acid binding protein in the metabolic channelling of retinoids. In: Livrea MA, Packer L (eds) Retinoids: progress in research and clinical applications. Marcel Dekker, New York, p 29–48

Ruberte E, Dollé P, Chambon P, Morriss-Kay G 1991 Retinoic acid receptors and cellular retinoid binding proteins. II. Their differential pattern of transcription during early morphogenesis in mouse embryos. Development 111:45–60

Ruberte E, Friederich V, Morriss-Kay G, Chambon P 1992 Differential distribution patterns of CRABP I and CRABP II transcripts during mouse embryogenesis. Development 115:973–987

Sharrard WJW 1964 The segmental innervation of the lower limb muscles in man. Ann R Coll Surg Engl 35:106–122

Simeone A, Acampora D, Arcioni L, Andrews PW, Boncinelli E, Mavilio F 1990 Sequential activation of *HOX2* homeobox genes by retinoic acid in human embryonal carcinoma cells. Nature 346:763–766

Teillet M-A, Kalcheim C, Le Douarin N 1987 Formation of the dorsal root ganglia in the avian embryo: segmental origin and migratory behavior of neural crest progenitor cells. Dev Biol 120:329–347

Wilkinson DG, Bhatt S, Chavrier P, Bravo R, Charnay P 1989a Segment-specific expression of a zinc-finger gene in the developing nervous system of the mouse. Nature 337:461–464

Wilkinson DG, Bhatt S, Cook M, Boncinelli E, Krumlauf R 1989b Segmental expression of *Hox-2* homeobox-containing genes in the developing mouse hindbrain. Nature 341:405–409

Neurulation in the normal human embryo

Ronan O'Rahilly and Fabiola Müller

Institut für Anatomie und Spezielle Embryologie, Universität Freiburg, Freiburg, Switzerland

Abstract. The neural groove and folds are first seen during stage 8 (about 18 postovulatory days). Two days later (stage 9) the three main divisions of the brain, which are not cerebral vesicles, can be distinguished while the neural groove is still completely open. Two days later (stage 10) the neural folds begin to fuse near the junction between brain and spinal cord, when neural crest cells are arising mainly from the neural ectoderm. The rostral (or cephalic) neuropore closes within a few hours during stage 11 (about 24 days). The closure is bidirectional; it takes place from the dorsal and terminal lips and may occur in several areas simultaneously. The two lips, however, behave differently. The caudal neuropore takes a day to close during stage 12 (about 26 days) and the level of final closure is approximately at future somitic pair 31, which corresponds to the level of sacral vertebra 2. At stage 13 (4 weeks) the neural tube is normally completely closed. Secondary neurulation, which begins at stage 12, is the differentiation of the caudal part of the neural tube from the caudal eminence (or end-bud) without the intermediate phase of a neural plate.

1994 Neural tube defects. Wiley, Chichester (Ciba Foundation Symposium 181) p 70–89

The significance of the neural folds and neural tube has been clarified gradually during the past few centuries, as recounted by Adelmann (1966). Although Malpighi discovered the neural folds of the chick embryo in the 17th century, their import escaped him. He thought that they 'guarded' the beginnings of the brain and spinal cord, which lay between them. In addition, he recognized the rostral neuropore as an 'apical opening' for entrance of fluid. It was Rusconi in 1826 who appreciated that the neural folds of the frog actually represent the future brain and spinal cord. Von Baer understood that the neural folds form a tube, but, in contrast to Rusconi, he believed that the nervous system is delaminated from the neural folds. For the chick embryo, this view was corrected by Reichert in 1840 and Remak showed conclusively in 1855 that the neural plate is the actual primordium of the central nervous system. The stage was then set for a more complete study of neurulation.

Neurulation is the formation of the neural tube in the embryo. The rostral portion of the tube then develops into the brain, whereas the caudal portion becomes the spinal cord. It is to be stressed at the outset that the terms anterior and posterior are not appropriate during early development. Neurulation is of two types: primary, in which the neural plate is a major feature, and secondary, which occurs within the caudal eminence.

Primary neurulation

Primary neurulation (Fig. 1), the folding of the neural plate to form the neural tube, has been studied extensively in the human embryo. The essentials were outlined by Streeter (1912), who later (1927) expressed a justifiable doubt concerning the existence of the so-called cerebral vesicles. The following account is arranged according to the internationally accepted system of staging for the human embryo (O'Rahilly & Müller 1987a). The ages given are those commonly used in human embryology and are postovulatory, that is, taken from the last ovulation, which is very close to fertilization. They are approximate, however, and those for some of the earlier stages may be slightly on the low side (Dickey & Gasser 1993). Moreover, variability may be greater than has previously been recognized (Shiota 1991).

Stage 7 (approximately 16 days). Although no histological or other distinctions have so far been recorded, it is possible, on the basis of autoradiographic studies of the chick embryo, to estimate the general position of the neural plate on the dorsal aspect of the human embryonic disc (O'Rahilly & Müller 1987b).

Stage 8 (approximately 18 days). The neural groove, which is the first morphological manifestation of the nervous system, is present in one quarter of embryos of this stage (O'Rahilly & Müller 1981). Such embryonic discs are 1 mm or more in length and 0.6 mm or more in width; they possess a primitive streak of nearly 0.3 mm or more in length and a notochordal process of 0.4 mm or greater. In other words, the neural groove seems to be present only when a certain degree of size and maturity has been attained. The future dorsal lamina of the neural tube (the alar plate) is situated more laterally in the neural plate, whereas the future ventral lamina (the basal plate) is placed more medially (Fig. 1). The neural folds at this time are very largely cerebral rather than spinal.

Stage 9 (approximately 20 days). Somites first become visible at this stage, although they belong entirely to the occipital region. The embryo is longer and the neural groove is deeper. It needs to be stressed that the three major divisions of the brain (prosencephalic, mesencephalic and rhombencephalic), which are not so-called cerebral vesicles, can be identified in the completely unfused neural folds before any portion of the neural tube has formed (Müller & O'Rahilly 1983).

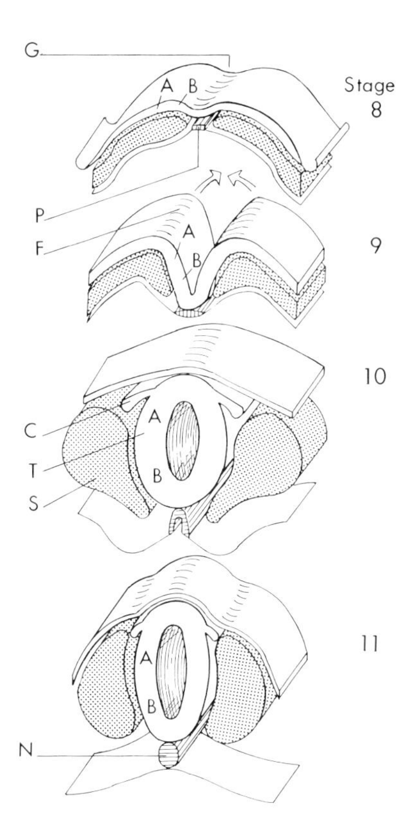

FIG. 1. Schematic representation of primary neurulation from 2½ to 3½ weeks (stages 8 to 11). These transverse sections, which are based on photomicrographs, show the conversion of the neural groove (G) (stage 8) and neural folds (F) (stage 9) into the neural tube (T) (stage 10). A indicates the future alar plate (the dorsal lamina) and B the future basal plate (the ventral lamina) of the neural tube. The neural crest (C) is shown arising from neural ectoderm at stages 10 and 11, at which stages somites (S) are visible on each side of the tube. N, notochord; P, notochordal plate. From O'Rahilly & Müller (1994), by permission of Wiley-Liss, New York. Copyright ©1993.

Neural crest has been recorded at the neurosomatic junction and rostral to the otic disc (in the region of the midbrain) in two embryos of this stage (Müller & O'Rahilly 1983), but its characteristic development occurs in the following stage.

Inductional processes between the future notochord and the neural tube would likely be greatest during stages 8 and 9, when the future notochord is present as a plate and hence presents a relatively wide surface. The contact is particularly close at stages 8–10 because, as has been shown in the macaque, a basement membrane is lacking between the two tissues.

Stage 10 (approximately 22 days). When about six somitic pairs are present, the neural tube begins to form by fusion of the neural folds (O'Rahilly & Gardner 1979). The site of the initial fusion is rhombencephalic, upper cervical or both, that is, at the cerebrospinal junction. The forebrain now consists largely of two diencephalic subdivisions: D1, which is the site of the optic primordia, and D2, which is the future thalamic region (Fig. 2). It is not generally appreciated, however, that, in embryos of 10 somitic pairs, the forebrain already begins to include a telencephalic portion (Müller & O'Rahilly 1985), which can be discerned more clearly within a few days (Fig. 2C).

The neural crest continues to form and, in the head, probably reaches its maximum. The preotic (mesencephalic and trigeminal) crest develops from the neural folds bounding the rostral neuropore. Further caudally, however, where the neural tube has formed, crest (associated with cranial nerves 7–8, 9, 10, 11, 12 and with future spinal ganglia) also arises, probably mainly, if not entirely, from the neural ectoderm.

Stage 11 (approximately 24 days). The two ends of the neural groove that still remain open are termed the rostral (or cephalic) and the caudal neuropore (Nomina embryologica 1989). Towards the end of this stage, when some 20 somitic pairs are present, the rostral neuropore closes (O'Rahilly & Gardner 1979) within a few hours. The rostral end of the neural tube is completed (from

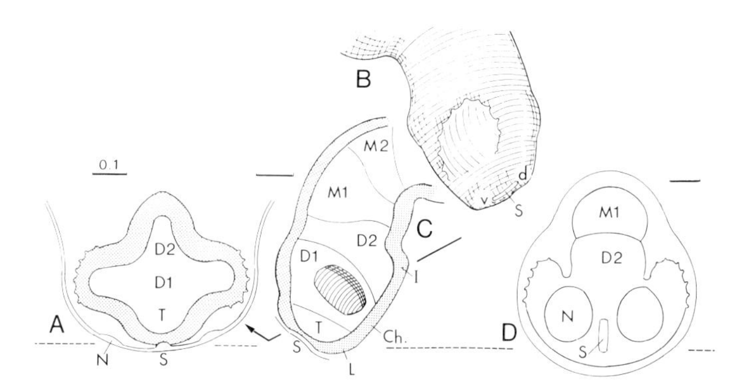

FIG. 2. The rostral end of the neural tube at 3½ weeks (stage 12, 22–23 somitic pairs, embryo No. 8943)). A is an oblique section through the optic vesicles and the situs neuroporicus (S). B is a right lateral view showing the situs. The optic vesicle resembles a frightened hedgehog because of its surrounding optic neural crest. C is the right half of the brain as seen on median section. The opening leading to the optic ventricle is shaded. The oblique line indicates the plane of section A. D is a 'head-on' view showing the situs neuroporicus between the nasal discs. Ch., chiasmatic plate; d, dorsal lip of neuropore; D1 and D2, diencephalic neuromeres; I, infundibular region; L, lamina terminalis; M1 and M2, mesencephalic neuromeres; N, nasal disc; T, telencephalon medium; v, ventral lip of neuropore. The bars represent 0.1 mm.

the right and left neural folds) by fusion that proceeds from the region of the midbrain and diencephalon (D2) and simultaneously from the telencephalic region immediately adjacent to the chiasmatic plate (Müller & O'Rahilly 1986). These two sites of bidirectional closure of the neural tube are termed, respectively, the dorsal and the terminal lip of the neuropore (O'Rahilly & Müller 1989a). It is entirely possible, however, that in the region of the dorsal lip, more than two areas of the neural folds are involved.

The two lips (Fig. 2B) behave differently. At the dorsal lip, fusion of the surface epithelium seems to precede that of the neural epithelium. At the terminal lip, however, fusion of the surface epithelium and fusion of the neural epithelium seem to take place simultaneously. The surface epithelium is also slightly thicker than that of the dorsal lip, so that the seam is believed to be stronger at the terminal lip. It is of interest that, in an example of complete dysraphia in an embryo of stage 11 (Dekaban & Bartelmez 1964), closure had occurred at the terminal, but not at the dorsal, lip.

The median fusion at the terminal lip results in the appearance of the embryonic lamina terminalis, accounting at least in part for the increase in length of the developing prosencephalic floor. The fusion takes place between the right and left neural folds, as is characteristic of primary neurulation elsewhere. Because this fusion begins immediately rostral to the chiasmatic plate, the rostral end of the neural plate in the median plane is, in the human, at the preoptic recess.

It is worth noting that, in future anencephaly, fusion at the terminal lip of the rostral neuropore to form the lamina terminalis may not be disturbed (Müller & O'Rahilly 1984), so that the abnormal opening would be in the region of the weaker dorsal lip. It is possible, therefore, that the histological differences between the two lips during fusion is associated with a difference in teratogenic susceptibility.

Occlusion of the lumen of the neural tube has been recorded in the chick embryo; it is thought to have a possible role in enlargement of the brain. The situation in the human, however, is different (Müller & O'Rahilly 1986). In stage 11 embryos with an open rostral neuropore, occlusion (at the somitic level) is the exception. In stage 11 embryos with a closed rostral neuropore, no occlusion is found. Moreover, far from showing an enlargement, the brain is not growing at stage 11.

Neural crest is still being given off from rhombomeres 4, 6 and 7, but these now belong to the closed portion of the neural tube. These crest cells are derivatives of the neural ectoderm. Optic crest and spinal crest are also evident.

The neurenteric canal, which appears first during stage 8, is a more or less vertical (perpendicular to the embryonic disc) passage between the amniotic cavity and the umbilical vesicle (or so-called yolk sac) (Fig. 3). It may be regarded as the remains of the notochordal canal at the level of the primitive node. It persists in some embryos of stages 9 and 10. Even after it becomes closed,

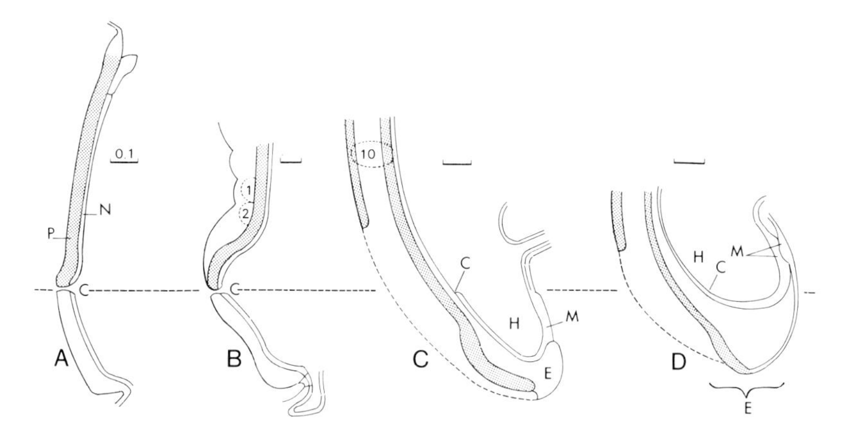

FIG. 3. Median sections of the caudal end of the embryo from 2½ to 3½ weeks (stages 8 to 11). A (stage 8, embryo No. 10 157, which possesses a neural groove) shows the neurenteric canal (C) in the region of the primitive node. The area rostral to the notochordal process is the prechordal plate. B (stage 9, embryo Da 1, two somitic pairs) includes the position of somites 1 and 2. The primitive streak is the wider white band below the dashed line in A and B. C (stage 10, embryo No. 5074, 10 somitic pairs) shows the caudal neuropore (dashed line). The site of the neurenteric canal (C) can be discerned. The caudal eminence (E) has already replaced the primitive streak. In D (stage 11, embryo No. 7611, 16 somitic pairs) the caudal neuropore is shorter. The wall of the neural fold or of the neural tube is shaded in all the drawings. C, neurenteric canal or its site. H, hindgut; M, cloacal membrane; N, notochordal process; P, neural plate. 1, 2, 10 indicate somites. The bars represent 0.1 mm.

its site is generally revealed (Fig. 3C,D) by an abrupt increase in thickness of the endoderm of the roof of the alimentary canal, as can be seen in some embryos of stage 11 (Müller & O'Rahilly 1986). Its level varies from somitic pair 12 to somitic pair 21.

The closed neural tube is surrounded by a basement membrane, which has been shown in the macaque to be rich in fibronectin and laminin, except in areas of active neural crest emigration. A complete basement membrane does not form over the dorsal portion of the neural tube until neural crest emigration has ended. The notochord *sensu stricto* appears first during this stage and its basement membrane is in contact with that of the neural tube.

Stage 12 (*approximately 26 days*). Early in stage 12 the site of the former rostral neuropore, the situs neuroporicus, may still be visible because of a connection between the surface ectoderm and the neural ectoderm. This bridge indicates the commissural plate (Fig. 4).

The optic neural crest is now at its maximum and covers the optic vesicle, which, in the felicitous phrase of Bartelmez & Blount (1954), resembles a

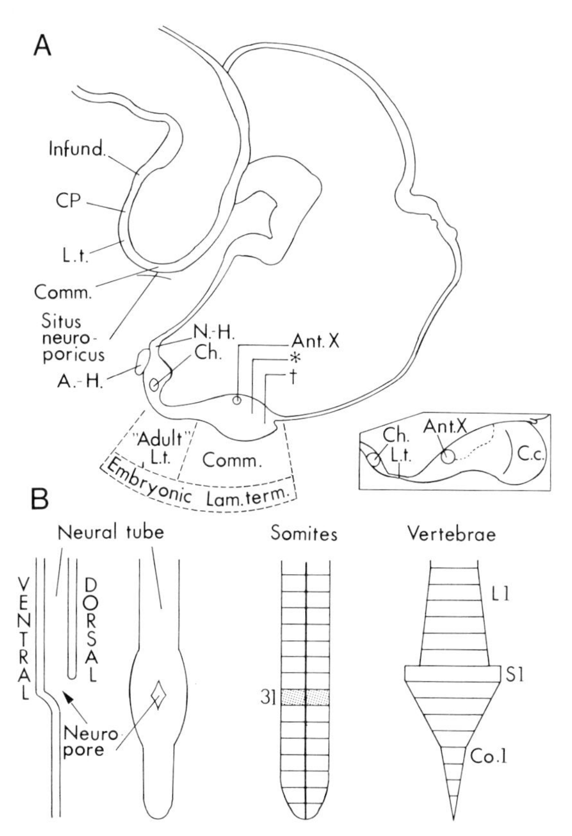

FIG. 4. The sites of the neuropores. A, median sections of the brain at 3½ (stage 12, embryo No. 7852) and 8 (stage 23, embryo No. 9226) postovulatory weeks. The site of the rostral neuropore is still visible in some embryos of stage 12. In later stages it probably corresponds to an area (asterisk) within the commissural plate, between the anterior commissure, present in anencephaly, and the future corpus callosum (dagger), absent in anencephaly. The embryonic lamina terminalis includes more than does the adult lamina. The inset (based on Hochstetter 1919) shows the main commissural region at 13 postovulatory weeks. B, schematic median and dorsal views of the caudal neuropore at 3½ weeks, and schemes to show the somites and the adult vertebrae. The neuropore closes at the level of somitic pair 31 (stippled), which corresponds approximately to sacral vertebra 2. A.-H., adenohypophysis; Ant. X, anterior commissure; C.c., corpus callosum; Ch. optic chiasma; Co. 1, coccygeal vertebra 1; Comm., commissural plate; CP, chiasmatic plate; Infund., infundibular region; L1, lumbar vertebra 1; L.t., lamina terminalis; N.-H., neurohypophysis; S1, sacral vertebra 1.

'frightened hedgehog'. Mesencephalic crest is generally not as clear at this stage because the cells have already migrated.

The caudal neuropore is clearly visible during stages 10 and 11 (Fig. 3C,D). During stage 12, generally when about 25 somitic pairs are present, the caudal neuropore closes (O'Rahilly & Gardner 1979) in a rostrocaudal direction. This process takes approximately one day to be completed. Once the neuropore is closed, the entire caudal end of the embryo is covered by surface ectoderm. The level of final closure of the caudal neuropore can be assessed by allowing space for additional somites in embryos of 25 pairs (Müller & O'Rahilly 1987). The result (Figs. 4B, 5A) is future somitic pair 31 (or 30/31), which corresponds to future sacral vertebra 2. This level is probably equivalent to that of the junction between Holmdahl's (1926, 1933, 1935) *primäre Körperentwicklung* und *sekundäre Körperentwicklung*.

Stage 13 (approximately 28–32 days). The embryo now possesses 30 or more somitic pairs and is about 5 mm in length. The neural tube is normally a closed system and it is filled with 'ependymal fluid' (Müller & O'Rahilly 1988) because the choroid plexuses will not develop for at least another fortnight. Once the neural tube has closed, its walls are subject to the pressure of the contained fluid (provided that formation of fluid is greater than absorption). The ependymal fluid, which is no longer in continuity with that in the amniotic cavity, may result in rostrocaudal enlargement and widening of the brain. Such a mechanism would be expected to contribute to the shaping of the neural tube and to preventing it from collapsing, although the chief factor in its growth is mitotic activity. Because of the absence of communication with the amniotic cavity, diffusion of α-fetoprotein into the liquor amnii ceases, except in such conditions as spina bifida aperta and anencephaly.

The spinal neural crest is beginning to be segregated and the spinal ganglia are clearly in series with the somites. In addition, terminal-vomeronasal crest is arising from the nasal discs.

Secondary neurulation

Secondary neurulation is the continuing formation of the spinal cord without direct involvement of the ectoderm, without the intermediate phase of a neural plate. It begins once the caudal neuropore has closed during stage 12.

The caudal eminence (or end-bud), which is already recognizable at stages 9 and 10, is an ectoderm-covered mass of pluripotent tissue (Fig. 3D). During stages 12 and 13 (Fig. 5), it gradually replaces the primitive streak, which appeared during stage 6. It is situated between the site of the neurenteric canal and the cloacal membrane at stage 11 (Fig. 3D) or between the caudal border of the caudal neuropore and the cloacal membrane at stage 12 (Fig. 5A). The caudal eminence provides structures comparable to those formed more rostrally from the three germ layers. Its derivatives include the caudal portions of the digestive tube, coelom, blood vessels, notochord, somites (represented at first by paraxial condensations) and spinal cord.

At stage 12 the caudal eminence gives rise to a solid cellular mass known as the neural cord (Fig. 5A); this forms the nervous system of the caudal part of the body. The cavity (central canal) of the spinal cord that is already present at more rostral levels extends into the neural cord in continuity. Isolated spaces have not been seen during cavitation within the neural cord.

At stage 13 the caudal eminence continues as the site of secondary neurulation (Fig. 5B). The neural cord extends to the caudal tip of the eminence and is in contact with the clearly delineated surface ectoderm. After stage 13 the caudal eminence is less noticeable, but its activity continues for at least several more stages.

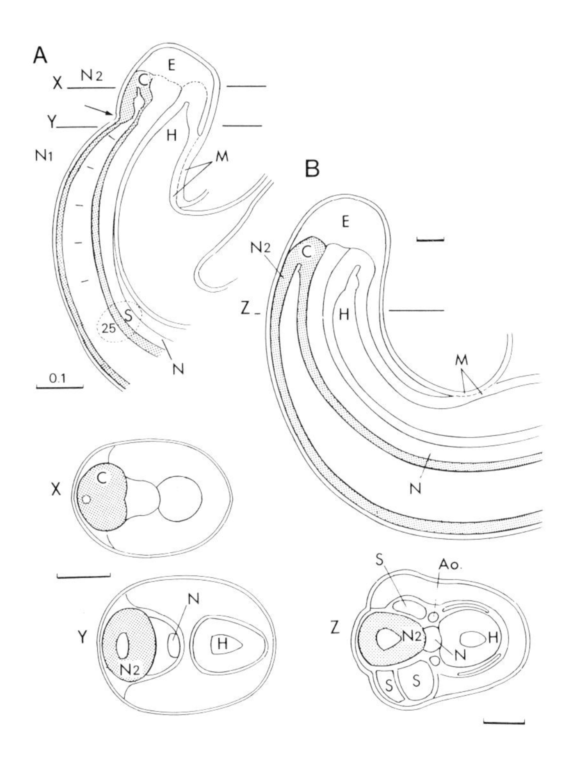

FIG. 5. Median and transverse sections of the caudal end of the embryo at 3½ and 4 weeks (stages 12 and 13). A (embryo No. 1062, 29 somitic pairs) shows the site of closure of the caudal neuropore, marked by an arrow. Somite 25 is identified and the positions of more caudal somites (including future No. 30) are indicated. The neuropore has closed opposite future somitic pair 31. Section X (embryo No. 8505a, 24 somitic pairs) is through the neural cord (C), which shows a small canal. Section Y shows spinal cord being formed by secondary neurulation, and also notochord (N) and hindgut (H) developing from the caudal eminence (E). B (embryo No. 9297, 33 somitic pairs) shows the caudal region at stage 13. The somites have not been included. Section Z (No. 8967) shows spinal cord, notochord and hindgut, and also right and left somitic plates that unite ventral to the hindgut. The neural cord and the wall of the neural tube are shaded. Ao., caudal aorta; M, cloacal membrane; N1 and N2, primary and secondary neurulation; S, somite or somitic plate.

The caudal eminence gives rise to at least somitic pair 32 and the following pairs. The mesenchyme for somitic pairs 30–34 is the material for sacral vertebrae 1–5; it is laid down during stages 12 and 13. A localized defect within the caudal eminence may result in sacral agenesis, although a later exaggeration of the normal process of caudal regression is another possible cause. Complete deficiency of the caudal eminence would result in cloacal deficiency and a consequent approximation of the two lower limb buds, characteristically symmelia. Such so-called fusions of the lower limbs are, in a certain sense, comparable to cyclopia (O'Rahilly & Müller 1989b).

In summary, primary neurulation is normally complete at 4–5 postovulatory weeks, whereas secondary neurulation probably continues until at least 7 weeks.

At the end of the embryonic period proper the spinal cord still reaches to the caudal end of the vertebral column (O'Rahilly et al 1980) (Fig. 4). Some dedifferentiation then takes place caudally and the spinal cord 'ascends' to a lumbar level during the first half of prenatal life.

The coverings of the neural tube

Because of the importance of the meningeal and skeletal coverings of the neural tube in various defects, their development will be summarized here.

The development of the cranial meninges was investigated by Hochstetter (1939) among others. Their complicated origin from several sources (such as prechordal plate, parachordal mesoderm, neural crest) was studied by O'Rahilly & Müller (1986). The loose mesenchyme around most of the brain at 5 weeks (stage 15) can be termed primary meninx. At 6 weeks (stage 17), the dural limiting layer begins to form basally and the skeletogenous layer of the head becomes visible. At 7 weeks (stage 19), pachymeninx and leptomeninx are distinguishable in the head.

The spinal meninges have been studied particularly by Hochstetter (1934) and Sensenig (1951). The future pia mater appears as neural crest cells by stage 11 and at 5 weeks (stage 15) the primary meninx is represented by a loose zone between the commencing vertebrae and the neural tube. After 6 weeks (stage 18) the mesenchyme adjacent to the vertebrae becomes condensed to form the dural lamella. At 8 weeks (stage 23) the dura completely lines the wall of the vertebral canal and the future subarachnoid space is becoming cell free. The arachnoid, however, for which an artifact is frequently mistaken, does not appear until either the third trimester or postnatally.

The development of the vertebrae has been studied extensively. Sensenig (1949) has shown that the first indication of clearly defined vertebrae is found at 5 weeks (stage 15) and chondrification can be observed at 6 weeks (stage 17). At that time a condensation for the base of the skull (chondrocranium) also appears, and both the vertebrae and the cranial base develop in close relationship with the notochord. At 8 weeks (stage 23) some 33 cartilaginous vertebrae are arranged in a flexed column approximately 33 mm in length. They lack spinous processes, however, and hence show a normal rhachischisis totalis (O'Rahilly et al 1980).

Ossification is detectable at about 9 postovulatory weeks. The neural processes (of the future vertebral arch) begin to unite and a cartilaginous spinous process begins to appear during the second trimester.

A concluding note on neuroteratology

A detailed study of normal development in staged human embryos enables some tentative comments to be made concerning the timing of the first morphological appearance of several abnormalities (Müller et al 1986). Thus, myelomeningocele involves the neural tube and probably appears at stages 10 or 11 in the cervical and thoracic regions. Lumbosacral myeloschisis, which is regarded by many as a precursor of myelomeningocele, begins also during primary neurulation (stages 8 to 12), although it is not a simple failure of neural closure. An example of caudal myeloschisis at stage 14 has been described by Lemire et al (1965). The lesion occurred before closure of the neural tube, defects in the basement membrane of the neural ectoderm were noted, and the condition was considered to be the forerunner of lumbosacral myelomeningocele.

It has been emphasized by Lemire et al (1975) that lumbosacral lesions covered by intact skin appear during secondary neurulation (stages 12 to 20). However, the conditions that occur in the transitional area of primary and secondary neurulation are not well understood.

The appearance of spina bifida occulta extends from the first sign of vertebrae (stage 15) until completion of the neural arches during fetal life. However, a radiologically detectable failure in the completion of one or more neural arches can be merely a dorsomedian lack of ossification in the lumbosacral region, which is a normal finding in children of two years of age and is found in the sacrum in one-fifth of the adult population. Meningocele probably occurs late in the embryonic (stages 18 to 23) or early in the fetal period.

It is to be stressed that the stages suggested above as being important in the origin of various abnormalities are tentative. The retrospective transfer of embryological timetables to teratogenesis is, as frequently stressed by Josef Warkany, hazardous.

A final point of embryological interest is the estimation that the prevalence of neural tube defects decreases from 2.5% of embryos at stage 12 to 0.06% at term, so that most embryos with neural tube defects are lost before birth and probably mainly within the embryonic period proper, that is, during the first 8 postovulatory weeks (Shiota 1991).

References

Adelmann HB 1966 Marcello Malpighi and the evolution of embryology, vol 2. Cornell University Press, Ithaca, NY

Bartelmez GW, Blount MP 1954 The formation of neural crest from the primary optic vesicle in man. Contrib Embryol Carnegie Inst 35:55–71

Dekaban AS, Bartelmez GW 1964 Complete dysraphism in a 14 somite human embryo. A contribution to normal and abnormal morphogenesis. J Neuropathol & Exp Neurol 22:533–548

Dickey RP, Gasser RF 1993 Ultrasound evidence for variability in the size and development of normal human embryos before the tenth post-insemination week after assisted reproductive technologies. Hum Reprod 8:331–337

Hochstetter F 1919 Beiträge zur Entwicklungsgeschichte des menschlichen Gehirns. I. Deuticke, Vienna

Hochstetter F 1934 Über die Entwicklung und Differenzierung der Hüllen des Rückenmarkes beim Menschen. Morphol Jahrb 74:1–104

Hochstetter F 1939 Über die Entwicklung and Differenzierung der Hüllen des menschlichen Gehirnes. Morphol Jahrb 83:359–494

Holmdahl E 1926 Die erste Entwicklung des Körpers bei den Vögeln und Säugetieren, inkl. dem Menschen, besonders mit Rücksicht auf die Bildung des Rückenmarks, des Zöloms und der entodermalen Kloake nebst einem Exkurs über die Entstehung der Spina bifida in der Lumbosakralregion. II–V. Morphol Jahrb 55:112–208

Holmdahl E 1933 Die zweifache Bildungsweise des zentralen Nervensystems bei den Wirbeltieren. Eine formgeschichtliche und materialgeschichtliche Analyse. Arch Entwicklungsmech Org (Wilhelm Roux) 129:206–254

Holmdahl E 1935 Primitivstreifen bzw. Rumpfschwanzknospe im Verhältnis zur Körperentwicklung. Z Mikrosk Anat Forsch 38:409–440

Lemire RJ, Shepard TH, Alvord EC 1965 Caudal myeloschisis (lumbo-sacral spinal bifida cystica) in a five millimeter (horizon XIV) human embryo. Anat Rec 152:9–16

Lemire RJ, Loeser JD, Leech RW, Alvord EC 1975 Normal and abnormal development of the human nervous system. Harper & Row, Hagerstown, MD

Müller F, O'Rahilly R 1983 The first appearance of the major divisions of the human brain at stage 9. Anat Embryol 168:419–432

Müller F, O'Rahilly R 1984 Cerebral dysraphia (future anencephaly) in a human twin embryo at stage 13. Teratology 30:167–177

Müller F, O'Rahilly R 1985 The first appearance of the neural tube and optic primordium in the human embryo at stage 10. Anat Embryol 172:157–169

Müller F, O'Rahilly R 1986 The development of the human brain and the closure of the rostral neuropore at stage 11. Anat Embryol 175:205–222

Müller F, O'Rahilly R 1987 The development of the human brain, the closure of the caudal neuropore, and the beginning of secondary neurulation at stage 12. Anat Embryol 176:413–430

Müller F, O'Rahilly R 1988 The development of the human brain from a closed neural tube at stage 13. Anat Embryol 177:203–224

Müller F, O'Rahilly R, Benson DR 1986 The early origin of vertebral anomalies, as illustrated by a 'butterfly vertebra'. J Anat 149:157–169

Nomina embryologica 1989 In: Nomina anatomica, 6th edn. Churchill Livingstone, Edinburgh

O'Rahilly R, Gardner E 1979 The initial development of the human brain. Acta Anat 104:123–133

O'Rahilly R, Müller F 1981 The first appearance of the human nervous system at stage 8. Anat Embryol 163:1–13

O'Rahilly R, Müller F 1986 The meninges in human development. J Neuropathol & Exp Neurol 45:588–608

O'Rahilly R, Müller F 1987a Developmental stages in human embryos including a revision of Streeter's 'horizons' and a survey of the Carnegie collection. Carnegie Institution of Washington, Washington, DC (publ 637)
O'Rahilly R, Müller F 1987b The developmental anatomy and histology of the human central nervous system. In: Vinken PJ, Bruyn GW, Klawans HL (eds) Handbook of clinical neurology. Elsevier Science Publishers, Amsterdam, p 1–17
O'Rahilly R, Müller F 1989a Bidirectional closure of the rostral neuropore in the human embryo. Am J Anat 184:259–268
O'Rahilly R, Müller F 1989b Interpretation of some median anomalies as illustrated by cyclopia and symmelia. Teratology 40:409–421
O'Rahilly R, Müller F 1994 The embryonic human brain: an atlas of developmental stages. Wiley, New York
O'Rahilly R, Müller F, Meyer DB 1980 The human vertebral column at the end of the embryonic period proper. I. The column as a whole. J Anat 131:565–575
Sensenig EC 1949 The early development of the human vertebral column. Contrib Embryol Carnegie Inst 33:21–41
Sensenig EC 1951 The early development of the meninges of the spinal cord in human embryos. Contrib Embryol Carnegie Inst 34:145–157
Shiota K 1991 Development and intrauterine fate of normal and abnormal human conceptuses. Congenital Anom 31:67–80
Streeter GL 1912 The development of the nervous system. In: Keibel F, Mall FP (eds) Manual of human embryology. Lippincott, Philadelphia, PA, vol 2:1–156
Streeter GL 1927 Archetypes and symbolism. Science 65:405–412

DISCUSSION

Shum: You have shown that in human embryos the lumina formed by primary and secondary neurulation join end to end. In chick embryos (Schoenwolf 1978), there is an overlap zone where primary neural tube formation occurs dorsally while the secondary neural tube forms ventrally. Lemire (1969) showed some human embryos that had more than one neural lumen in the transitional area. Does this suggest that an overlap zone may be present in human embryos as in the chick?

O'Rahilly: We have never seen an overlap zone. We are amazed at the way two such different processes unite the way they do. I think the presence of the junctional area, which is essentially lumbosacral, and the fact that these two processes join in the lumbosacral region might have something to do with the frequency of lesions in that region.

Stanley: Are there any sex-specific differences in neurulation? There is a widely and consistently observed higher frequency of upper spinal cord lesions in females, which may have resulted from sex-specific differences in primary or secondary neurulation.

O'Rahilly: There is no work on the human embryo that would address this. We have found slight differences in the embryonic length (nothing to do with neurulation) at about 7–8 postovulatory weeks: female embryos were about 1 mm shorter than male embryos (O'Rahilly & Muller 1984).

Hall: Was the sex determined externally or chromosomally?

O'Rahilly: By this time you can tell from the gonads.

Morriss-Kay: I have never sexed mouse embryos at that stage, apart from some we have been studying that have chromosomal abnormalities—trisomy 12 and 19 (Putz & Morriss-Kay 1981). We didn't see any differences in those.

Seller: In the mouse, female embryos grow more slowly at this stage than do male ones (Seller & Perkins-Cole 1987). This is true from cleavage onwards. There is no reason humans should be different.

Copp: I agree that male embryos are larger at this stage, but if you look at the rate of development or the rate of growth during neurulation, there is no difference between the sexes (F. A. Brook & A. J. Copp, unpublished). The growth advantage of the male embryos originates earlier, probably before or soon after implantation. From beginning to end of neurulation in the mouse, there is no difference in rate of development or growth between the sexes. So this cannot account for the predisposition of females to anencephaly that is seen in the mouse and in humans.

We have done preliminary studies on the stages of head closure, using the same staging system as Juriloff et al (1991). We found a difference between males and females in the latest stages of head closure. Females become delayed so that the final closure of the caudal part of the midbrain and hindbrain (closure 2 back towards closure 4) is retarded relative to the rest of their development (F. A. Brook & A. J. Copp, unpublished). This may or may not be related to their predisposition to anencephaly.

Hall: Doesn't closure 4 have a very different mechanism? Isn't it a membranous closure?

Juriloff: It looks quite different.

Hall: Is the sex-specific difference at that point?

Copp: In the *curly tail* mouse the exencephaly affects midbrain and hindbrain, predominantly. It is in this region that we find the difference between the sexes.

Cuckle: There have been seven epidemiological studies of spina bifida reporting on the relationship between sex ratio and position of the lesion on the spine (summarized in Cuckle 1993). The studies were all rather small and so the results may be affected by reporting bias, but taken together they indicate that for higher lesions a greater proportion of the affected children is female.

Hall: In the Sikh population in British Columbia, it looks as though there is no difference between the sexes. We have to be very careful which ethnic group we are talking about.

Mills: In clinical terms, the latest stage at which an event could occur is very important. When does secondary neurulation occur in humans? What do you think about the theory that a neural tube defect can occur after the neural tube is closed?

O'Rahilly: Secondary neurulation begins at about 3½ postovulatory weeks. It's more or less finished before the end of the embryonic period proper, which is 8 weeks.

With regard to the opening of a closed tube, we don't really see any evidence for this, in the human. We are aware that it can be produced experimentally in animals.

Hall: Isn't there evidence from Hook's article (1992) that vitamins do some good after they shouldn't?

Mills: This has been debated. In the Milunsky et al (1989) study there were many women who supposedly took vitamins very early in pregnancy who were not taking them prior to pregnancy. It is hard to imagine how that could have an effect. If these women actually began taking vitamins later in pregnancy (after 26 to 28 days post-conception when the neural tube closes), yet they had a reduced risk of having a child with a neural tube defect, it suggests that there was a beneficial effect beyond 28 days. From what you say about secondary neurulation, there might be an effect throughout the embryonic period.

O'Rahilly: We could not eliminate that as a possibility.

Morriss-Kay: Could it be that a delay in neural tube closure normally remains as an open spina bifida, but if the mother takes vitamins then there could be completion of that delayed neurulation after the normal time of neural tube closure?

O'Rahilly: It is certainly a reasonable hypothesis.

Hall: Could you say a little more about the timing? Is there a broader time frame during which neural tube defects may arise than we thought in the past? Or is there variability between embryos?

O'Rahilly: We always said that the human embryo at 8 weeks was some 30 mm in length; this has been known for a very long time (O'Rahilly & Müller 1987). Streeter (1912) based his timing (7 weeks) on the rhesus monkey; that was definitely incorrect. This issue was resolved in the 1960s when ultrasonography was developed, which showed quite clearly that at 8 weeks the embryo is about 30 mm long.

The recent introduction of a more sophisticated method, namely transvaginal sonography, has made quite a difference. We have now seen pictures that are claimed to be of a stage 6 or a stage 8 embryo. A stage 8 embryo is only 1 mm in total length; we do not believe that such embryos can be staged by ultrasonography, we have trouble staging them with the embryo in our hand. According to some authors (e.g. Dickey & Gasser 1993), however, what used to be described as a 4 week embryo may in fact be 5 weeks. For early stages, it seems that we may have been underestimating their age. Some defects, including anencephaly, probably arise extremely early, possibly at or even before stage 8.

Hall: Is there any chance that's the difference between males and females?

O'Rahilly: It's a possibility; it hasn't been considered.

van Straaten: Nevertheless, Dickey & Gasser (1993) stated that a 2 mm embryo seen by sonography was 28 days post-fertilization. They suggested that histologically treated embryos were swollen, resulting in the larger embryos we know from the Carnegie stages. This seems unlikely to me.

O'Rahilly: H. P. Robinson in the 1970s was one of the first to do these ultrasonographic studies. He spoke about shrunken embryos and swollen embryos treated histologically. I sent him a photo of a first-class 30 mm embryo.

van Straaten: What else could explain that at 28 days post-fertilization an embryo as seen by ultrasonography measures only 2 mm?

O'Rahilly: Shiota (1991) based his study on instances of a single isolated coitus. A great range in age was found for a given stage, e.g. 36–54 days for stage 18. In a better controlled study, based on *in vitro* fertilization or gamete intratubal transfer, Dickey & Gasser (1993) found less variation in age, e.g. 40–50 days for embryos of 13–17 mm (comparable to stage 18). We have been using approximately 44 days for stage 18. According to studies with transvaginal sonography (Dr Josef Wisser, personal communication 1992), embryos of stages 6 to 13 may commonly be 3–6 days older than the embryological norms (O'Rahilly & Müller 1987).

Lindhout: If you want to know the normal time schedule of these embryonic developments, it's very important to know how normal your embryos are. What was the source of your embryos? If you are studying very early embryos from miscarriages, we know that more than 50–60% of very early embryos have chromosomal abnormalities.

O'Rahilly: The embryos came from spontaneous abortions and from hysterectomies.

Lindhout: If you compare these two groups, do you find differences?

O'Rahilly: We have not compared the two groups as such, but we find no differences at all. We are less concerned with the source than with the histological quality. When we see embryos that are histologically superior to most published photomicrographs I have seen of chick and mouse, we know we have excellent material. We have no information about the chromosomes.

Lindhout: If the embryos come from hysterectomies, how can you say that they are so many days postovulatory? If there was an hysterectomy, presumably it was accepted that the woman was not pregnant, implying an unknown duration of pregnancy.

O'Rahilly: We no longer date individual embryos, we stage them with the Carnegie morphological staging system (O'Rahilly & Müller 1987). Then, on the assumption that the embryo is normal (which can be disputed, but if we can find nothing that looks abnormal, we call it normal), we stage it and say it should be of a certain age (O'Rahilly & Müller 1987). Now we have the ultrasonographic evidence, in general, the data have held up very well. A 30 mm embryo has been shown by various techniques (including ultrasound) to be commonly 8 weeks after fertilization.

Lindhout: So if there is discrepancy of 2–3 days between groups and your timing is based on morphology, conclusions can only be drawn safely when an embryo has only a few abnormalities with otherwise proportionate morphology.

Opitz: Usually, aneuploidy or polyploidy in human embryos, where you have the opportunity of sectioning, produce dissynchronies. The systems don't all go through their appointed morphological stages at synchronous times. That's the easiest way of telling whether or not an embryo is aneuploid.

O'Rahilly: George Streeter took embryos of the Carnegie collection at stage 18 (about 44 days) and measured the length of the paramesonephric ducts (future uterine tubes). He was able to assess the number of semicircular ducts that had been freed in the internal ear. If the paramesonephric duct was 0.2 to 0.4 mm long, one semicircular duct was established, if 0.4 to 0.7 mm, two ducts, and if 0.8 to 1.1 mm, all three were formed. Hence, from the status of the paramesonephric ducts, the degree of development of an organ as distant as the internal ear can be estimated.

Hall: It would be possible now with *in situ* hybridization to go back and look at a cell and tell whether it was trisomic. That would be very reassuring.

Scott: I have a question related to nutrient supply during the whole process of development. Is there a difference during this period of development in the embryonic circulation? I ask because it is important to know how folic acid affects the embryo.

O'Rahilly: Yes, there is. When speaking of the supply of nutrients to the human embryo, one has to specify the stage. For the zygote, the only source is the liquid in the uterine tube and then that in the uterine cavity. During implantation the trophoblastic lacunae communicate with the endometrial vessels (stage 5) and this is the basis of the uteroplacental (haemochorial) circulation. Blood islands and blood vessels appear in the wall of the umbilical vesicle (so-called yolk sac) at about 18 days, the heart and embryonic vessels begin to develop within another two days, and cardiac contraction is believed to start at 3 weeks. Coelomic fluid provides a nutrient supply and an ebb-and-flow movement occurs in the blood vessels. Finally, at about 26 days (stage 12), a connected circulatory system with a unidirectional flow is present. The functional activity of the future placenta is well established by 4 weeks.

Do we know where folic acid acts? Could there be folic acid in the liquid in the uterine cavity? Does it get into everything?

Scott: One would anticipate it would.

O'Rahilly: It could affect the embryo even before there is any real circulation.

Seller: During neurulation, the embryonic nutrition is histiotrophic rather than haemotrophic. In rabbit, vitamin B_{12} does get into the uterine secretions (Jacobson & Lutwak-Mann 1956), but I don't know whether any studies have been done on humans.

Hall: There were some studies on amniotic fluid.

Seller: That's later.

Morriss-Kay: I learned my embryology from Hamilton (Boyd) and Mossman (Hamilton & Mossman 1972). They have a very clear picture of a 7-somite-stage embryo (22 days postovulation), which you would call stage 10, which has a complete circulation with a single aortic arch continuous with two dorsal aortae running down to supply the yolk sac and via the umbilical arteries to the placenta (Fig. 174, p 231). They state (p 99) that there is a simple circulatory system within the embryo, yolk sac, connecting stalk (primitive umbilical cord) and chorion at about 20 days (see also Fig. 97, p 105), i.e. before neural tube closure starts.

O'Rahilly: It's possible, but the vascular system is incomplete at stage 10, and it is believed that any movement of blood would be at most ebb-and-flow.

Morriss-Kay: How do you know that? At that stage in rat and mouse embryos, there is exactly the same blood system and it is circulating, you can see it in embryos in culture.

Copp: The circulation is going through the yolk sac and histiotrophic nutrition is via the yolk sac. Is the secondary yolk sac an organ of exchange at that time in humans?

O'Rahilly: I would think so. That's where the first vessels appear.

Morriss-Kay: And this is just prior to the beginning of neural tube closure. So, if we are talking about the access of folate to the embryo, this correlation is crucial. There may be a functional blood circulation enabling uptake of folate by the embryo directly, not only by histiotrophic nutrition but via the haemochorial placenta, at a fairly early stage.

Opitz: I have always marvelled at how short a process neurulation actually is. You put it down to almost eight days, between 18 and 26 days. Most text books say that it lasts about 14 days, beginning at about the 16th and going to the 30th day.

O'Rahilly: Most text books are wrong! I would say that in the human, primary neurulation begins at 18 days and should be ended at about 28 days. So that is 10 days for primary neurulation. Secondary neurulation takes much longer.

Copp: Is the dorsal lip of the rostral neuropore equivalent to the midbrain/forebrain apposition point that we see in rodents? Is there an open region behind?

O'Rahilly: Yes.

Copp: From the literature, one has the impression that the point of closure of the caudal neuropore has shifted position over the years. You find it at somite 31 in human embryos. We find it, by experimental marking, at somite 33 in the mouse (Copp & Brook 1989). Thus, the neuropore closes in the upper sacral region in both human and mouse. Why have other authors claimed the site of closure is in the upper lumbar region (e.g. Lemire et al 1975)? Are they just not allowing for the unsegmented mesoderm that intervenes between the last formed somite and the caudal neuropore?

O'Rahilly: It's very difficult to count the somites. Unless one reconstructs them, one can never be sure about the number. Most people don't count the occipital somites. For example, in the interesting paper on the spinal cord by Kunitomo (1918), the numbering of the somites is incorrect because only three (instead of four) occipital somites were acknowledged, and even these were not included from 7 weeks onwards, leading to an erroneous conclusion about reduction of segments (Müller & O'Rahilly 1986).

Shurtleff: You didn't mention where the cells that are involved in the secondary neurulation come from. It sounded as if the neural tube suddenly appears, rather than going through the cystic multiple canalization stage.

O'Rahilly: The cells are from the caudal eminence; the caudal eminence is originally really from the primitive streak, which produces vertebrae, hindgut, blood vessels and somites.

Shurtleff: Is there evidence of extension? Do stem cells migrate and then change to become recognizable as neural cells? Do neural cells migrate into the area of canalization or do they differentiate locally from primitive mesoderm?

Schoenwolf: In the chick some of the cells that give rise to the medullary cord, the structure that cavitates and forms secondary neural tube, arise very close to the caudal neural plate wings that we have mapped. Whether they are precursor cells that originate from the neural plate and migrate away, we don't know, but they do arise in close proximity to neural plate.

Goulding: Is the notochord required for secondary neurulation or what are the signals that actually drive secondary neurulation? The mouse mutants *Brachyury* and *Danforth short tail* both have a truncated tail and a truncated nervous system. Do they undergo secondary neurulation?

Shum: Brachyury homozygotes (*T/T*) die on Day 11 of gestation when secondary neurulation has normally just begun.

Copp: Mice homozygous for *Danforth short tail* have a notochord that degenerates later.

Opitz: The same must be true of Manx cats, especially for 'rumpies', the most extreme form. They have a short spinal cord and deficiencies of the coccygeal vertebrae.

Do marsupials accelerate their neurulation? Because in metamorphosis, which in humans would be the end of the eighth week, most marsupials are born, so they had better have the neural tube closed.

Stanley: They certainly can alter their development drastically, depending on the external conditions. They can arrest it well before neurulation. Marilyn Renfree's work shows that the Australian honey possum can stop at the blastocyst stage for months when environmental conditions are harsh. Once these improve, development continues normally.

References

Copp AJ, Brook FA 1989 Does lumbosacral spina bifida arise by failure of neural folding or by defective canalisation? J Med Genet 26:160–166

Cuckle H 1993 Sex differences in the location of a spina bifida lesion. J Med Genet 30:262–264

Dickey RP, Gasser RF 1993 Ultrasound evidence for variability in the size and development of normal human embryos before the tenth post-insemination week after assisted reproductive technologies. Hum Reprod 8:331–337
Hamilton WJ, Mossman HW 1972 Hamilton, Boyd and Mossman's human embryology. W Heffer, Cambridge
Hook EB 1992 Neural tube rupture as a cause of neural tube defects. Lancet 2:1000
Jacobson W, Lutwak-Mann C 1956 The vitamin B_{12} content of the early rabbit embryo. J Endocrinol 14:19–20
Juriloff DM, Harris MJ, Tom C, Macdonald KB 1991 Normal mouse strains differ in the site of initiation of closure of the cranial neural tube. Teratology 44:225–233
Kunitomo K 1918 The development and reduction of the tail, and of the caudal end of the spinal cord. Contrib Embryol Carnegie Inst 8:161–198
Lemire RJ 1969 Variations in development of the caudal neural tube in human embryos (horizons XIV–XXI). Teratology 2:361–370
Lemire RJ, Loeser JD, Leech RW, Alvord EC 1975 Normal and abnormal development of the human nervous system. Harper & Row, Hagerstown, MD
Milunsky A, Jick H, Jick SS et al 1989 Multivitamin/folic acid supplementation in early pregnancy reduces the prevalence of neural tube defects. JAMA (J Am Med Assoc) 262:2847–2852
Müller F, O'Rahilly R 1986 Somitic-vertebral correlation and vertebral levels in the human embryo. Am J Anat 177:3–19
O'Rahilly R, Müller F 1984 Embryonic length and cerebral landmarks in staged human embryos. Anat Rec 209:265–271
O'Rahilly R, Müller F 1987 Developmental stages in human embryos including a revision of Streeter's 'horizons' and a survey of the Carnegie collection. Carnegie Institution of Washington, Washington, DC (publ 637)
Putz B, Morriss-Kay GM 1981 Abnormal neural fold development in trisomy 12 and trisomy 14 mouse embryos. I. Scanning electron microscopy. J Embryol Exp Morphol 66:141–158
Schoenwolf GC 1978 An SEM study of posterior spinal cord development in the chick embryo. Scanning Electron Microsc 2:739–746
Seller MJ, Perkins-Cole KJ 1987 Sex difference in mouse embryonic development at neurulation. J Reprod Fertil 79:159–161
Shiota K 1991 Development and intrauterine fate of normal and abnormal human conceptuses. Congenital Anom 31:67–80
Streeter GL 1912 The development of the nervous system. In: Keibel F, Mall FP (eds) Manual of human embryology. Lippincott, Philadelphia, PA, vol 2:1–156

Molecular genetics of neurulation

Nancy Papalopulu and Chris R. Kintner

Molecular Neurobiological Laboratory, The Salk Institute for Biological Studies, 10010 North Torrey Pines Road, PO Box 85800, San Diego, CA 92186-5800, USA

Abstract. The formation of the neural tube begins during gastrulation when ectoderm, an epithelial sheet on the outside of the embryo, is induced to form the neural plate. During the process of neural induction, the epithelium of the neural plate is regionalized along both the dorsoventral and anteroposterior axes of the embryo; this regionalization is likely to contribute to the cellular processes underlying neurulation. Genes whose expression marks the formation and regionalization of the neural plate and which encode cell adhesion molecules or putative transcription factors have been recently identified. The differential expression of these genes apparently subdivides the epithelium of the neural plate into small regions. Evidence from transgenic embryo experiments supports the idea that the differential expression of these genes in the neural plate plays a role in neural tube formation.

1994 Neural tube defects. Wiley, Chichester (Ciba Foundation Symposium 181) p 90–102

Neural tube defects appear to have a number of different multifactorial aetiologies. One important goal towards a basic understanding of these defects is to identify and characterize animal models that recapitulate the diversity of the defects that arise in humans. In the long term, however, our ability to understand the faults underlying neural tube defects will depend on the extent to which we understand the normal processes that take place during neurulation. Unfortunately, our current knowledge of these processes is severely limited, particularly in those aspects of neurulation involving processes such as tissue patterning and morphogenesis. Fortunately, this situation is changing rapidly because of the attention that is now being given to early neural development at a number of different levels of analysis. These include cellular studies of neurulation (Shoenwolf, this volume, Jacobson, this volume), studies of single gene mutations in the mouse that lead to neural tube defects (Goulding & Paquette, this volume, Copp, this volume), the study of the tissue interactions that are responsible for inducing and patterning the neural plate and tube (Jessell & Dodd 1992) and molecular studies of neural genes (Goulding et al 1991). The hope is that these studies will dissect neurulation into distinct cellular processes that, in turn, can be further dissected at the molecular level.

In this paper, we summarize recent work on the analysis of neurulation as a problem in gene expression (Kintner 1992a). This analysis is based on the idea that at least some of the morphological changes underlying neurulation will depend on gene products that are newly expressed in ectoderm during the formation of the neural plate and tube. Using the techniques of gene cloning, considerable progress has been made in recent years in identifying genes that are expressed in ectoderm following induction. We describe the properties of such genes in *Xenopus* embryos, many of which encode cell adhesion molecules and homeodomain proteins. Because the ultimate goal of this analysis is to determine the functional contribution of these genes to neurulation, we also describe the use of transgenic *Xenopus* embryos to study gene function. Although this approach is still in the early stages, such analyses are likely to be important in the future for understanding the molecular basis of neurulation.

Results and discussion

Characterization of gene expression during neurulation may identify molecules that mediate the specialized properties of neuroepithelial cells during neural tube formation. Towards this end, the search for genes has centred on those whose expression distinguishes the ectoderm of the neural plate from surrounding ectoderm, or different regions of the neural plate from each other. The rationale behind this strategy is that the cellular properties required for neurulation are likely to arise first during gastrulation when ectoderm is induced to form the neural plate. At this stage, the neural plate consists of columnar epithelial cells that can be distinguished morphologically from the epidermis that is formed by the surrounding ectoderm. These cells show specialized properties, or behaviours, in subregions of the neural plate that lead to the complex movements of neurulation. One such behaviour is seen at the dorsal midline, where cells produce an intercalary movement called convergent extension that appears to underlie the axial elongation and closure of the neural tube. Specialized cell behaviour patterns around the edges of the neural plate are thought to contribute to the elevation and fusion of the neural folds. Finally, the shaping of the neural tube differs markedly along the anteroposterior axis of the embryo, suggesting that there are distinctive features of the anterior and posterior neural plate which allow these different shapes to be achieved during neurulation. The idea is to identify genes whose expression correlates with these specialized features of the neural plate.

Expression patterns of cell adhesion molecules mark the separation of the ectoderm into neural and epidermal tissue

One class of genes expressed differentially in ectoderm during gastrulation encodes cell adhesion molecules. The neural cell adhesion molecule (N-CAM) is the prototypic immunoglobulin-like cell adhesion molecule and was one of the first examples of a gene whose expression appeared to be regulated in ectoderm during gastrulation (Edelman et al 1983). Studies in chick and *Xenopus* embryos showed that N-CAM is expressed in the neural plate but not in the surrounding ectoderm (Edelman et al 1983, Jacobson & Rutishauser 1986, Kintner & Melton 1987). Subsequently, two members of another cell adhesion molecule family, the cadherins, were also found to be expressed differentially during neural plate formation. N-cadherin is expressed in the neural plate but not in the surrounding ectoderm (Hatta & Takeichi 1986, Detrick et al 1990), while E-cadherin shows the inverse pattern of expression (Levi et al 1991). The cadherins (reviewed in Takeichi 1991) are notable because they appear to play an important role in the morphological integrity of epithelial cells through the formation of adherens junctions. In addition, fibroblasts transfected with the cadherins not only adhere strongly to each other, but undergo cell sorting when they express different cadherin types, indicating that the cadherins mediate specific, homotypic cell adhesion. Embryological studies show that the expression of N-CAM and N-cadherin in ectoderm requires neural induction and that their expression in the neural plate coincides with morphological changes that take place during neural tube formation (Kintner & Melton 1987, Detrick et al 1990). Taken together, these studies indicate that the differential expression of cell adhesion molecules is one of the first detectable changes in ectoderm as it proceeds along two very different pathways of epithelial differentiation (see Fig. 1A).

How might cell adhesion molecules contribute to neurulation? One possibility is that they generate the differential adhesion between neuroepithelial cells of the neural tube and the surrounding non-neural epithelium of the skin that was first demonstrated by Townes & Holtfreter (1955). In this classic study, epidermal and neural tube cells were isolated from early embryos, dissociated, mixed together and allowed to reassociate in culture. Upon reassociation, the epidermal cells segregated and reformed an epidermis with the correct polarity on the outside, while the neural cells formed a tube of neuroepithelial cells on the inside. From this result, the idea arose that epidermal and neuroepithelial cells express a distinctive class of cell surface molecules, members of which allow the cells to adhere to each other and to form tissues with the correct morphology. The implication is that the differential expression of such molecules during embryogenesis would be one change underlying neural tube formation (Edelman 1986).

Functional studies of cell adhesion molecules in neural tube formation

If the differential expression of cell adhesion molecules is responsible for certain aspects of neural tube morphogenesis, then embryos in which the expression of these molecules is altered should fail to neurulate normally. One alteration

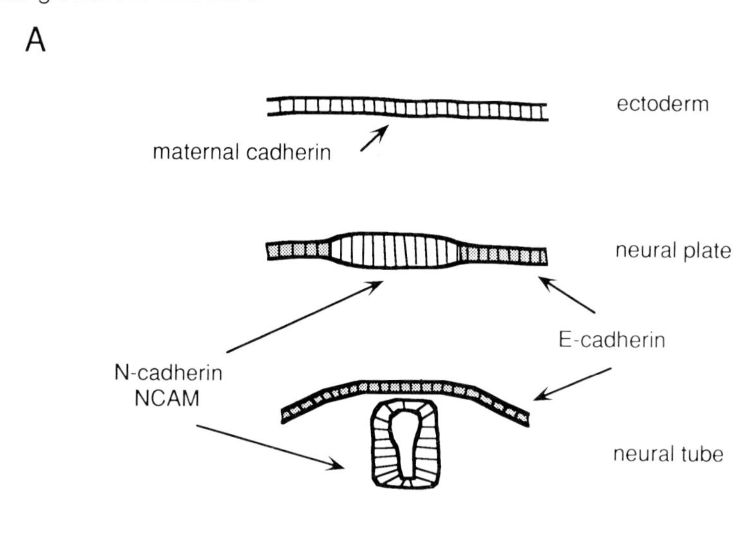

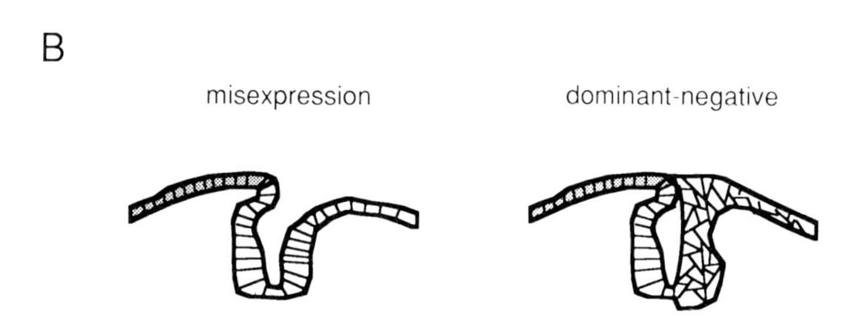

FIG. 1. Expression and function of cell adhesion molecules during neural tube formation in *Xenopus* embryos. (A) Summary of the expression of N-CAM and two members of the cadherin family in ectoderm during the formation of the neural plate and tube (Edelman et al 1983, Hatta & Takeichi 1986, Kintner & Melton 1987, Levi et al 1987, Detrick et al 1990). The expression of N-CAM and N-cadherin RNAs (light stipple) appear to mark the formation of neural tissue while the expression of E-cadherin RNA (dark stipple) appears to mark the formation of epidermis. (B) Diagrams illustrating the experimental outcome of manipulating the expression of N-cadherin in transgenic *Xenopus* embryos. Each experiment is illustrated by a tissue section through an embryo where RNA encoding N-cadherin (misexpression) or a truncated N-cadherin (dominant-negative) has been injected on just the right side, while the left side serves as an internal control. Misexpression of N-cadherin by ectopic injection of its RNA results in a failure to form a boundary between the neural and epidermal epithelium (left panel, for experimental details see Detrick et al 1990). Expression of mutant N-cadherin lacking extracellular sequences at moderate levels in embryos results in a putative dominant-negative effect where the epithelium in both the neural tube and epidermis is morphologically disorganized (right panel, see Kintner 1992b).

is the misexpression of molecules such as N-CAM or N-cadherin in ectoderm to determine whether this has a dominant effect on morphology, leading to changes that would be consistent with a role in cell segregation and the formation of tissue boundaries. Towards this end, we have injected frog embryos with RNA encoding N-CAM or N-cadherin in order to generate embryos in which the ectopic expression of these cell adhesion molecules occurs in a small region of ectoderm. When these embryos were analysed at later stages, we found that the expression of N-cadherin in ectodermal cells increases cell adhesion, leading to the formation of cell boundaries across which epithelial cell mixing does not occur (Detrick et al 1990). In contrast, N-CAM expression in a subpopulation of ectodermal cells does not generate cell boundaries (Detrick et al 1990).

These dominant effects of N-cadherin expression on ectodermal cell mixing suggested that the tissue boundary between the neural plate and surrounding non-neural ectoderm may be due to a boundary of N-cadherin expression. This hypothesis predicted that if the normal boundary of N-cadherin expression were to be erased in embryos by N-cadherin misexpression, then the segregation of neural from non-neural ectoderm should fail to occur. This prediction was fulfilled when embryos injected with RNA encoding N-cadherin were examined for morphological defects after neural tube formation (Fig. 1B). Neural tube defects were evident in all embryos in which RNA injection had resulted in the misexpression of N-cadherin across the boundary between neural and non-neural ectoderm (Detrick et al 1990). In a large fraction of the embryos, the morphological defect consisted of a continuous cell layer between epithelium in the neural tube and surrounding ectoderm, suggesting that ectoderm had failed to segregate into neural and non-neural tissue. In contrast, the morphology of the neural tube was normal in embryos injected with N-CAM RNA as well as in embryos injected with a variety of control RNAs (Kintner 1988). The results from this analysis indicated that the differential expression of N-cadherin, but not N-CAM, in the neural plate is essential for normal neurulation and suggested that the cadherins could play an important role in the normal process of morphogenesis.

These misexpression experiments with N-CAM and N-cadherin illustrate how the dominant effects of a gene product can be analysed in terms of neurulation. A complementary approach is to ask whether normal neurulation requires the action of a particular molecule. One way to address this experimentally is to express mutated forms of the molecule that are not only non-functional but are designed to interfere with the function of an endogenous molecule (Herskowitz 1987). For example, the results of misexpression experiments described above indicate that cadherin expression can have a dominant effect on morphogenesis of epithelial tissues such as the neural tube. To determine what occurs in epithelial tissues when cadherin function is blocked, we introduced an altered form of N-cadherin into *Xenopus* embryos, which lacked the extracellular domain (Kintner 1992b). This mutant protein should lack extracellular sequences required for function, but retain intracellular sequences known to bind cytoplasmic components required for cadherin activity. By competing for these intracellular components, the mutant cadherin could inactivate endogenous cadherins, producing a dominant-negative phenotype. Indeed, when this mutant protein was introduced into embryos, it proved to be a very potent inhibitor of cell adhesion. At high levels of expression, the mutant cadherin dissociates cells in the embryo completely, while at moderate levels it disrupts the normal architecture of the neural tube epithelium (see Fig. 2B). This result supports the idea that cadherins play a pivotal role in the morphogenesis of epithelial tissues, such as that which occurs during neural tube formation.

Expression of genes that divide the neural plate into different regions

The analysis of cell adhesion molecules illustrates how the characterization of genes expressed in the neural plate can identify the molecules that mediate

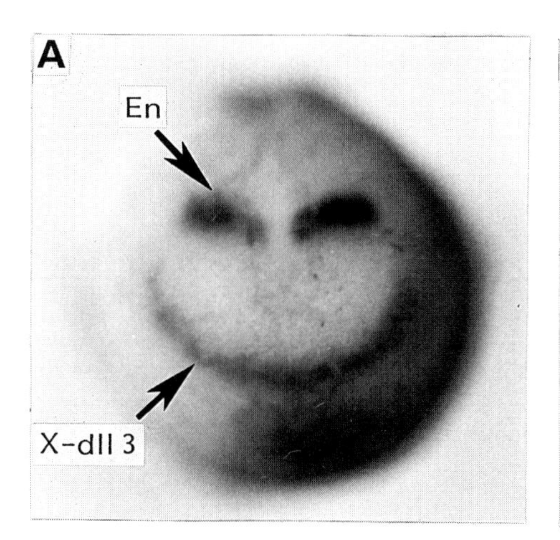

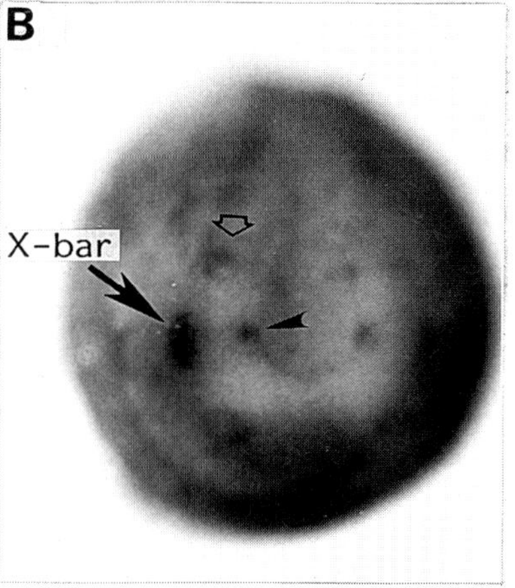

FIG. 2. Gene expression regionally restricted before closure of the neural tube. (A) An albino *Xenopus* embryo at the neural plate stage of development (stage 16) that was reacted in whole-mount with probes for *Xenopus Engrailed-2* (*En*) (Hemmati-Brivanlou et al 1991) and *Xenopus distal-less 3* (*X-dll3*) (for experimental details see Papalopulu & Kintner 1993). The area of the neural plate marked as expressing *En* corresponds to the prospective midbrain/hindbrain boundary (Hemmati-Brivanlou et al 1991). The area marked as expressing *X-dll3* corresponds to the prospective ventral forebrain/olfactory placodes. (B) Anterior view of a neural plate stage embryo hybridized with a probe for *X-bar* as in panel A. Expression occurs in the portion of the neural plate that gives rise to the dorsal retina. An arrowhead designates additional expression in the prospective optic stalk and an open arrow points to *X-bar* expression in an area that appears to overlap with the area that expresses *En-2* and may correspond to part of the midbrain.

the cellular processes underlying neurulation. This approach is likely to become increasingly important, because recent results have revealed the extent to which the neuroepithelium of the neural plate can be subdivided into a number of different regions according to differential gene expression. For example, several putative transcription factors have been isolated from vertebrate embryos on the basis of their homology to structural motifs such as the homeobox (reviewed in Kessel & Gruss 1990, Papalopulu & Kintner 1992). It has been known for several years now that the expression of these genes can be strikingly localized to different regions of the vertebrate nervous system, as exemplified by the expression of the *Hox* gene cluster along the anteroposterior axis of vertebrate embryos (reviewed in McGinnis & Krumlauf 1992). Perhaps more relevant to the problem of neurulation are the observations which show that the differential expression of these genes in subdomains of the neuroepithelium can occur as early as the neural plate stage of development. Fig. 2 presents several examples of such results obtained by localizing the expression of *Xenopus* homeobox genes at the neural plate stage of development by whole-mount *in situ* hybridization.

The first example is the localization of transcripts encoding a *Xenopus* homeobox protein, *X-dll3*, which is related in sequence to the *Drosophila* homeobox gene, *Distal-less*. *X-dll3* belongs to a family of *Distal-less*-related homeobox genes identified in vertebrates (see Papalopulu & Kintner 1993 and references therein, also Beauchemin & Savard 1992, Dirksen et al 1993). Vertebrate *distal-less* genes fall into at least four subfamilies in terms of sequence relatedness and show a diverse pattern of expression during embryogenesis. All vertebrate *distal-less* genes described so far are expressed only in ectodermal derivatives and most, including *X-dll3*, show a restricted pattern of expression in the anterior end of the neural tube, the ventral forebrain. *X-dll3* transcripts have also been detected in *Xenopus* embryos at the neural plate stage by whole-mount *in situ* hybridization, where they are expressed in the anterior ridge of the neural plate (Papalopulu & Kintner 1993, Fig. 2A). The portion of the anterior neural plate expressing *X-dll3* has been shown by fate mapping to give rise to the anterior neural tube and the olfactory placodes, sites where *X-dll3* expression is localized after neural tube closure (Papalopulu & Kintner 1993). Thus, these results suggest that the neuroepithelium at the anterior ridge of the neural plate is already specified to generate a particular portion of the neural tube, as marked by the expression of *X-dll3*, a putative transcription factor.

Similar results have been obtained using another *Xenopus* homeobox gene, called *X-bar*. The *X-bar* homeobox is most closely related to the homeobox found in the *Drosophila BarH1* and *BarH2* genes, which are required for proper eye development (N. Papalopulu & C. R. Kintner, unpublished observations, Higashijima et al 1992 and references therein). Localization of *X-bar* expression by whole-mount *in situ* hybridization to *Xenopus* embryos indicates that it is found predominantly in the developing eye. The striking feature of this expression is that *X-bar* RNA appears to be restricted to the progenitor cells in just the dorsal half of the eye at all stages of development (data not shown). Importantly, *X-bar* RNA can also be detected in the anterior neural plate (Fig. 2B), where it is localized to the portion of the neural plate fate map that later gives rise to the dorsal retina (Eagleson & Harris 1989). Thus, these results further support the notion that very small regions of the neural plate, in this case the dorsal half of the eye neuroepithelium, are specified before neural tube closure. The picture emerging from these observations is that the neuroepithelium prior to neurulation can be subdivided into very small domains on the basis of gene expression. This extends previous studies that showed that the regional properties of the neuroepithelium become established at the neural plate stage (Saha & Grainger 1992). One scenario is that the products of these regionally expressed genes specify the diverse properties of cells in the neural plate that have been shown by the cellular studies of neurulation.

One model suggested by these studies is that the cellular diversity of the neural plate is brought about by the regulation of downstream gene targets by the transcription factors localized to different regions of the neural plate. These gene targets presumably include molecules underlying morphogenesis, such as cell adhesion molecules. This expectation appears to be borne out by the analysis of a new cadherin that we have recently identified in the developing *Xenopus* nervous system. This cadherin, called F-cadherin, is localized to boundaries that subdivide the neuroepithelium into different regions and is detected at the neural plate stage of development. We are presently investigating the contribution of F-cadherin to neurulation and the role of transcription factors expressed in the neural plate in establishing the expression of F-cadherin at neural boundaries.

Concluding remarks

The characterization of mouse mutants with neural tube defects is one approach that can be taken to identify genes whose activity is critical to neurulation. This approach is severely limited, however, by the fact that these mutants are identified only on the basis of dominant phenotypes in heterozygous mice. Molecular genetics is a complementary approach that is likely to be increasingly important, particularly with the advent of technology to generate mice homozygous for a particular gene mutation by homologous recombination. Gene expression studies suggest a model in which the neuroepithelial cells in the neural plate are specialized by the differential expression of putative regulatory genes, for example the homeobox genes, and that one activity of these genes may be to turn on the expression of molecules such as F-cadherin which in turn confer specialized behaviours on cells within the neural plate during neurulation. The goal in the future will be to test this model by a genetic analysis of neural plate genes combined with detailed analyses of the cellular changes underlying neurulation.

Acknowledgements

We wish to thank Drs Oliver Bögler and Martyn Goulding for critical reading of the manuscript. This work was supported by a grant from the NIH to C. R. K. and by a long-term HFSPO post-doctoral fellowship to N. P.

References

Beauchemin M, Savard P 1992 Two *Distal-less* related homeobox-containing genes expressed in regeneration blastemas of the newt. Dev Biol 154:55–65
Copp AJ 1994 Genetic models of mammalian neural tube defects. In: Neural tube defects. Wiley, Chichester (Ciba Found Symp 181) p 118–143
Detrick RJ, Dickey D, Kintner CR 1990 The effect of N-cadherin misexpression on morphogenesis in *Xenopus* embryos. Neuron 4:493–506
Dirksen M-L, Mathers P, Jamrich M 1993 Expression of a *Xenopus Distal-less* homeobox gene involved in forebrain and cranio-facial development. Mech Dev 41:121–128
Eagleson GW, Harris WA 1989 Mapping of the presumptive brain regions in the neural plate of *Xenopus laevis*. J Neurobiol 21:427–440
Edelman GM 1986 Cell adhesion molecules in the regulation of animal form and tissue pattern. Annu Rev Cell Biol 2:81–116
Edelman GM, Gallin WJ, Delovee A, Cunningham BA 1983 Early epochal maps of two different cell adhesion molecules. Proc Natl Acad Sci USA 80:4384–4388
Goulding MD, Paquette A 1994 *Pax* genes and neural tube defects in the mouse. In: Neural tube defects. Wiley, Chichester (Ciba Found Symp 181) p 103–117
Goulding MD, Chalepakis G, Deutsch U, Erselius JR, Gruss P 1991 Expression of developmentally regulated genes in the embryonic mouse nervous system. Discuss Neurosci 1:17–24
Hatta K, Takeichi M 1986 Expression of N-cadherin adhesion molecules associated with early morphogenetic events in chick development. Nature 320:447–449
Hemmati-Brivanlou A, de la Torre JR, Holt C, Harland RM 1991 Cephalic expression and molecular characterization of *Xenopus EN-2*. Development 111:715–724
Herskowitz I 1987 Functional inactivation of genes by dominant negative mutations. Nature 329:219–222
Higashijima S-I, Kojima T, Michiue T, Ishimaru S, Emori Y, Saigo K 1992 Dual *Bar* homeo box genes of *Drosophila* required in two photoreceptor cells, R1 and R6, and primary pigment cells for normal eye development. Genes & Dev 6:50–60
Jacobson AG 1994 Normal neurulation in amphibians. In: Neural tube defects. Wiley, Chichester (Ciba Found Symp 181) p 6–24
Jacobson M, Rutishauser U 1986 Induction of neural cell adhesion molecule (N-CAM) in *Xenopus* embryos. Dev Biol 116:524–531
Jessell TM, Dodd J 1992 Midline signals that control the dorso-ventral polarity of the neural tube. Semin Neurol 4:317–325
Kessel M, Gruss P 1990 Murine developmental control genes. Science 249:374–379
Kintner CR 1988 Effects of altered expression of the neural cell adhesion molecule, NCAM, on early neural development in Xenopus embryos. Neuron 1:545–555
Kintner CR 1992a Molecular bases of early neural development in *Xenopus* embryos. Annu Rev Neurosci 15:251–284
Kintner C 1992b Regulation of embryonic cell adhesion by the cadherin cytoplasmic domain. Cell 69:225–236
Kintner CR, Melton DM 1987 Expression of Xenopus N-CAM RNA is an early response of ectoderm to induction. Development 99:311–325

Levi G, Crossin KL, Edelman G 1987 Expression sequences and distribution of two primary cell adhesion molecules during embryonic development of *Xenopus laevis*. J Cell Biol 105:2359–2372
Levi G, Gumbiner B, Thiery JP 1991 The distribution of E-cadherin during *Xenopus laevis* development. Development 111:159–169
McGinnis W, Krumlauf R 1992 Homeobox genes and axial patterning. Cell 68:283–302
Papalopulu N, Kintner CR 1992 Induction and patterning the neural plate. Semin Neurol 4:295–306
Papalopulu N, Kintner CR 1993 *Xenopus Distal-less* related homeobox genes are expressed in the developing forebrain and are induced by planar signals. Development 117:961–975
Saha M, Grainger R 1992 A labile period in the determination of the anterior–posterior axis during early neural development in *Xenopus*. Neuron 8:1003–1014
Schoenwolf GC 1994 Formation and patterning of the avian neuraxis: one dozen hypotheses. In: Neural tube defects. Wiley, Chichester (Ciba Found Symp 181) p 25–50
Takeichi M 1991 Cadherin cell adhesion receptors as a morphogenetic regulator. Science 251:1451–1455
Townes PL, Holtfreter J 1955 Directed movements and selective adhesion of embryonic amphibian cells. J Exp Zool 128:53–120

DISCUSSION

Shum: Have you looked at the interactions between different cell adhesion molecules? For example, in the embryos expressing the dominant-negative N-cadherin, have you looked at the expression of other cell adhesion molecules to see whether they are perturbed as well?

Kintner: That's a good experiment; we haven't done it. Epithelial cell layers such as the neural plate are held together by a collection of different junctions. There is good evidence that the cadherin junction is pivotal in that if cadherin function is perturbed, all the other junctions fail. This suggests that the cadherin junction is providing mechanical stability that is required for the integrity of the other junctions. Alternatively, perturbing the cadherin junction might generate an intracellular signal that feeds back on and disrupts other junctional molecules. The major problem in testing this idea is the lack of good probes for other junctional molecules.

Shum: Have you transplanted, for example, some of the ventral epidermal cells to the dorsal midline, where usually the neuroepithelium is formed? This would test whether these cells have the potential to express N-cadherin instead of E-cadherin.

Kintner: There is a critical time after which the ectoderm will no longer respond to inducing signals and become neural. The time when ectoderm loses the ability to become neural correlates well with when it turns on the expression of other cell adhesion molecules. For example, E-cadherin begins to be expressed in the ectoderm at the time this is no longer able to respond to inductive signals and become neural and express N-cadherin.

In general, there is a very tight correlation between when cell fate decisions can no longer be made in the embryo and when differential cell adhesion takes place. These observations have led to the idea that cell adhesion may contribute to the ability of cells to respond to inductive signals and change cell fate. Ectoderm isolated from late-stage embryos expresses E-cadherin and cannot be induced to express N-cadherin. If taken early enough, ectoderm does not express N- or E-cadherin but can be induced to express N-cadherin. Whether or not the expression of E-cadherin at later stages prevents ectoderm from responding and becoming neural is not yet known.

Shum: Is N-cadherin or E-cadherin expressed in the mouse embryo?

Kintner: Yes. There are a few oddities about *Xenopus*, but all the molecules I have talked about have homologues in mammals and show similar expression patterns.

Shum: Have you looked at N-cadherin expression in the neural tube formed by secondary neurulation, because in the tail bud the secondary neural tube is not formed by the ectoderm?

Kintner: No, I haven't looked at the tail bud.

van Straaten: Do you know if any of these adhesion molecules are involved in the process of fusion of the neural folds?

Kintner: One can make a very strong case that differential expression of molecules like the cadherins could be important when epithelial sheets segregate or fuse. The problem is how to devise experiments to address this hypothesis. When we perturb the expression of these molecules, we do it in a very global way and we do it at the very start of the process. A better experiment would be to manipulate the expression of these molecules at different stages during neurulation and in specific regions of the neural plate. In addition, these experiments need to be done under circumstances where one can actually watch the behaviour of cells in the neural plate in order to make conclusions about the role of cell adhesion molecules in a particular aspect of neurulation.

Opitz: Have you looked at N-CAM later during morphogenesis, specifically in the mesonephros? G. Edelman has shown expression of N-CAM in the mesonephros during development. I'm trying to look for a developmental connection between correlated kidney and central nervous system abnormalities. They were first described by J. Bernstein many years ago and then again by Steve Kornguth and his colleagues at the University of Wisconsin. They mostly involve just histogenetic abnormalities of the cerebellum and renal dysplasias. The only thing that's ever come to my attention that might connect these two is that apparently the same N-CAM is involved in morphogenesis of mesonephros and metencephalon.

Kintner: The role of N-CAM as a cell adhesion molecule is somewhat controversial. N-CAM was originally described as a cell adhesion molecule on the basis of cell aggregation assays. These assays have their limitations as a test for cell adhesion molecules. Molecules that one does not normally think of as cell adhesion molecules, such as receptor-configured tyrosine kinases, can cause cells to aggregate *in vitro*. In the experiments we have done in embryos, we have no hint of N-CAM promoting cell adhesion. One caveat is that our results are negative and so one could always argue that we haven't provided something that N-CAM needs in order to work.

I should mention that there is considerably better evidence that N-CAM is an anti-cell adhesion molecule. It bears a very large, negatively charged polysialic acid modification. Rutishauser (1992) has shown in developing embryos that this modification acts to push cells apart. If anything, the effects of N-CAM that we have seen in our experiments are consistent with this idea. The point that I am trying to make is that in order to know what these molecules are really doing, one needs to do experiments in the embryo rather than extrapolate from *in vitro* aggregation assays and expression studies.

The answer to your question will become apparent when the mice in which the N-CAM gene has been mutated by homologous recombination have been analysed in detail. Embryos lacking N-CAM function will presumably tell us whether N-CAM is required for the morphogenesis of several tissues.

Hall: Are those mice viable?

Kintner: I believe the mutation is lethal. If we are unlucky, the embryos may die at too young an age to analyse.

Hall: Cell surface molecules in general are used and re-used several times during development. For any one of those, I would expect knocking out the gene to be lethal.

Kintner: This may become a problem in the future in analysing knock-out mice: in cases where molecules are re-used in embryogenesis, the knock-out phenotype will just tell you about the first step of embryogenesis for which the molecule is required.

Trasler: In *splotch* mouse embryos, immunofluorescence studies showed much brighter staining for N-CAM than in normal littermates. When we looked for the protein on Western blots (with monoclonal antibody H28123), we found that the embryos at the time of neural tube closure had a heavier form of N-CAM. It seems that the molecule becomes sialylated earlier than normal: in wild-type mice, N-CAM becomes sialylated from Day 11 to Day 19 of gestation; in *splotch* embryos this occurs at Day 9.5, at the time of neural tube closure (Moase & Trasler 1991).

Kintner: Our experiments do not address the role of sialylation. Most of the N-CAM that we have expressed by RNA injections is not sialylated.

Trasler: We have been using a monoclonal antibody specific to N-CAM containing moderate to long chains of polysialic acid (mAB5A5(IgM)), given to us by Urs Rutishauser. We find that N-CAM from *splotch* mice is sialylated as early as the 12-somite stage, before anterior neural tube closure. In the wild-type, N-CAM (depending on strain) only becomes sialylated at the 20-somite stage or as late as Day 11.

Kintner: To relate that directly to the neural tube defect, one could artificially increase sialylation of N-CAM to see what effects that has.

Opitz: Human X-linked hydrocephalus now turns out to be a defect in a neural cell adhesion molecule (Rosenthal et al 1992). Does X-linked hydrocephalus occur in mice?

Kintner: I don't know.

References

Moase CE, Trasler DG 1991 N-CAM alterations in splotch neural tube defect mouse embryos. Development 113:1049–1058

Rosenthal A, Jouet M, Kenwrick S 1992 Aberrant splicing of neural cell adhesion molecule L1 messenger RNA in a family with X-linked hydrocephalus. Nat Genet 2:107–112

Rutishauser U 1992 NCAM and its polysialic acid moiety: a mechanism for pull/push regulation of cell interactions during development? Dev Suppl (Gastrul) p 99–104

Pax genes and neural tube defects in the mouse

Martyn Goulding and Alice Paquette

Molecular Neurobiology Laboratory, The Salk Institute for Biological Studies, 10010 North Torrey Pines Road, PO Box 85800, San Diego, CA 92186-5800, USA

Abstract. The *Pax* genes encode a family of transcription factors that are expressed in restricted regions of the developing embryo. Several *Pax* genes are expressed in the developing nervous system where they are believed to regulate the morphogenesis of neural structures. Loss-of-function mutations in the *Pax-3* gene have been identified in a number of alleles of the mouse mutant *splotch*. In homozygous *splotch* embryos closure of the neural tube is defective with embryos exhibiting spina bifida and/or exencephaly. Other structures in which *Pax-3* is expressed are also affected, most notably those tissues derived from the neural crest and somites.

1994 Neural tube defects. Wiley, Chichester (Ciba Foundation Symposium 181) p 103–117

The vertebrate nervous system develops from a dorsal sheet of epithelial cells that are induced to form neural plate tissue in response to signals from the mesoderm (Mangold 1931, Slack & Tannahill 1991). The primordium of the adult nervous system, the neural tube, is generated by the progressive folding of the neural plate along the ventral midline, followed by the fusion of the dorsal neural folds. At first the neural tube consists of a simple pseudostratified neuroepithelium containing neural progenitor cells. These progenitor cells are mitotically active. After their final cell division, postmitotic cells migrate radially then differentiate into a variety of glial and neuronal cell types.

The formation of the neural tube depends on morphological changes in cells of the neural plate. Many of these changes involve specialized cell behaviour that depends in part on the position of the cell in the neuraxis. As such, genes that play a role in specifying position or cell type in the early nervous system may regulate aspects of cellular behaviour that are important for neurulation.

The molecular events that regulate neural induction and the subsequent patterning of the vertebrate nervous system are largely unknown. One important aspect of regionalization in the early nervous system is the expression of different transcription factors and cell adhesion molecules in discrete regions of the neural plate prior to neurulation (see Papalopulu & Kintner, this volume). Many of the transcription factors present in the developing nervous system contain conserved protein domains that mediate DNA binding, such as the helix-loop-helix motif, zinc-finger motif, fork-head domain, homeodomain and paired domain. The *Pax* genes encode a family of transcription factors containing the paired domain; some of them also contain a homeodomain. They are expressed in the neuroepithelium before and during neural tube closure and are therefore of particular interest since they may regulate cellular events associated with neurulation.

Evidence that the *Pax* genes regulate embryonic development has come from studies on invertebrates and vertebrates. In the *Drosophila* embryo, proteins encoded by paired box genes regulate the segmentation of the blastoderm and the development of the nervous system. The vertebrate *Pax* genes also appear to regulate embryogenesis. All of the mouse *Pax* genes isolated so far exhibit spatially restricted patterns of gene expression in the embryo consistent with a regulatory role in development. Confirmation of their importance for development has come from the discovery that mutations in three mouse *Pax* genes have dramatic effects on embryonic development.

Pax gene expression in the developing nervous system

Six of the *Pax* genes are expressed in the developing nervous system in distinct subsets of neural progenitor cells and differentiating neurons. The early expression of *Pax-2, Pax-3, Pax-6* and *Pax-7* in discrete regions of the neural plate and neural tube suggests these genes may regulate cellular events important for neurulation. *Pax-3, Pax-6* and *Pax-7* are expressed in discrete regions of the mitotic neuroepithelium in both the hindbrain and spinal cord. All three genes are expressed as the neural tube is closing. *Pax-3* and *Pax-7* are expressed in dorsal regions of the spinal cord. *Pax-6* is initially expressed throughout the spinal cord, with the exception of ventral and dorsal midline cells. Expression of *Pax-6* appears then to be down-regulated in ventral and dorsal regions of the spinal cord until it is eventually restricted to midlateral regions of the neuroepithelium. While loss-of-function mutations in *Pax-3* are known to cause neural tube defects in the mouse, mutations in *Pax-6* do not affect neural tube closure. The role of *Pax-7* in neurulation has not been determined.

Pax-2 and *Pax-6*, along with a number of other homeobox genes, are expressed at the earliest stages of brain development (Krauss et al 1991a,b, Walther & Gruss 1991, Kessel & Gruss 1991, Püschel et al 1992). Their expression in the cephalic neuroepithelium appears to coincide with the partitioning of the neuroepithelium into distinct regions that later will give rise to discrete anatomical structures in the brain. For example, the expression of *Pax-2* in the caudal mesencephalon appears to be important for the formation of the midbrain/hindbrain boundary, since this rhombencephalic isthmus fails to form in zebrafish embryos injected with antibodies specific for the *Pax-2/Pax* [*zf-b*] protein (Krauss et al 1992). These and other studies suggest that the activity of *Pax-2* and *Pax-6* in the early brain may regulate some of the morphological changes that are necessary for the regionalization of the brain.

In the mouse spinal cord, *Pax-2* and *Pax-8* are transcribed in two dorso-ventrally restricted populations of postmitotic neuroblasts. The expression of both genes in the spinal cord occurs too late to have any role in neurulation. Instead, these genes probably regulate the differentiation of a subset of spinal cord neurons. In zebrafish, the homologue of *Pax-2*, *Pax* [*zf-b*], is initially expressed in two columns of spinal cord interneurons (Krauss et al 1991b, Püschel et al 1992). These interneurons may be functionally equivalent to the two populations of *Pax-2*-expressing cells in the mouse.

Two lines of evidence suggest that *Pax-3* may be important for neurulation in vertebrates. In the mouse, expression of *Pax-3* is first seen in the neural plate and head fold region at eight days post coitum, which coincides with the closure of the neural tube. Second, loss-of-function mutations in the *Pax-3* gene can cause neural tube defects such as exencephaly and spina bifida. Expression of *Pax-3* in the spinal cord and hindbrain is initially restricted to the neural folds. As the neural tube closes, *Pax-3* expression expands in the dorsal region of the spinal cord, and, by Day 10, cells throughout the dorsal half of the spinal cord neuroepithelium express *Pax-3* (Fig. 1). The exact role, if any, of *Pax-3* in neurulation remains unclear, in spite of the close spatial and temporal correlation between *Pax-3* expression in the developing nervous system and the closure of the neural tube.

Transcripts of *Pax-3* are found in tissues other than the neural tube. *Pax-3* is expressed in cephalic and trunk neural crest cells prior to migration and then later in neural crest-derived structures, including the cranial ganglia, dorsal root ganglia and facial mesectoderm (Goulding et al 1991). It is unclear whether *Pax-3* is expressed in all neural crest cell types, such as the sympatho-adrenal lineage and melanocytes. The fact that melanocytes fail to develop and sympathetic ganglia are often missing in caudal regions of homozygous *Pax-3* mutant mice suggests these cell types and/or their progenitors could express *Pax-3*. Cells in the somitic mesoderm also express it. Expression of *Pax-3* in somites appears to coincide with the earliest stages of somitogenesis, with cells positive for *Pax-3* present in dorsal regions of the newly formed somite. As the somite differentiates into sclerotome and dermomyotome, *Pax-3* expression becomes restricted to dermomyotomal cells. Transcripts of *Pax-3* are also present in the developing limb bud. Analysis of *Pax-3* expression in the limb bud indicates that *Pax-3* is expressed in the dorsal and ventral muscle masses. These cells are probably derived from *Pax-3*-expressing cells located at the ventrolateral edge of the adjacent somites that migrate into the limb bud.

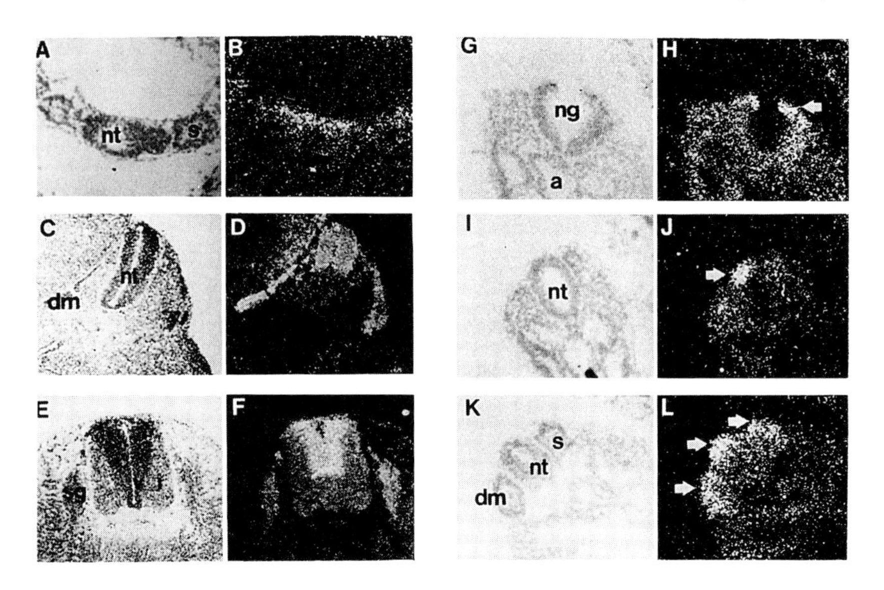

FIG. 1. Expression of *Pax-3* in the mouse. Left panel: transverse sections through (A,B) Day 8.5, (C,D) Day 10.5 and (E,F) Day 12.5 embryos showing expression of *Pax-3* transcripts in the dorsal neural tube. In the neural tube it is restricted to the ventricular zone in the alar and roof plates (D,E). Right panel: transverse sections through a Day 8.5 embryo showing expression of *Pax-3* in the closing neural tube. It is first seen in the dorsal neural folds (G,H); after closure of the neural tube it is present in the roof plate and adjacent cells (I–L). a, aorta; dm, dermomyotome; i, intermediate zone; ng, neural groove; nt, neural tube; s, somite; sg, spinal ganglia; v, ventricular zone.

The *splotch* phenotype is caused by mutations in the *Pax-3* gene

Six alleles of *splotch* have been described and are currently available. Two of these alleles arose spontaneously, *splotch* (*Sp*), the original allele (Russell 1947), and *splotch-delayed* (Sp^{d}). Four other alleles of *splotch* were derived by X-irradiation, Sp^{1H}, Sp^{2H}, Sp^{4H} and *splotch retarded* (Sp^{r}). Heterozygous *splotch* animals carry a white belly spot, sometimes accompanied by spotting on the limbs and tail. Embryos homozygous for *splotch* die prenatally and have characteristic neural tube defects that include exencephaly and spina bifida (see Table 1). Tissues derived from the neural crest are also severely affected in *splotch* homozygous embryos.

An interspecific backcross analysis mapped the *Pax-3* gene to mouse chromosome 1 between the *leaden* (*ln*) and *villin* (*Vil*) loci (Walther et al 1991), placing *Pax-3* close to the known mouse mutant allele, *splotch*. The mapping data, together with the close correlation between the expression pattern of *Pax-3* in mouse embryos and tissues affected in *splotch* embryos, was a further indication that *Pax-3* and *splotch* might be allelic.

TABLE 1 Summary of *splotch* allele phenotypes and mutations in *Pax-3*

Splotch allele	*Heterozygote phenotype*	*Homozygote phenotype*	*Mutation in Pax-3*
Sp^r	White belly spot Retarded growth	Preimplantation lethal	Large deletion including *Pax-3 Vil, Des, Inha, Akp-3*
Sp	White belly spot	Extensive neural tube and neural crest defects Die at Days 13 to 15 of gestation (E13 to E15)	Splice acceptor site mutation at the intron 3/exon 4 boundary giving a truncated *Pax-3* transcript
Sp^1H	White belly spot	Extensive neural tube and neural crest defects Die at E13 to E15	Not known—may be the same as *Sp^2H*
Sp^2H	White belly spot	Extensive neural tube and neural crest defects Die at E13 to E15	32 bp deletion within the *Pax-3* homeobox, giving a predicted truncated protein
Sp^4H	White belly spot	Not known	Deletion of entire *Pax-3* gene but not *Vil, Tp-1, Inha* and *Acrg*
Sp^d	White belly spot	Moderate neural tube defects Small dorsal root ganglia Die at birth	Mutation in *Pax-3* paired domain and reduced levels of *Pax-3* transcript in homozygotes

Acrg, acetylcholine receptor γ; *Akp-3*, intestinal alkaline phosphatase; *Des*, desmin; *Inha*, inhibin A; *Tp-1*, transition protein 1; *Vil*, villin.

The initial mapping data for *Pax-3* were used in an effort to determine whether *Pax-3* and *splotch* were the same gene. In two separate linkage analyses, no recombination events between *Pax-3* and *Sp* were detected (Mancino et al 1992, Goulding et al 1993a). The first clear demonstration that mutation of *Pax-3* might be responsible for the *splotch* phenotype came from the analysis of the *Sp^2H* allele by the group of Philippe Gros (Epstein et al 1991a). They were able to show that the *Sp^2H* allele contained a 32 bp deletion in the homeobox of *Pax-3*. This mutation causes a frame shift and introduces a stop codon into the open reading frame. As a result, the *Pax-3* protein produced by the *Sp^2H* allele lacks a functional homeodomain and downstream sequences known to be necessary for transcriptional activation by *Pax-3*.

The *Pax-3* gene has now been examined in four other alleles of *splotch*. In the original *splotch* allele, *Sp*, a mutation in the splice acceptor site of exon 4 results in the abnormal splicing of exon 3 to exon 5. The consequence is production of a *Pax-3* protein that lacks the 45 amino acids encoded by exon 4 (Fig. 2).

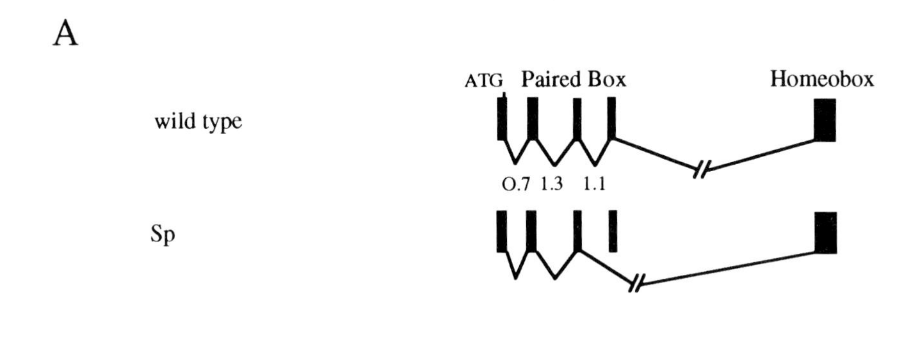

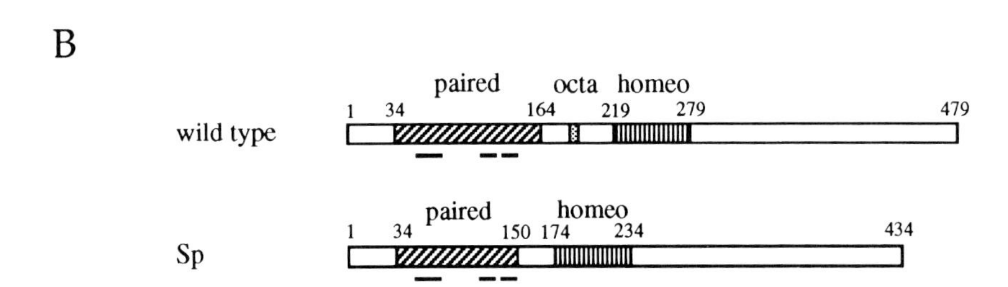

FIG. 2. Scheme of the mutation in the *Sp* allele. (A) In *Sp* mice exon 3 is spliced directly to exon 5. Exons are shown as black bars. (B) The protein produced by this shorter transcript lacks some of the paired box and the octapeptide sequence.

Interestingly, two conserved domains are partially encoded by the fourth exon, namely the last 14 amino acids of the paired domain and the octapeptide sequence. The paired domain is important for the activity of the *Pax* proteins; it mediates DNA binding by this family of proteins both *in vitro* (Treisman et al 1991, Goulding et al 1991, Chalepakis et al 1991) and *in vivo* (Zannini et al 1992, Adams et al 1992). The *Pax-1* gene is mutated in the mouse mutant *undulated*. In experiments analysing this mutant, it has been shown that a single glycine to serine change within the first helix of the paired domain of *Pax-1* results in a 10-fold reduction in DNA binding to the e5 sequence (Balling et al 1988, Chalepakis et al 1991). Deletion of the first helix of the paired domain in the *Drosophila paired* protein also abolishes DNA binding activity (Treisman et al 1991). As no function has yet been ascribed to the octapeptide sequence, the consequences of its deletion are unknown.

The entire coding sequence of the *Pax-3* gene is deleted in both the Sp^r and Sp^{4H} alleles. The Sp^r allele contains a large deletion in band C4 on chromosome 1 that is cytogenetically visible (Evans et al 1988). This deletion

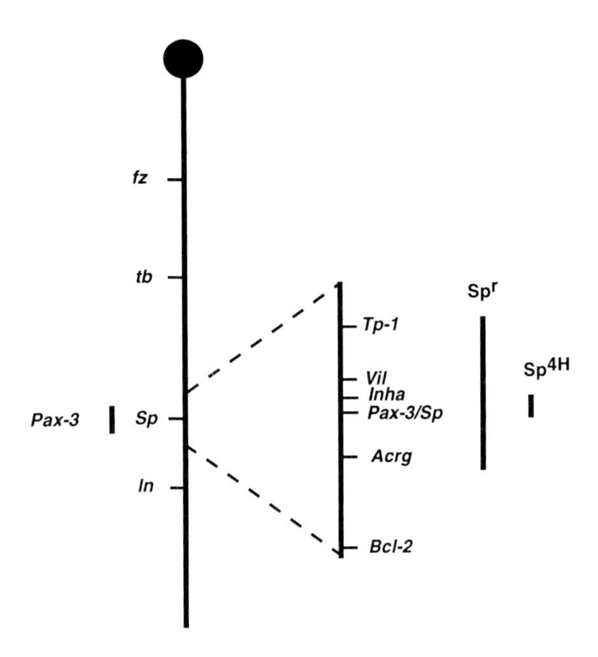

FIG. 3. Chromosomal mapping of *Pax-3* and deletions in *Sp*^{r} and *Sp*^{4H} alleles. *Pax-3* was mapped to the long arm of chromosome 1, distal between *tumbler* (*tb*) and *leaden* (*ln*) loci. The position of the *Sp* locus is also shown. *Pax-3* and a number of flanking genes are shown together with the extent of the deletions present in the *Sp*^{r} and *Sp*^{4H} alleles. Acrg, acetylcholine receptor γ; fz, fuzzy locus; Inha, inhibin A; Tp-1, transition protein 1; Vil, villin.

encompasses a number of flanking markers on chromosome 1 in addition to the *Pax-3* gene (Epstein et al 1991b). The deletion of *Sp*^{4H} is smaller than that of *Sp*^{r}. While the entire coding region of *Pax-3* is deleted, two closely flanking markers, the genes for inhibin A and the acetylcholine receptor γ subunit, are still present (Fig. 3). Both of these *Sp* alleles are clear loss-of-function alleles. It is thought that loss of genes flanking *Pax-3* in the *Sp*^{r} allele is responsible for the death of *Sp*^{r} homozygotes before implantation. The *Pax-3* gene has also been examined in a fifth allele of *splotch, Sp*^{d}. In *Sp*^{d} mice, the abundance of *Pax-3* transcripts in Day 12 embryos is approximately fivefold lower than in normal wild-type embryos. A point mutation resulting in a glycine to arginine change at the ninth amino acid of the paired domain has recently been identified in *Sp*^{d} mice (Vogan et al 1993). This finding is consistent with the *Sp*^{d} allele being the least severe of all *splotch* alleles.

Analysis of the *splotch* phenotype

Considerable effort has gone into describing the defects that occur in homozygous *splotch* embryos. Failure of the neural tube to close in lumbosacral regions is the most conspicuous malformation that occurs during development (see Copp, this volume). While it is clear that mutations in *Pax-3* predispose the mouse to closure defects, this trait is somewhat variable and is also dependent on genetic background. The basis of the neural tube defects seen in *splotch* mice is not known. Differences in rates of cell division (Wilson 1974) and disorganization of the neuroepithelium have been observed within the neural tube of *splotch* embryos. Defects in the extracellular matrix, the basal lamina and the extracellular space surrounding the neural tube have also been proposed as candidate mechanisms for lack of neural tube closure (O'Shea & Liu 1987, Kapron-Bras & Trasler 1988, Moase & Trasler 1991).

Neural crest-derived cell types are severely affected in homozygous *splotch* embryos (Auerbach 1954, Moase & Trasler 1989). A delay in neural crest cell emigration has been observed both *in vivo* (Kapron-Bras & Trasler 1988) and *in vitro* (Moase & Trasler 1990). Neural crest derivatives are affected in a rostrocaudal gradient. Dorsal root ganglia are reduced in size at cervical levels and often missing in thoracic or lumbar regions. The sympathetic ganglia in *splotch* mice are similarly affected. Interestingly, the cranial ganglia in *splotch* mice appear not to be significantly altered; this observation is particularly striking for the trigeminal ganglion, which has a significant neural crest contribution. In some *Sp* homozygous embryos the superior ganglia are diminished in size, consistent with a defect in the migration of hindbrain neural crest cells.

A conspicuous feature of *Pax-3* expression in the spinal cord is the establishment of a sharp ventral boundary that coincides with the border between the alar and basal plates, the sulcus limitans. Signals emanating from the ventral midline appear to play a major role in establishing this boundary. Transplantation studies in the chick have demonstrated that signals from the notochord and floor plate regulate the ventral boundaries of expression of *Pax-3* and *Pax-6* genes in the spinal cord. Implantation of a supernumerary notochord lateral to the spinal cord causes a dorsal shift in the expression domains of both genes, while removal of the notochord causes the domain of *Pax-3* expressing cells to expand into ventral regions of the spinal cord (Goulding et al 1993b).

In light of experiments in the chick showing that changes in *Pax* gene expression induced by notochord closely match subsequent changes in neuronal differentiation (Goulding et al 1993b), we assessed the development of various classes of neuron in the spinal cord of *splotch* mice. Analysis was confined to levels of the cord where tube closure had occurred successfully. Using acetylcholinesterase activity as a marker of motor neurons, we detected no significant differences in the position or size of the ventral motor pools. Intermediately situated sympathetic preganglionic neurons were also present in similar numbers in both wild-type and *splotch* embryos. In addition, no difference was seen in the antigenicity of cellular retinoic acid binding protein (CRABP) in *splotch* spinal cords, indicating that CRABP-positive commissural neurons differentiate normally in the absence of a functional *Pax-3* gene.

Waardenburg syndrome

The *splotch* phenotype in the mouse is believed to be analogous to the human disease, Waardenburg syndrome. This syndrome maps to chromosome 2q37, a region that is syntenic with the *Pax-3* locus on mouse chromosome 1. Many of the defects in patients with Waardenburg syndrome are similar to the defects observed in *splotch* mice. A variety of tissues that are derived from the neural crest are affected. The phenotype includes abnormal pigmentation of the eyes and a white forelock, Hirshsprung's disease and broadened facial features. In addition, many individuals with this syndrome exhibit sensorineural deafness. Recently, a number of mutations have been identified within the human *PAX-3/HuP2* gene in families with the syndrome (Burri et al 1989, Baldwin et al 1992, Tassabehji et al 1992, 1993). Many of these mutations are found in the paired domain and are known to affect DNA binding by the *Pax-3* protein (Chalepakis et al 1994). It is unclear at this stage whether patients with Waardenburg syndrome have increased susceptibility to neural tube defects. In this respect it should be noted that these patients carry one normal allele of *Pax-3* and in *splotch* mice neural tube defects are seen only in homozygous embryos.

Acknowledgements

I would like to thank Peter Gruss and Karen Steel who contributed to the work described in this paper, Elise Lamar and Susan Koester for critical reading of this manuscript, and the Weingart Foundation and NIH for financial support for this work.

References

Adams B, Dorfler P, Aguzzi et al 1992 Pax-5 encodes the transcription factor BSAP and is expressed in B lymphocytes, the developing CNS, and adult testis. Genes & Dev 6:1589–1607

Auerbach R 1954 Analysis of the developmental effects of a lethal mutation in the house mouse. J Exp Zool 127:305–329

Baldwin CT, Hoth CF, Amos JA, da-Silva EO, Mulinsky A 1992 An exonic mutation in the *HuP2* paired domain gene causes Waardenburg's syndrome. Nature 355:637–638

Balling R, Deutsch U, Gruss P 1988 *Undulated*, a mutation affecting the development of the mouse skeleton, has a point mutation in the paired box of *Pax-1*. Cell 55:531–535
Burri M, Tromvoukis Y, Bopp D, Frigerio G, Noll M 1989 Conservation of the paired domain in metazoans and its structure in three isolated human genes. EMBO (Eur Mol Biol Organ) J 8:1183–1190
Chalepakis G, Fritsch R, Fickenscher H, Deutsch U, Goulding M, Gruss P 1991 The molecular basis of the *undulated/Pax-1* mutation. Cell 66:873–884
Chalepakis G, Goulding M, Tassabehji M, Read A, Gruss P 1994 Functional analysis of Pax-3 in Splotch and Waardenburg syndrome alleles. Submitted
Copp AJ 1994 Genetic models of mammalian neural tube defects. In: Neural tube defects. Wiley, Chichester (Ciba Found Symp 181) p 118–143
Epstein DJ, Vekemans M, Gross P 1991a *Splotch* (Sp^{2H}), a mutation affecting development of the mouse neural tube, shows a deletion within the paired homeodomain of *Pax-3*. Cell 67:767–774
Epstein DJ, Malo D, Vekemans M, Gros P 1991b Molecular characterization of a deletion encompassing the Splotch mutation on mouse chromosome 1. Genomics 10:89–93
Evans EP, Burtenshaw M, Beechey CV, Searle A 1988 A splotch locus deletion visible by Giemsa banding. Mouse News Lett 81:66
Goulding MD, Chalepakis G, Deutsch U, Erselius JR, Gruss P 1991 Pax-3, a novel murine DNA binding protein expressed during early neurogenesis. EMBO (Eur Mol Biol Organ) J 10:1135–1147
Goulding M, Sterrer S, Fleming J et al 1993a Analysis of the *Pax-3* gene in the mouse mutant *splotch*. Genomics 17:355–363
Goulding MD, Lumsden A, Gruss P 1993b Signals from the notochord and floor plate regulate the region-specific expression of two *Pax* genes in the developing spinal cord. Development 117:1001–1016
Kapron-Bras CM, Trasler DG 1988 Histological comparison of the effects of the splotch gene and retinoic acid on the closure of the mouse neural tube. Teratology 37:389–399
Kessel M, Gruss P 1991 Homeotic transformations of murine vertebrae and concomitant alteration of *Hox* codes induced by retinoic acid. Cell 67:89–104
Krauss S, Johansen T, Korzh V, Fjose A 1991a Expression of the zebrafish paired box gene *pax* [*zf-b*] during early neurogenesis. Development 113:1193–1206
Krauss S, Johansen T, Korzh V, Fjose A 1991b Expression pattern of zebrafish *pax* genes suggests a role in early brain regionalization. Nature 353:267–270
Krauss S, Maden M, Holder N, Wilson SW 1992 Zebrafish *pax*[*6*] is involved in the formation of the midbrain–hindbrain boundary. Nature 360:87–89
Mancino F, Vekemans M, Trasler DG, Gros P 1992 Segregation analysis reveals tight genetic linkage between the spontaneously arising neural tube defect *splotch* (*Sp*) and Pax-3 in an intraspecific mouse backcross. Cytogenet Cell Genet 61:143–145
Mangold O 1931 Über die induktionsfahigkeit der verschiedenen Bezirk der Neurula von Urodelen. Naturwissenschaften 21:394–397
Moase CE, Trasler DG 1989 Spinal ganglia reduction in splotch delayed neural tube defect mutants. Teratology 40:67–75
Moase CE, Trasler DG 1990 Delayed neural crest cell emigration from *Sp* and *Sp delayed* mouse neural tube explants. Teratology 42:171–182
Moase CE, Trasler DG 1991 N-CAM alterations in splotch neural tube defect mouse embryos. Development 113:1049–1058
O'Shea KS, Liu LH 1987 Basal lamina and extracellular matrix alterations in the caudal neural tube of the delayed Splotch embryo. Brain Res 465:11–20
Papalopulu N, Kintner CR 1994 Molecular genetics of neurulation. In: Neural tube defects. Wiley, Chichester (Ciba Found Symp 181) p 90–102

Püschel AW, Westerfield M, Dressler GR 1992 Comparative analysis of Pax-2 protein distributions during neurulation in mice and zebrafish. Mech Dev 38:197–208

Russell WL 1947 *Splotch*, a new mutation in the house mouse *Mus musculus*. Genetics 32:102

Slack JMW, Tannahill D 1991 Mechanisms of anteroposterior axis specification in vertebrates: lessons from the amphibians. Development 114:285–302

Tassabehji M, Read AP, Newton VE et al 1992 Waardenburg's syndrome patients have mutations in the human homologue of the *Pax-3* paired box gene. Nature 355:635–636

Tassabehji M, Read AP, Newton VE et al 1993 Mutations in the *PAX3* gene causing Waardenburg syndrome type 1 and type 2. Nat Genet 3:26–30

Treisman J, Harris E, Desplan C 1991 The paired box encodes a second DNA-binding domain in the paired homeodomain protein. Genes & Dev 5:594–604

Vogan KJ, Epstein DJ, Trasler DG, Gros P 1993 The *splotch-delayed* (*Sp^d*) mouse mutant carries a point mutation within the paired box of the *Pax-3* gene. Genomics 17:364–369

Walther C, Gruss P 1991 *Pax-6*, a murine paired box gene is expressed in the developing CNS. Development 113:1435–1449

Walther C, Guenet J-L, Simon D et al 1991 Pax: a multigene family of paired box containing genes. Genomics 11:424–434

Wilson DB 1974 Proliferation in the neural tube of the splotch (*Sp*) mutant mouse. J Comp Neurol 154:249–256

Zannini M, Francis-Lang H, Plachov D, Di Lauro R 1992 Pax-8, a paired domain-containing protein, binds to a sequence overlapping the recognition site of a homeodomain and activates transcription from two thyroid-specific promoters. Mol Cell Biol 12:4230–4241

DISCUSSION

Shum: In the chick the notochord is involved in restricting the expression of *Pax-3* to the dorsal part of the neural tube. Have you looked in *Danforth short tail* (*Sd*) mice or *Brachyury* mice, in which there is no notochord, to see whether *Pax-3* expression is evenly distributed throughout the neural tube or not?

Goulding: We looked at *Pax-3* expression in *Sd* mice. Rudi Balling has done the same. We see expression in ventral regions of the spinal cord. We have a lot of data from studies on the chick, which show that if you take away the notochord, *Pax-3* is expressed throughout the spinal cord, even in very ventral regions. The correlation between the expression of these genes and the cell types that are generated subsequent to their expression is pretty tight.

Shum: Does *Sd* share some of the phenotypic characteristics of *splotch*—since the pattern of *Pax-3* expression is altered in both mutants?

Goulding: The problem is that in *Sd*, these defects occur in the lumbosacral region, where there is not much neural crest. In these mice we see an opposite effect: cells are expressing *Pax-3* where they normally wouldn't. In *splotch* mice there is a loss of functional *Pax-3* mRNA from the embryo.

Hall: Is it clear how notochord affects expression of *Pax-3*?

Goulding: We don't know what the signals are, but we have some ideas. The basic model is that there are signals emanating from the neural fold that are required for activation of *Pax-3* expression. Normally, *Pax-3* is switched on progressively in more ventral cells and there is a wave of expression that moves from dorsal to ventral cells. If there is a notochord there, the cells in the ventral part of the spinal cord are unable to express *Pax-3*. If you place a notochord laterally, cells adjacent to the notochord fail to switch on *Pax-3* and as a result *Pax-3* is expressed only in cells located very dorsally on the side of the graft. If you take the notochord away, expression of *Pax-3* extends down into the very ventral part of the cord.

Hall: Is there a migratory effect or is it something that expresses because of the 'previous history' of the cells?

Goulding: We don't think cells are migrating. We think the signals may be diffusible factors that switch on *Pax-3* or prevent *Pax-3* from being switched on in these cells.

Schoenwolf: Is there any evidence for regulation of *Pax-3* expression in the mitotic cycle?

Goulding: We haven't looked at that in detail. Katharina Mullbacher in Peter Gruss' lab took a *Pax-3* expression construct that we made driven by the CMV (cytomegalovirus) promoter and put it into NIH/3T3 cells. The original assay was for oncogenes. Genes are introduced into these cells and if they have any transformation potential or if they can change the cell cycle in any way, the transformed cells tend to form foci. They are no longer contact inhibited and they pile up. Under those conditions, Katharina found that *Pax-3* seems to be quite a strong 'oncogene' in that it is able to induce foci in 3T3 cells. This goes quite nicely with the recent study that has shown that a human rhabdomyosarcoma is caused by a chromosomal translocation that involves the *Pax-3* gene (Barr et al 1993). This gives rise to an aberrant transcript that contains *Pax-3* sequences and sequences from some other gene. This transcript may be involved in generating these rhabdomyosarcomas.

Schoenwolf: Are there any early expressed genes having a dorsoventral sequence of expression in the neuraxis that could account for the migratory pathways that occur down the length of the axis? Is there anything that imparts positional information early in the dorsoventral plane? One would like to see a gene turned on very early in the floor plate, for example.

Goulding: Pax-3 is expressed very early—at the neural plate stage and in the primitive streak. You see a ring of cells around the neural folds in the chick expressing *Pax-3*; then, as the neural tube closes, expression of *Pax-3* shifts ventrally.

Morriss-Kay: Several genes are expressed differentially dorsoventrally at quite early stages, e.g. *HNF-3β* (Sasaki & Hogan 1993) and the gene for CRBPI (Ruberte et al 1993).

Hall: But do they actually do something to make cells stream down the chord?

Morriss-Kay: I don't know. These are just descriptive studies at the moment; obviously that's the kind of thing that we would like to study. With Gary Schoenwolf's notochord rearrangements, we are a bit closer to an experimental manipulative system that might give some of those answers (Schoenwolf, this volume).

Scott: Is anything known about the level of methylation of the *Pax-3* genes?

Goulding: We have not looked at it.

Scott: It is known that genes are up-regulated or down-regulated by methylation (Bird 1986). Folate has a role in methylation, so it is a possible site of action, if there were differences in methylation.

Hall: What about the other *Pax* genes?

Goulding: Most of them are expressed in the developing nervous system. *Pax-1* and *Pax-9* are not, they are expressed in the mesoderm. They seem to have a specific role in the segmentation of the somitic mesoderm.

Oakley: Were you showing us heterozygotes?

Goulding: No, those were homozygous *splotch* embryos. Only homozygous *splotch* embryos have exencephaly. In patients with Waardenburg syndrome, one sees several of the defects that are seen in *splotch* mice, for instance, Hirshprung's disease, which would correlate with a loss of enteric ganglia. They also have craniofacial abnormalities: they have the white forelock, some have mental retardation. These are all closely correlated with the phenotype of *splotch* mice. But the neural tube defects that you see in the homozygous *splotch* embryos are not seen in patients with Waardenburg syndrome, who are presumably heterozygous for the mutant *Pax-3*.

Copp: A *Pax-6* mutation in the rat has also been claimed to cause a neural crest abnormality (Matsuo et al 1993). *Pax-6* is correlated with the *small-eye* phenotype. The *Pax-6* mutation is similar to *Pax-3* mutations in that the neural crest is affected, but there is no neural tube defect.

Goulding: The paper reports that neural crest cells migrating into the region of the optic vesicle are affected. Whether this is the primary defect in these mice remains to be determined. In *small-eye* mice, the optic placode doesn't form. I believe this defect arises before the crest cells have migrated into the region of the optic placode.

Morriss-Kay: The crest cells migrate into the forebrain region long before there is an optic placode there (Tan & Morriss-Kay 1985). There are also crest cells that migrate from the forebrain around the developing optic vesicle; that's after optic placode formation (Bartelmez & Blount 1954). So crest cells derived from the midbrain neural folds are covering the basal surface of the forebrain neuroepithelium before there is an optic placode in the surface ectoderm.

O'Rahilly: The forebrain-derived crest is formed directly from the optic vesicle.

Lindhout: Are there phenotypic differences among the *splotch* mutants? Do they all show, for example, telencephalic or diencephalic disorganization?

Trasler: For the more extreme mutants, that have the large deletions in the gene, you only see the heterozygote, you don't see the homozygote. *splotch retarded* dies before implantation. Concerning the brain defect, I would say that once there is an exencephaly, they all look alike. The spina bifida is slightly larger or slightly smaller, whether it's the *splotch delayed* (Sp^d) or the *splotch* mutation. Sp^d was called that because it only has spina bifida; *splotch* has spina bifida with or without exencephaly.

Lindhout: Are there also heterozygotes that show epilepsy, as a result of the cerebral disorganization?

Trasler: I have never seen a *splotch* mouse behaving in a peculiar manner, however, my students have seen *splotch* heterozygotes have convulsions for short periods.

Hall: Do any of the *splotch* mutants have limb defects? Patients with Waardenburg type III syndrome have limb defects. Is there a correlation between the severity of the gene mutation and whether or not there is limb involvement or how severely the patients with Waardenburg syndrome are affected?

Trasler: Thomas Franz has shown changes in the muscle of the limb in two mutants, the Sp^{1H} and Sp^d.

Goulding: Pax-3 is normally expressed in the fore limb and in the hind limb.

Czeizel: Do mutations in *Pax-3* cause heart defects?

Goulding: Yes, truncus arteriosus.

Trasler: The *Sp* mouse is not deaf, whereas patients with Waardenburg syndrome are. What kind of deafness do they have? Karen Steel (personal communication) said it could be melanocytes missing in the ear, leading to sensorineural deafness.

Lindhout: As far as I know, it's nerve deafness and it occurs, depending on the type of Waardenburg syndrome, in 25–50% of family members.

Morriss-Kay: Melanin is missing from the stria vascularis of the cochlea as well as from all or part of the skin in a number of mutant animals, including the white cat (Mair 1973).

Trasler: The white cat is deaf.

Hall: The cats that are deaf are easy to tell because they have a blue eye on the deaf side.

Opitz: The white cat also has polydactyly.

References

Barr FG, Galili N, Holick J, Biegel JA, Rovera G, Emanuel BS 1993 Rearrangement of the PAX3 paired box gene in the paediatric solid tumour alveolar rhabdomyosarcoma. Nat Genet 3:113–117

Bartelmez GW, Blount MP 1954 The formation of neural crest from the primary optic vesicle in man. Contrib Embryol Carnegie Inst 35:55–71

Bird AP 1986 CpG-rich islands and the function of DNA methylation. Nature 321:209–213

Mair IWS 1973 Hereditary deafness in the white cat. Parts I–IV. Acta Oto-laryngol Suppl 314:1–48

Matsuo T, Osumi-Yamashita N, Noji S et al 1993 A mutation in the Pax-6 gene in rat *small eye* is associated with impaired migration of midbrain crest cells. Nat Genet 3:299–304

Ruberte E, Friederich V, Chambon P, Morriss-Kay GM 1993 Retinoic acid receptors and binding proteins. III. Their differential transcript distribution during mouse nervous system development. Development 118:267–282

Sasaki H, Hogan BLM 1993 Differential expression of multiple fork head related genes during gastrulation and axial pattern formation in the mouse embryo. Development 118:47–59

Schoenwolf GC 1994 Formation and patterning of the avian neuraxis: one dozen hypotheses. In: Neural tube defects. Wiley, Chichester (Ciba Found Symp 181) p 25–50

Tan SS, Morriss-Kay GM 1985 The development and distribution of the cranial neural crest in the rat embryo. Cell Tissue Res 240:403–416

Genetic models of mammalian neural tube defects

Andrew J. Copp

Developmental Biology Unit, Division of Cell and Molecular Biology, Institute of Child Health, University of London, 30 Guilford Street, London WC1N 1EH, UK

Abstract. Several mouse mutations disturb the embryonic process of neurulation, yielding neural tube defects. Analysis of the mutations offers the most feasible approach to understanding the aetiology and pathogenesis of human neural tube defects. Interactions between the non-allelic mutant genes and between several of the mutant genes and modifying genes in the genetic background modulate the frequency and severity of the defects that develop. Environmental factors interact with the genetic predisposition either to increase or to decrease the incidence of defects. The gene loci corresponding to two of the mutations, *splotch* (*Sp*) and *extra toes* (*Xt*), have been identified as those encoding the transcription factors Pax-3 and Gli3, respectively; their human homologues are associated with Waardenburg type I syndrome and Greig's cephalopolysyndactyly. Embryological analysis reveals that several of the mutations disturb the process of neural tube closure at the posterior neuropore (in the lumbosacral region), yielding spina bifida and/or tail defects. The different mutations appear to achieve this developmental end-point by different underlying mechanisms. In *curly tail* (*ct*), non-neural tissues proliferate abnormally slowly causing ventral curvature of the neuropore region and inhibiting neural tube closure. Neural tube defects can be prevented in cultured *ct/ct* embryos by experimentally correcting either the proliferative imbalance or the ventral curvature. In *Sp* the primary defect appears to reside in the neuroepithelium. A combination of genetic analysis, gene cloning and experimental embryology is revealing that neural tube defects in mice and, by implication, in humans are a developmentally heterogeneous group of malformations.

1994 Neural tube defects. Wiley, Chichester (Ciba Foundation Symposium 181) p 118–143

Disturbance of the embryonic process of neurulation, which results in the development of neural tube defects, is a common abnormality of human embryogenesis. Despite extensive study over many years, our understanding of the aetiology and pathogenesis of such defects remains rudimentary. Epidemiological studies have indicated a multifactorial aetiology for most neural tube defects: the interaction of multiple genetic and environmental factors is seen as determining the propensity of an individual to undergo faulty neurulation (Carter 1974). But none of the hypothetical genetic factors, and few environmental factors, have been identified, so that the multifactorial aetiology of neural tube defects remains to be convincingly demonstrated. As regards the pathogenesis of the defects, numerous hypotheses have been, and continue to be, proposed on the basis of observations on affected human fetuses. None of these hypotheses has been subjected to rigorous testing, however, because this would require examination of human embryos around the time of neurulation, which occurs when the embryo is at its most inaccessible, during the fourth week of embryonic development (the sixth week after the last menstrual period).

One possible solution to the problem of understanding the aetiology and pathogenesis of human neural tube defects is to study the development of corresponding defects in experimental animals, preferably mammals, and then attempt to extrapolate to the human condition. Fundamental differences in maternal physiology, for instance between rodents and humans, make it potentially hazardous to extrapolate findings concerning the effect of environmental factors on the embryo (Fraser 1977). The same does not appear to be true of genetic lesions, however. Conservation of genetic structure and function is being found to extend far down the evolutionary tree, for instance to the lowest vertebrates and even to invertebrates (Holland 1992). Within the mammals, genetic homologies are proving to be present to a quite remarkable degree (Nadeau 1989). Thus, by studying genetic causes of neural tube defects in experimental mammals (effectively this means the mouse, which is the only non-human mammal with well studied genetics), we may be able to perform fruitful extrapolation to humans, and so begin to unravel the aetiology and pathogenesis of the human defects.

Genetic models of neural tube defects in the mouse

Several genetic mutations have been identified that disrupt neurulation and yield neural tube defects in the mouse (Table 1). The existence of so many neural tube defect-producing genes argues that a number of distinct gene products may be required for neurulation to occur normally. Although most of the mutants were described many years ago, their analysis has only recently begun to provide information on the aetiology and pathogenesis of the defects that may be of direct relevance to understanding the human malformations. In this review, three main questions will be addressed in relation to research with mouse genetic models of neural tube defects.

(i) The role of interactions, both gene–gene and gene–environment, in the generation of neural tube defects. Is there evidence for a multifactorial aetiology of these defects in the mouse, as suggested for humans?

(ii) The nature of the genes disrupted in the neural tube defect-producing mutants. What is the function of the gene products and do these genes have human homologues?

TABLE 1 The principal mutant genes causing neural tube defects in the mouse[a]

Locus name	Alleles	Chromosome	Neural tube defect in homozygotes		Gene product	
			Cranial	Spinal	Name[b]	Type
Loop-tail	*Lp*	1	Craniorachischisis		nd	nd
cranioschisis	*crn*	nd	EX	—	nd	nd
exencephaly	*xn*	nd	EX	—	nd	nd
extra-toes	Xt, Xt^{bph}	13	EX	—	Gli3, GLI3	zinc finger, DNA-binding
SELH[c]	—	nd	EX	—	nd	nd
Bent-tail	*Bn*	X	EX	SB	nd	nd
curly tail	*ct*	4	EX	SB	nd	nd
splotch	Sp, Sp^{d}, Sp^{r}, Sp^{1H}, Sp^{2H}, Sp^{4H}	1	EX	SB	Pax-3, PAX-3	homeodomain, DNA-binding
Axial deformity	*Axd*	nd	—	SB	nd	nd
vacuolated lens	*vl*	1	—	SB	nd	nd

nd, not determined; EX, exencephaly; SB, spina bifida ± tail defect.
[a]References to individual mutants are given either in the text or in Copp et al (1990).
[b]Name of gene product is given for mouse (lower case letters) and human (upper case letters).
[c]SELH is a mouse strain carrying several polymorphic loci that interact additively to produce neural tube defects (MacDonald et al 1989).

(iii) The embryonic mechanisms that underlie the development of the defects. Do they all result from a uniform pathogenetic process or are multiple developmental mechanisms involved?

Gene–gene and gene–environment interactions in the aetiology of neural tube defects

Gene–gene interactions

Interactions between genes are being studied in two ways. First, interactions between the mutant genes themselves can be identified when the mutant genes are expressed in paired combinations in the same individual. Second, the expression of a single mutant gene can be compared when placed on different genetic backgrounds (i.e. inbred mouse strains) to discern differences in the 'modifier' gene composition of inbred strains.

There have been only a few studies, so far, in which mice have been bred carrying two or more of the neural tube defect-producing mutations, but the preliminary findings suggest that interactions between these non-allelic mutations may be a common occurrence. For instance, an interaction is known to occur between the *splotch* (*Sp*) and *curly tail* (*ct*) mutations (Estibeiro et al 1993). Around 10% of mice with genotype *Sp/+*, *ct/+* have been found to develop tail defects that resemble those occurring in the single homozygote *ct/ct* but not in either of the heterozygotes *Sp/+* or *ct/+*. Moreover, mice with genotype *Sp/+*, *ct/ct* were found to develop severe spinal neural tube defects, comprising lumbosacral spina bifida and tail defects, resembling those seen in *Sp/Sp* mice but more frequent and severe than those seen in *ct/ct* mice; few of the *Sp/+*, *ct/ct* mice survived beyond birth. Thus, defects in the mice carrying two mutant genes exceeded those in the corresponding single mutants in both incidence and severity.

It is a frequent finding that mutant genes are expressed variably on different inbred backgrounds in the mouse. This had led to the idea that the mouse genome contains 'modifying' genes that, while not producing a specific mutant phenotype by themselves, are capable of modulating the incidence and severity of mutant phenotypes. These modifying genes can be recognized by virtue of their polymorphisms among inbred strains. Expression of the *ct* mutation has been studied after matings of *ct/ct* mice to mice of a number of different inbred backgrounds, followed by backcrossing to the *ct/ct* parent: widely varying incidences of defects are seen between the different matings, with BALB/c and C57BL/6 promoting the highest penetrance and DBA/2 the lowest (Table 2). In a recent study (Neumann et al 1994), *ct/ct* mice were mated to each component strain of the BxD series of recombinant inbred (RI) strains. Each RI strain in the series is derived from brother–sister matings between a pair of F1 hybrids of the two parental strains, in this case C57BL/6 and DBA/2. The frequency

TABLE 2 Effect of genetic background on the incidence of neural tube defects in *curly tail* mutant mice

Strain[a]	*% offspring affected*[b] *(total no. mice scored)*	*Presumed penetrance (%) in ct/ct genotype*
ct stock matings	45.2 (2282)	45.2
BALB/c	23.9 (117)	47.8
C57BL/6	12.0 (1305)	24.0
A/Strong	8.0 (53)	16.0
PO	5.1 (970)	10.2
DBA/2	0.4 (256)	0.8

Data are from Embury et al (1979) (for the BALB/c and A/Strong strains) and A. J. Copp, unpublished (for the other strains).
[a]Mice of each inbred (BALB/c, C57BL/6 and DBA/2) and outbred (A/Strong and PO) strain were mated to *ct/ct* mice and then the F1 hybrids backcrossed to *ct/ct*. The backcross litters comprise *ct/ct* and *ct/+* individuals in a 1:1 ratio; the incidence of defects in these mice is shown in the second column, while the penetrance in *ct/ct* homozygotes is shown in the third column, calculated on the assumption that half the backcross offspring are of genotype *ct/ct*. In the case of the *ct* stock matings (first row of table), both parents were *ct/ct* homozygotes, so all offspring were also homozygous.
[b]Mice were scored within three days of birth. Affected individuals had either a spina bifida (± tail defect) or a tail defect alone.

of defects observed when *ct/ct* was mated to each RI strain was found to vary in a continuous fashion, from the maximum seen with the C57BL/6 strain to a minimum value, as seen with DBA/2. The distribution of defect frequencies among the RI strains suggested that at least three modifier genes vary between the parental strains. Comparison of the distribution of *ct* defect frequencies with the strain distribution pattern for a large number of mapped genes indicates that two of the modifiers map to chromosomes 3 and 5 (Neumann et al 1994). Thus, use of the RI strain approach permits the localization of genes that, while not causing neural tube defects by themselves, can modify the expression of defective phenotypes.

Gene–environment interactions

A great many teratogenic agents have been identified as causing neural tube defects in rodent embryos (Copp et al 1990), but in only a few cases have such teratogens been administered to embryos carrying neural tube defect-producing mutations. These studies show that gene–teratogen interactions can either increase or decrease the frequency of defects. It is perhaps most instructive to consider the interaction of a single teratogen with several different mutants. For instance, retinoic acid has an interesting spectrum of interactions with the mutants *Axial deformity* (*Axd*), *ct*, *Loop tail* (*Lp*), *Sp* and with strain SELH (Table 3). Treatment at 8.5 days of gestation affects mainly closure

TABLE 3 Effect of treatment with retinoic acid at 8.5 or 9.5 days of gestation on the incidence of neural tube defects in litters containing mutant embryos, by comparison with litters containing wild-type control embryos

		8.5 days of gestation		*9.5 days of gestation*		
Mutant	*Genotype*	*Exencephaly*	*SB*	*Exencephaly*	*SB*	*Ref*[a]
Axial deformity	*Axd/+*	nd	nd	No effect	No effect	1
curly tail	*ct/ct*	Increase	No effect	No effect	Decrease	2
Loop tail	*Lp/+*	Increase	No effect	nd	nd	3
Splotch	*Sp/+*	Increase	Increase	No effect	Increase	4
	Sp/Sp	Increase	Decrease[b]	Decrease[b]	Decrease[b]	5
SELH	—	Increase	nd	No effect	nd	6

SB, spina bifida ± tail defect; nd, not determined or not applicable.
[a]References: 1, Haviland & Essien (1990); 2, Seller et al (1979); 3, Wilson et al (1990); 4, Kapron-Bras & Trasler (1984; 1988); 5, Kapron-Bras & Trasler (1985); 6, Tom et al (1991).
[b]Decrease in incidence may be due to selective loss of affected embryos.

of the cranial neural tube and the effect of retinoic acid seems to be uniformly exacerbating: the incidence of exencephaly, the developmental precursor of anencephaly, is increased in all cases studied. In the lower spinal region, by contrast, retinoic acid has varying interactions with the different mutants: it exacerbates the development of spinal neural tube defects in *Sp/+*, prevents the malformations in *ct/ct* and has no apparent effect on the incidence of defects in *Lp/+* or *Axd/+*. This variability of effect is consistent with an idea, developed later in this review, that the cellular/molecular mechanisms underlying closure of the posterior neuropore are diverse and, therefore, likely to be affected in different ways by administration of excess retinoic acid during development.

Of particular relevance to human studies of neural tube defects is the possibility of an interaction between folic acid and such defects in animals. Wild-type mice when rendered folate deficient exhibit a diminished reproductive performance, but neural tube defects are not seen among their offspring (Heid et al 1992). Similarly, rat embryos develop more slowly *in vitro* in the absence of exogenous folate but do not develop neural tube defects (Cockroft 1988). There is some evidence for a role of folate in neural tube defects resulting from the action of valproic acid in the mouse, but this result is controversial (see Nau, this volume, Hansen & Grafton 1991). In the case of the *ct* (Seller & Adinolfi 1981) and *Axd* (Essien & Wannberg 1993) mutants, administration of folate during pregnancy has no effect on either the incidence or severity of the defects. These largely negative results with respect to the effect of folate on animal neural tube defects may indicate that folate has an effect on these defects which is based on a particular genetic predisposition that has not yet been studied in the animal models.

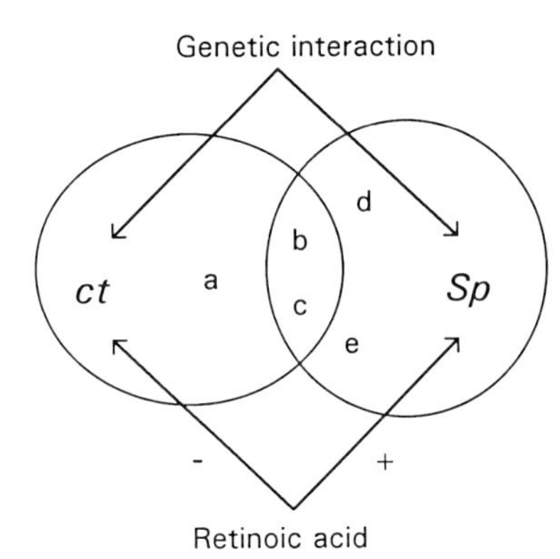

FIG. 1. Diagram illustrating the interaction of genetic and environmental factors in the aetiology of mouse neural tube defects. Mutant genes such as *curly tail* (*ct*) and *splotch* (*Sp*) interact with each other either to increase or decrease the incidence and severity of the neural tube defect phenotype. The expression of each mutant is also affected by a set of polymorphic modifying genes (depicted by a, b, c, d and e) that have no neural tube defect-producing activity by themselves. Environmental factors such as retinoic acid further interact with the mutant and/or modifying genes to determine the overall predisposition of each individual to develop a defect.

In conclusion, a picture is emerging from studies of the mouse mutants of a complex interacting network of neural tube defect-producing genes, some of large and some of small effect, whose expression is modulated by environmental teratogens (Fig. 1). Extrapolated to a randomly breeding human population, this can be seen to generate exactly the aetiological framework proposed by Carter (1974). The chance association of one or more mutated genes, together with a suitable combination of modifying genes and environmental factors, could be expected to lead to apparently 'sporadic' cases of neural tube defects. The genetic component will be shared with first-degree relatives giving a higher rate of recurrence in family members.

Gene products encoded by the mouse neural tube defect mutations and their human homologues

Work is in progress in many laboratories worldwide towards cloning genes in the mouse that have likely relevance to human disease states. The neural tube defect-producing mutations are examples of this group of interesting mouse genes. To date, the gene sequences of two of the mutations, *Sp* and *extra toes* (*Xt*), have been isolated (Table 1). The *Sp* locus encodes the Pax-3 transcription factor which is a member of the family of *Pax* genes (see Goulding & Paquette, this volume) that were isolated by virtue of their homology to the *Drosophila* gene *paired*. The human homologue of *Pax-3* (formerly named *HuP2*, now called *PAX-3*) is mutated in patients with Waardenburg syndrome type I (Tassabehji et al 1992, Baldwin et al 1992). These patients share with the *Sp/+* mouse the feature of disturbed neural crest migration that produces reduced pigmentation in both species and craniofacial defects and sensorineural deafness in humans. Since most Waardenburg patients are heterozygous for the mutation, the features characterizing the *Sp/Sp* phenotype (severe spina bifida and exencephaly) are not expected to be widespread in humans. Nevertheless, there has been a recent report of spina bifida in a family with Waardenburg syndrome (Chatkupt & Johnson 1993).

The gene disrupted by the *Xt* mutation was found recently as a result of analysis of a transgenic mouse strain in which an insertional event had occurred, producing a phenotype similar to that seen in *Xt* (Pohl et al 1990). The *Xt* mutation disrupts the *Gli3* gene (Schimmang et al 1992, Hui & Joyner 1993), which encodes a transcription factor and is a member of a gene family with homology to the *Drosophila* gene *Krüppel*. The human *GLI3* locus is disrupted in some patients with Greig's cephalopolysyndactyly (Vortkamp et al 1991), confirming previous suggestions of homology between patients with this syndrome and the *Xt* mouse (Winter & Huson 1988). Patients with Greig's cephalopolysyndactyly are characterized mainly by craniofacial and limb malformations: neural tube defects are not a recognized feature of the syndrome. This is not surprising, however, since the neural tube defects in *Xt* are of low penetrance, occur only in homozygotes and are expected to be perinatally lethal because they comprise exencephaly/anencephaly, not spina bifida.

The prediction, stated earlier, that the mouse neural tube defect mutations will have important human homologues is proving to be true for the loci that have been isolated so far. *Sp* and *Xt* may not be the most likely candidates for genes that control the majority of human neural tube defects because each mutation produces multisystem defects that are not present in most affected humans. Genes such as *Axd*, *cranioschisis* (*crn*), *ct*, *Lp*, *exencephaly* (*xn*) and those mutated in strain SELH that produce more isolated defects, will perhaps prove to have human homologues of more general importance in the aetiology of human neural tube defects.

Embryonic mechanisms of neural tube defects

Apart from determining the aetiological factors responsible for neural tube defects, the other main aim of research in this field is to determine the chain of embryonic events that intervenes between the action of the causative factor, whether genetic or environmental, and the development of the birth defect (i.e. the pathogenesis of these defects). In the case of genetically determined defects, the possibility arises of relating the pathogenetic events to the action of a single gene product, whereas purely environmentally induced defects may have their origin in multisystem effects. For this reason, we have chosen to concentrate on genetic models of neural tube defects in the mouse embryo. This section will deal, firstly, with some methodological aspects of applying an experimental embryological approach to the development of the defects and, secondly, with the current view of the pathogenesis of lower spinal neural tube defects in several of the mouse mutants.

Culture and identification of genotype in mutant embryos

Experimental studies of mouse and rat neurulation are made possible by the use of whole embryo culture (New 1978). Prospective analysis can be performed in which embryos are either observed at several times during the culture period or are manipulated before culture and then observed to determine the effect of the manipulation. When using experimental manipulations *in vitro* that might alter an embryo's eventual phenotype, it is important to be able to establish the genotype of the embryos independently of their morphology. For this purpose, we are using the polymerase chain reaction (PCR) to amplify specific nucleotide sequences from the very small amounts of DNA that are available in the early embryo (Fig. 2). In cases where a gene has been cloned, it is possible to determine genotype directly: for instance, embryos carrying the Sp^{2H} mutation can be recognized (Fig. 3a) by virtue of a 32 bp deletion present within the coding sequence of the gene (Epstein et al 1991). In cases where the gene is not yet cloned, we identify embryo genotype using closely linked sequences that define a polymorphism detectable by PCR, so-called microsatellite sequences (Hearne et al 1992). Since microsatellites are plentiful and widespread throughout the genome, this method is generally applicable to the majority of mutant genes. We have used the *Crp* polymorphic microsatellite on distal chromosome 1 to 'mark' embryos segregating for the *Lp* mutation (Fig. 3b) (Copp & Checiu 1994).

Embryonic mechanisms underlying the development of lower spinal neural tube defects

The best understood category of neural tube defects, in terms of the underlying embryonic basis, comprises spina bifida and associated caudal defects. Although the morphology of exencephaly/anencephaly has been described in several genetic models, there has not yet been an experimental analysis of the embryonic mechanisms underlying this type of defect. A critical event in neural tube development in the lower spinal region is the transition between primary neurulation (where the neural tube is formed by elevation and fusion of neural folds) and secondary neurulation (where the neural tube forms by cavitation of a solid cord of cells). The transition between the two modes of neurulation occurs at the site of posterior neuropore closure in the upper sacral region of mouse (Copp & Brook 1989) and, probably, human (Müller & O'Rahilly 1987)

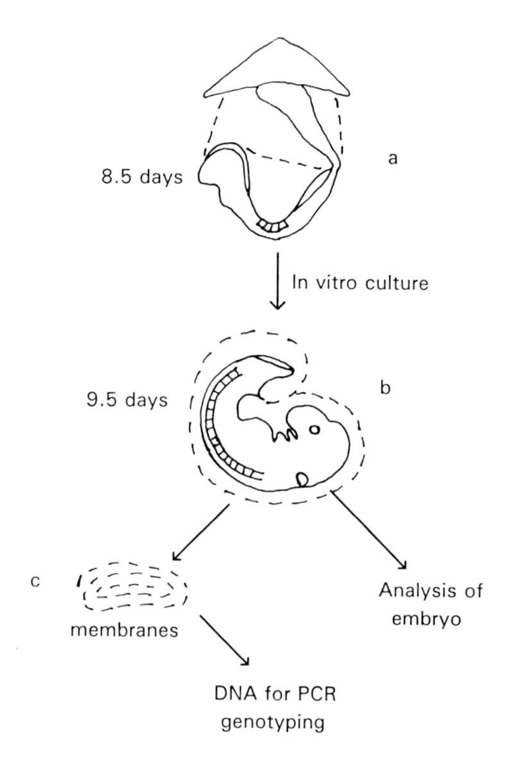

FIG. 2. Experimental methodology for studying the embryonic mechanisms of genetically determined mouse neural tube defects. (a) Embryos can be explanted from the uterus, maintaining the extraembryonic membranes intact (shown as dashed lines) and cultured by the method of New (1978). In the case illustrated, a 24 h culture period, from 8.5 to 9.5 days, is shown but other embryonic stages can also be grown *in vitro*. (b) At the end of the culture period, the extraembryonic membranes (dashed lines) are removed and processed for DNA extraction (c), to permit genotype identification. The embryo is then available for a variety of types of analysis, e.g. *in situ* hybridization to determine the pattern of gene expression.

embryos. This transition appears to be a point of susceptibility to neural tube defects, perhaps because of mechanical instability at the interface between the two modes of neurulation or perhaps because closure of the posterior neuropore is an event with a strict 'time window': that is, delay in neuropore closure, which is caused by a number of the mutations, can be tolerated by the embryo only within fairly narrow limits, because of the need to initiate secondary neurulation in a caudal direction from this level.

In most of the well-studied mouse mutants, open spina bifida is known to develop from failure of closure of the posterior neuropore. Moreover, even when the neuropore completes closure in a delayed fashion, the development of

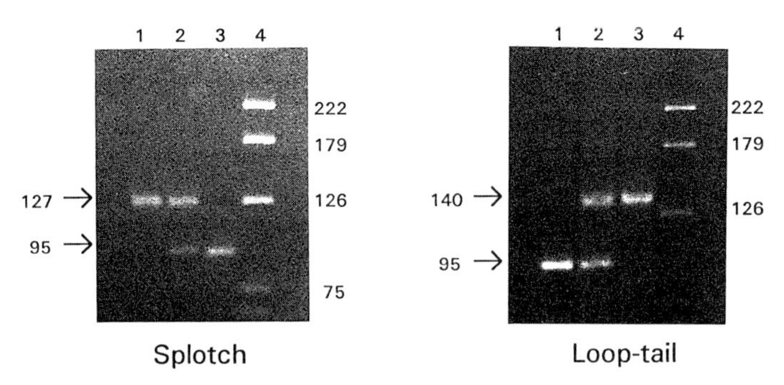

FIG. 3. Agarose gels (4%) stained with ethidium bromide to illustrate the use of PCR for identifying the genotype of embryos in litters segregating lethal mutations. For the Sp^{2H} mutation (left-hand panel), the embryo genotype can be determined directly since a 32 bp deletion in the *Pax-3* gene can be identified in a DNA fragment amplified by primers that flank the deletion (Epstein et al 1991). The +/+ embryo (lane 1) exhibits only the intact 127 bp PCR product, the Sp^{2H}/Sp^{2H} (lane 3) which develops neural tube defects has only the 95 bp fragment and the $Sp^{2H}/+$ heterozygote (lane 2) has both bands. Lane 4 shows marker DNA fragments with sizes as shown on the right side of the gel. For the *Lp* mutation (right-hand panel), embryonic genotype can be determined using a microsatellite polymorphism at the *Crp* locus, which is closely linked to *Lp* on distal chromosome 1. Wild-type embryos (lane 1) are homozygous for a 95 bp PCR product, *Lp/Lp* embryos (lane 3) are homozygous for a 140 bp variant and *Lp/+* embryos (lane 2) exhibit both fragments. Lane 4 shows DNA markers as for the other gel. The accuracy of genotyping using linked markers is inversely proportional to the genetic distance between the mutant and marker loci.

the tail is disturbed, resulting in a dorsal curl/loop/kink malformation. Thus, it is clear that spina bifida and a tail defect can be manifestations of the same developmental lesion—disturbed closure of the posterior neuropore. As studies progress on the embryonic mechanisms of posterior neuropore closure, it is becoming clear that several distinct processes can yield the same morphogenetic lesion (Fig. 4).

In the *ct* mutant, posterior neuropore closure is delayed as a result of a defect expressed in non-neural tissues, the notochord and hindgut endoderm. The neuroepithelium of *ct/ct* embryos is not compromised in its ability to undergo posterior neuropore closure, as was shown in a recent experiment (van Straaten et al 1993) in which the neuroepithelium was grown in isolation *in vitro*. While neuropore closure was delayed in *ct/ct* explants containing both neuroepithelium and non-neural tissues, the mutant microsurgically isolated neuroepithelium closed its neuropore at a rate comparable to wild-type explants. The defect in *ct/ct* notochord and hindgut endoderm is expressed as a reduced rate of cell proliferation, specifically in embryos with 27–29 somites, just before the normal time of neuropore closure (Copp et al 1988a). Experimental correction of the cell proliferation imbalance can prevent the development of neural tube defects

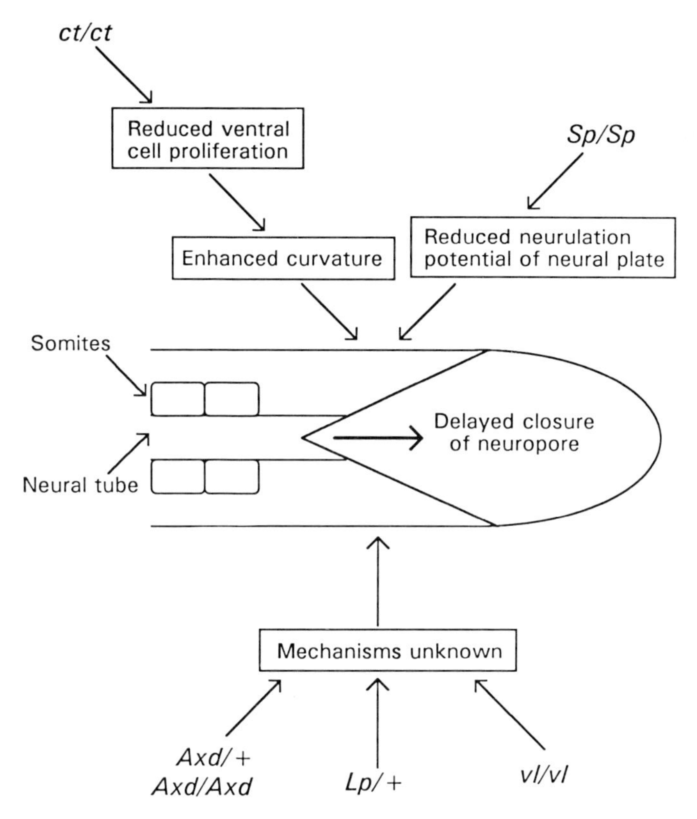

FIG. 4. Diagram illustrating the pathogenesis of spina bifida in the mouse mutants. A final common pathway for all of the mutants is represented by delayed closure of the posterior neuropore in the lumbosacral region. Each mutant seems likely to produce this phenotype by a different cellular and molecular mechanism. The known mechanisms for *curly tail* (*ct*) and *splotch* (*Sp*) are shown in the upper part of the diagram. The mechanisms underlying delayed neuropore closure in the *Axial deformity* (*Axd*), *Loop-tail* (*Lp*) and *vacuolated lens* (*vl*) mutants are unknown.

(Copp et al 1988b). Owing to tethering of the notochord and hindgut to the overlying (normally proliferating) neuroepithelium, the entire caudal region is caused to undergo ventral curvature. This curvature is responsible for delaying or preventing closure of the neuropore, as has been shown in an experiment in which the caudal region was splinted to prevent curvature from developing by insertion of an eyelash tip into the hindgut lumen. Embryos with eyelash implants developed significantly less curvature than controls and closed their neuropores normally in a high proportion of cases (Brook et al 1991). The molecular basis of the *ct* defect is unclear, although it is known that accumulation of newly synthesized hyaluronan, a component of the extracellular matrix, is reduced at the basal surface of the neuroepithelium and around the notochord in affected embryos (Copp & Bernfield 1988). The precise relationship between this abnormal accumulation of hyaluronan and the cell proliferation defect in *ct* embryos remains to be determined.

Failure of neuropore closure in *Sp/Sp* embryos appears to differ from the mechanism in *ct/ct* in that the neuroepithelium itself is primarily affected. Although there is, as yet, no experimental evidence for this assertion, three lines of descriptive evidence do support the idea: (i) the *Pax-3* gene, which is mutated in *Sp*, is first expressed in the neuroepithelium and neural crest cell lineages (Goulding et al 1991). (ii) Migration of the neural crest is defective in *Sp/Sp* embryos (Auerbach 1954). Since the neural crest cell lineage arises from the junction of neuroepithelium and surface ectoderm, this finding is consistent with a general neuroepithelial defect in *Sp*. (iii) Measurements of the angle of ventral curvature indicate that, unlike in *ct/ct* embryos, the caudal region of *Sp/Sp* is less curved than normal (Estibeiro et al 1993). This is consistent with the idea that *Sp/Sp* embryos have a normal rate of cell proliferation in ventral tissues, together with a defective neuroepithelium, leading to a relative straightening of the caudal region. Pax-3 is a DNA-binding transcription factor (Goulding et al 1991) and its absence in a functional form in *Sp/Sp* embryos presumably leads to downstream misregulation of gene expression. Several potential downstream targets of Pax-3 are known to be misexpressed in *Sp/Sp* embryos, including the cell adhesion molecule N-CAM (Moase & Trasler 1991) and the extracellular matrix components, heparan sulphate and chondroitin sulphate proteoglycans (Trasler & Morriss-Kay 1991), but it is unclear whether these molecules are involved in the development of neural tube defects in *Sp* embryos.

In the mutant *Lp*, the homozygote develops a severe condition resembling craniorachischisis in humans, in which the neural tube is open throughout the brain and spine. This is due to failure of initiation of neural tube closure at the 6–7 somite stage (Copp & Checiu 1994). It has only recently emerged that *Lp/+* heterozygotes display delayed closure of the posterior neuropore and that this is responsible for development of tail defects and a low frequency of spina bifida in these mice (Copp & Checiu 1994). Thus, homozygosity and heterozygosity for this mutant gene lead to disturbance of two distinct neurulation events. Interestingly, a third event, apposition of the neural folds at the midbrain/forebrain boundary, occurs apparently normally in *Lp/Lp* and *Lp/+* embryos. Thus, the *Lp* mutant provides a good indication of the mechanistic heterogeneity responsible for development of neural tube defects at different levels of the body axis. The cellular and molecular mechanisms responsible for the neurulation defects in the *Lp* mutant are not yet understood. Similarly, although delayed neuropore closure has been demonstrated in *vacuolated lens* homozygotes (Wilson & Wyatt 1988), and probably in *Axd* heterozygotes and homozygotes (Essien et al 1990), there are no clues, as yet, to the pathogenetic mechanisms in these mutants.

Conclusions

It is now possible to offer tentative answers to the questions posed at the beginning of this review. The concept of a multifactorial aetiology of neural tube defects is borne out by the studies of the mouse mutants. Multiple genes, some with neural tube defect-inducing potential and some with apparently no effect by themselves, interact to determine the propensity of each individual to develop defects. Moreover, environmental perturbations further modulate the action of the genes, controlling the severity and incidence of the resulting defects. In terms of the pathogenesis of neural tube defects, the mutant mouse genes are proving to act through a small number of final common pathways of development that correspond to the major morphological transitions during neurulation. Posterior neuropore closure is one of these transitions that appears to be readily disturbable, leading to spina bifida and tail defects. Closure of the neural folds at the forebrain/midbrain boundary and within the hindbrain probably represents the corresponding events that, when disturbed, lead to exencephaly/anencephaly. For posterior neuropore closure, it is beginning to appear that a variety of cellular and molecular disturbances can independently yield the same morphological result. If this conclusion can be extrapolated to humans, it will have implications for the prevention of neural tube defects. Preventive measures may need to be specific for each of the different mechanisms of pathogenesis. For instance, we have shown that splinting the caudal region can prevent defects in *ct/ct* embryos, but this would not be expected to rescue *Sp/Sp* embryos which do not exhibit enhanced ventral curvature. There may be a need for a series of distinct preventive measures designed to match the variety of pathogenetic mechanisms that exist. In this context, it will be essential to predict the type of pathogenetic mechanism operating, by developing genetic diagnostic techniques for prospective parents. The finding of human homologues for the recently cloned mouse genes suggests that this goal may be realizable in the not too distant future.

Acknowledgements

I thank Peter Thorogood for reading the manuscript. I am indebted to my colleagues, Paul Neumann, Wayne Frankel and Merton Bernfield for allowing me to describe our unpublished data on the map position and genetic interactions of *ct*. The original research described in this review was supported by the Imperial Cancer Research Fund and as part of a Multicenter Agreement for Studying Neural Tube Defects in Mutant Mice funded by the National Institutes of Child Health and Human Development, USA, through Cooperative Agreement HD 28882-01.

References

Auerbach R 1954 Analysis of the development effects of a lethal mutation in the house mouse. J Exp Zool 127:305–329

Baldwin CT, Hoth CF, Amos JA, da-Silva EO, Milunsky A 1992 An exonic mutation in the *HuP2* paired domain gene causes Waardenburg's syndrome. Nature 355:637–638

Brook FA, Shum ASW, van Straaten HWM, Copp AJ 1991 Curvature of the caudal region is responsible for failure of neural tube closure in the curly tail (*ct*) mouse embryo. Development 113:671–678

Carter CO 1974 Clues to the aetiology of neural tube malformations. Dev Med Child Neurol Suppl 32:3–15

Chatkupt S, Johnson WG 1993 Waardenburg syndrome and myelomeningocele in a family. J Med Genet 30:83–84

Cockroft DL 1988 Changes with gestational age in the nutritional requirements of postimplantation rat embryos in culture. Teratology 38:281–290

Copp AJ, Bernfield M 1988 Accumulation of basement membrane-associated hyaluronate is reduced in the posterior neuropore region of mutant (curly tail) mouse embryos developing spinal neural tube defects. Dev Biol 130:583–590

Copp AJ, Brook FA 1989 Does lumbosacral spina bifida arise by failure of neural folding or by defective canalisation? J Med Genet 26:160–166

Copp AJ, Checiu I 1994 Microsatellite DNA markers for mouse developmental mutations: defining the embryonic defects in the loop-tail (*Lp*) mutant. Submitted

Copp AJ, Brook FA, Roberts HJ 1988a A cell-type-specific abnormality of cell proliferation in mutant (curly tail) mouse embryos developing spinal neural tube defects. Development 104:285–295

Copp AJ, Crolla JA, Brook FA 1988b Prevention of spinal neural tube defects in the mouse embryo by growth retardation during neurulation. Development 104:297–303

Copp AJ, Brook FA, Estibeiro JP, Shum ASW, Cockroft DL 1990 The embryonic development of mammalian neural tube defects. Prog Neurobiol 35:363–403

Embury S, Seller MJ, Adinolfi M, Polani PE 1979 Neural tube defects in curly-tail mice. I. Incidence and expression. Proc R Soc Lond B Biol Sci 206:85–94

Epstein DJ, Vekemans M, Gros P 1991 *splotch* (Sp^{2H}), a mutation affecting development of the mouse neural tube, shows a deletion within the paired homeodomain of *Pax-3*. Cell 67:767–774

Essien FB, Wannberg SL 1993 Methionine but not folinic acid or vitamin B-12 alters the frequency of neural tube defects in *Axd* mutant mice. J Nutr 123:27–34

Essien FB, Haviland MB, Naidoff AE 1990 Expression of a new mutation (*Axd*) causing axial defects in mice correlates with maternal phenotype and age. Teratology 42:183–194

Estibeiro JP, Brook FA, Copp AJ 1993 Interaction between *splotch* (*Sp*) and *curly tail* (*ct*) mouse mutants in the embryonic development of neural tube defects. Development 119:113–121

Fraser FC 1977 Relation of animal studies to the problem in man. In: Wilson JG, Fraser FC (eds) Handbook of teratology. Plenum Press, New York, vol 1:75–96

Goulding M, Paquette A 1994 *Pax* genes and neural tube defects in the mouse. In: Neural tube defects. Wiley, Chichester (Ciba Found Symp 181) p 103–117

Goulding MD, Chalepakis G, Deutsch U, Erselius JR, Gruss P 1991 Pax-3, a novel murine DNA binding protein expressed during early neurogenesis. EMBO (Eur Mol Biol Organ) J 10:1135–1147

Hansen DK, Grafton TF 1991 Lack of attenuation of valproic acid-induced effects by folinic acid in rat embryos in vitro. Teratology 43:575–582

Haviland MB, Essien FB 1990 Expression of the *Axd* (axial defects) mutation in the mouse is insensitive to retinoic acid at low dose. J Exp Zool 256:342–346
Hearne CM, Ghosh S, Todd JA 1992 Microsatellites for linkage analysis of genetic traits. Trends Genet 8:288–294
Heid MK, Bills ND, Hinrichs SH, Clifford AJ 1992 Folate deficiency alone does not produce neural tube defects in mice. J Nutr 122:888–894
Holland P 1992 Homeobox genes in vertebrate evolution. BioEssays 14:267–273
Hui C, Joyner AL 1993 A mouse model of Greig cephalopolysyndactyly syndrome: the *extra-toes^J* mutation contains an intragenic deletion of the *Gli3* gene. Nat Genet 3:241–246
Kapron-Bras CM, Trasler DG 1984 Gene–teratogen interaction and its morphological basis in retinoic acid-induced mouse spina bifida. Teratology 30:143–150
Kapron-Bras CM, Trasler DG 1985 Reduction in the frequency of neural tube defects in splotch mice by retinoic acid. Teratology 32:87–92
Kapron-Bras CM, Trasler DG 1988 Interaction between the splotch mutation and retinoic acid in mouse neural tube defects in vitro. Teratology 38:165–173
MacDonald KB, Juriloff DM, Harris MJ 1989 Developmental study of neural tube closure in a mouse stock with a high incidence of exencephaly. Teratology 39: 195–213
Moase CE, Trasler DG 1991 N-CAM alterations in splotch neural tube defect mouse embryos. Development 113:1049–1058
Müller F, O'Rahilly R 1987 The development of the human brain, the closure of the caudal neuropore, and the beginning of secondary neurulation at stage 12. Anat Embryol 176:413–430
Nadeau JH 1989 Maps of linkage and synteny homologies between mouse and man. Trends Genet 5:82–86
Nau H 1994 Valproic acid-induced neural tube defects. In: Neural tube defects. Wiley, Chichester (Ciba Found Symp 181) p 144–160
Neumann PE, Copp AJ, Frankel WN, Letts VA, Coffin JM, Bernfield M 1994 Multifactorial inheritance of neural tube defects: localization of the major gene and two modifiers in curly-tail mutant mice. Submitted
New DAT 1978 Whole-embryo culture and the study of mammalian embryos during organogenesis. Biol Rev 53:81–122
Pohl TM, Mattei M-G, Rüther U 1990 Evidence for allelism of the recessive insertional mutation *add* and the dominant mouse mutation *extra-toes* (*Xt*). Development 110:1153–1157
Schimmang T, Lemaistre M, Vortkamp A, Rüther U 1992 Expression of the zinc finger gene *Gli3* is affected in the morphogenetic mouse mutant *extra-toes* (*Xt*). Development 116:799–804
Seller MJ, Adinolfi M 1981 The curly-tail mouse: an experimental model for human neural tube defects. Life Sci 29:1607–1615
Seller MJ, Embury S, Polani PE, Adinolfi M 1979 Neural tube defects in curly-tail mice. II. Effect of maternal administration of vitamin A. Proc R Soc Lond Ser B Biol Sci 206:95–107
Tassabehji M, Read AP, Newton VE et al 1992 Waardenburg's syndrome patients have mutations in the human homologue of the *Pax-3* paired box gene. Nature 355:635–636
Tom C, Juriloff DM, Harris MJ 1991 Studies of the effect of retinoic acid on anterior neural tube closure in mice genetically liable to exencephaly. Teratology 43:27–40
Trasler DG, Morriss-Kay G 1991 Immunohistochemical localization of chondroitin and heparan sulfate proteoglycans in pre-spina bifida splotch mouse embryos. Teratology 44:571–579

van Straaten HWM, Hekking JWM, Consten C, Copp AJ 1993 Intrinsic and extrinsic factors in the mechanism of neurulation: effect of curvature of the body axis on closure of the posterior neuropore. Development 117:1163–1172
Vortkamp A, Gessler M, Grzeschik K-H 1991 GLI3 zinc-finger gene interrupted by translocations in Greig syndrome families. Nature 352:539–540
Wilson DB, Wyatt DP 1988 Closure of the posterior neuropore in the vl mutant mouse. Anat Embryol 178:559–563
Wilson DB, Wyatt DP, Gookin JL 1990 Cranial effects of retinoic acid in the loop-tail (*Lp*) mutant mouse. J Craniofacial Genet Dev Biol 10:75–81
Winter RM, Huson SM 1988 Greig cephalopolysyndactyly syndrome: a possible mouse homologue (Xt—extra toes). Am J Med Genet 31:793–798

DISCUSSION

Seller: There are many agents that induce neural tube defects. Trypan blue is one for which the mechanism was worked out a long time ago (Williams et al 1976). It seems particular relevant when we are considering a possible nutritional deficiency in humans as a cause of neural tube defects, because trypan blue works by depressing histiotrophic nutrition in rodents.

Hall: Are the neural tube defects similar to those caused by other agents?

Nau: Trypan blue is one of the rare examples which produces mostly spina bifida and not exencephaly.

Copp: Williams et al (1976) claimed that trypan blue exerted its effect via the yolk sac. But, Dencker (1977) showed that it does actually reach the embryo. The mechanism of trypan blue teratogenicity may not be as simple as first thought.

Nau: A Japanese group has compared the rat and the rabbit in this regard. In the rat, the yolk sac is more prominent during earlier gestation (early organogenesis) than in the rabbit. If one administers trypan blue at a time when the yolk sac is not fully established in the rabbit, but is present in the rat, you can differentiate whether or not the effect occurs through the yolk sac. Teratogenic effects of trypan blue were observed during that gestational period in the rat, but not the rabbit. These findings are compatible with the hypothesis that the neural tube defects induced by trypan blue in the rat are the result of interference with yolk sac function.

Seller: You mentioned that wild-type mice deficient in folate did not show an increased incidence of neural tube defects (Heid et al 1992). M. N. Nelson (1960) found that folic acid depletion does produce neural tube defects in wild-type mice. He and others also showed that there was an alteration of the mitotic cycle in neural plate cells (Johnson et al 1963).

Copp: Those studies used a cytotoxic drug that was anti-folate in its action, which produced neural tube defects. But a drug of that sort might have other effects.

Seller: How did Heid et al produce folate depletion?

Copp: They fed the mice a folate-deficient diet before and during pregnancy.

Seller: That could be more akin to chronic folate deficiency.

Morriss-Kay: Is folate stored? How long does it take to get folate-deficient mice?

Copp: These mice were assessed for their folate level and were deficient; there's no question about that.

Scott: It has always been surprising that the use of folate-deficient diets has not increased the incidence of neural tube defects in experimental animals. It is now clear that other studies using such diets failed to produce true folate deficiency because such diets contain appreciable amounts of folate. True folate deficiency in rats or mice can be achieved only by using the chemically defined diet of Walzem et al (1983). This diet is completely free of folate and also contains sulphonamide to prevent synthesis of the vitamin by the gut microflora. Unlike previous folate-deficient diets, this diet produces clear signs of folate deficiency, such as anaemia. This diet was used by Heid et al (1992) and failed to produce neural tube defects in mice. Perhaps folate-deficient embryos in such experiments do not survive and thus one sees no increase in incidence of neural tube defects.

Copp: With mice, we can look all the way through gestation, so survival is not an issue. Heid et al looked during gestation and did not see any evidence of neural tube defects.

Seller: Mice don't have any intestinal synthesis of folate, whereas humans do.

Scott: Mice do have intestinal synthesis but it can be prevented by the presence of sulphonamides in the diet.

Hall: When you kill off the bacteria, don't the mice become deficient in vitamin B_{12}?

Scott: The diet of Walzem et al (1983) contains vitamin B_{12}.

Juriloff: We should make a clear distinction between making genetically normal mice folate deficient and making mice that are genetically liable to neural tube defects folate deficient. In the human situation, the folate may not be interacting with the normal genotype, it may be interacting with a susceptible genotype.

van Straaten: Mooij et al (1992) found that in hamster development, maternal folate deficiency results in a decreased fertility rate and increased embryonic loss, but they saw no cases of spina bifida.

Trasler: Moffa & White (1983) reported that folic acid supplementation significantly reduced the frequency of neural tube defects in hamster embryos from 17% to 6%. This could not be confirmed as it was one particular strain, which they didn't keep.

Juriloff: We have developed an animal model for neural tube defects that is not due to a single gene mutation, so it differs from the mouse models Andrew Copp has described.

The model that is used to explain the majority of neural tube defects in human beings is a multifactorial threshold model. It basically says that for the human embryo population there is a normal (Gaussian) distribution of values for a quantitative embryonic trait, which is influenced by the cumulative effect of genetic and environmental variation. The majority of embryos fall within the normal range, but there is a minority at the extreme of the distribution that fall beyond a critical value. They fail to close their neural tubes. These individuals have accumulated the effects of several genes, each with small effect, which jointly create the embryonic deficiency or delay that leads to failure of closure. According to this model, the genetic variation leading to the distribution of values for the embryonic quantitative trait is *not* due to mutations, but to polymorphic alleles, that is, *normal* allelic variation at many genetic loci. This suggests that the type of developmental errors or the mechanisms causing neural tube defects in this multifactorial system may be different from the mechanism in single locus mutants.

About 12 years ago, the neural tube defect exencephaly arose spontaneously in a heterogeneous stock of mice in our laboratory. I selected for this trait. Exencephaly in mice is lethal. Therefore, when I obtained the first exencephalic newborn, I selected the normal siblings and bred them with each other. They produced litters, some of which had all normal pups and some of which had occasional exencephalic pups. I selected the siblings of the exencephalics again and repeated this process for about 30 generations. We now have a colony of inbred mice that produce about 20% exencephaly in their litters. We have named this strain SELH/Bc. Every single animal, when tested, produces exencephaly in its progeny. A further 7% of the surviving, non-exencephalic pups develop ataxia.

Having created the SELH/Bc stock, we investigated the genetic cause of the exencephaly. We did an extensive genetic study using standard classical genetic methods: making crosses to normal strains, going to the F1, F2, backcross 1, and backcross 2 generations and test-crossing individual backcross 1 animals (Juriloff et al 1989). We concluded that the defect in our strain is caused by the cumulative action of 2–3 gene loci. The risk of having exencephaly increases as more alleles are accumulated at these loci, that is, this is an 'additive' genetic trait. We are currently mapping these loci using QTL (quantitative trait loci) methods and microsatellite marker loci.

To find the developmental cause of the exencephaly, we have looked at the development of the neural tube in our SELH stock of mice. The closure of the cranial neural tube has several initiation sites in the mouse. The first is in the cervical (occipital) region. The second initiation ('closure 2') is at the mesencephalon/prosencephalon border and closure proceeds bidirectionally. The third ('closure 3') begins at the most rostral end of the neural tube and closure proceeds toward the posterior. The fourth closure site covers the rhombencephalon. Our SELH mice all leave out closure 2 and begin at the rostral

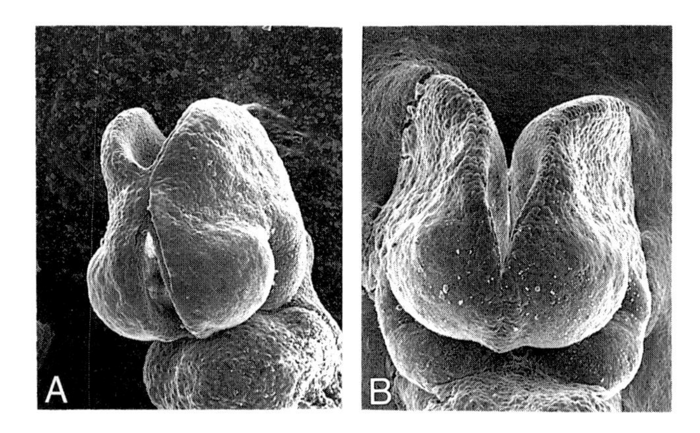

FIG. 1. (*Juriloff*) Scanning electron micrographs of cranial neural tube closure in mouse embryos on Day 9 of gestation. (A) Normal strain during initiation of closure 2. (B) SELH strain showing absence of closure 2 and posterior extension of closure from the more rostral closure 3 initiation site.

end, closure 3 (Fig. 1). From this initiation site closure proceeds toward the posterior to close the whole of the cranial neural tube except the rhombencephalon. This results in delayed closure of the mesencephalon. We believe that closure often stalls at the mesencephalon/prosencephalon boundary, resulting in failure of closure of the mesencephalon and subsequent exencephaly. If closure successfully closes that region and continues, the last region to close would be the junction with closure 4, an area which gives rise to the rhombic lip and is the origin of the cerebellum (Fig. 2). This points to the cause of ataxia in these mice. The ataxic mice in the SELH stock have a midline cleft of the cerebellum. We think this might be like Dandy–Walker syndrome in humans. The cleft cerebellum in SELH is due to an abnormality of neural tube closure at the rhombic lip. The surface epithelial layer successfully closes, but the neuroepithelial layer may not.

This SELH strain has taught us that there must be developmental redundancy of the closure initiation sites. It is possible for the most rostral initiation site, closure 3, to do the job of closure 2. In about 75% of SELH mice it does this successfully. In the rest it fails at the prosencephalon/mesencephalon border, leading to exencephaly, or in the neuroepithelial layer of the posterior mesencephalon, leading to ataxia.

Copp: You make the distinction between the aetiology of the defects in your mice and in those with single gene defects. I wonder whether there really is a distinction. I described the role of modifiers in *ct*. Our model is that you must have the *ct* mutation, plus some combination of modifiers to get a phenotype.

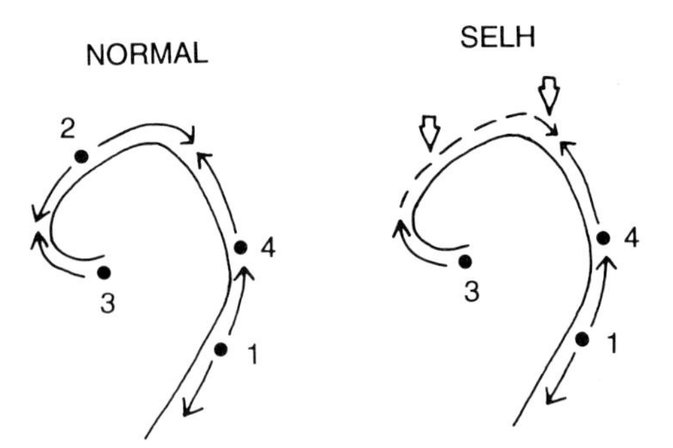

FIG. 2. (*Juriloff*) Diagram of the pattern of initiation sites and closure of the cranial neural tube in normal and SELH mouse embryos. The wide arrows denote sites of failure of extension of closure 3, leading to exencephaly or ataxia.

Can you be sure that you do not have an obligatory gene amongst your two or three loci? If one of the loci must always be mutant, the model would be no different from *ct*.

Juriloff: Our genetic analysis does not support that model for SELH. An obligatory homozygous recessive mutation with modifiers is expected to produce certain patterns of frequencies of exencephaly in classical crosses. We looked for those patterns and did not find them. The only model that the data did fit was additivity at 2–3 loci.

Lindhout: The analysis is based on segregation of the phenotype and not on the segregation of linked markers. So it is still possible that at least one factor preventing neural tube closure is necessary to get the defective phenotype and this factor is always present in all the other ones. So the distinction really is: can you have two factors within this model, and two completely different factors producing the same phenotype, or is the presence of one specific single gene obligatory?

Juriloff: This is a complex analysis and there are a lot of data. If we were dealing with recessive mutations, we would see no exencephaly in the F1 generation, but we do. On the other hand, it's not a simple dominant gene, because the frequency is so low in the F1. This is model fitting and one has to be careful not to overinterpret, but the data do not fit a major gene/modifier system or other epistatic systems, they fit only an additive model. In particular, there is no bimodality of backcross 1 individual breeding values; an obligatory recessive mutation would necessarily produce bimodality in this test. It is the classical genetic way to detect a single locus of major effect in a complex system and it indicated that there is not a single obligatory mutation in SELH.

Opitz: You suggested that your ataxic mice were similar to patients with Dandy–Walker syndrome. Is it sufficient to say absence of midline cerebellar structures without atresia of the foramina of Luschka and Magendie with a fourth ventricle cyst, and maybe hydrocephalus?

Juriloff: Our studies of this are just beginning. The mice have a midline cleft of the cerebellum and the vermis is absent.

Hall: John, do you think Diana's idea of redundancy fits with Waddington's buffering?

Opitz: Waddington's hypothesis of buffering and the canalization of development refers to quite a different concept, relating to the evolution of fail-safe mechanisms, which enable a reasonably normal morphogenetic outcome after a developmental system has been perturbed. Diana Juriloff was speaking of two or three alternative morphogenetic paths: if one fails, another takes over and allows completion. If this system exists and there are dyssynchronies in the closure of the anterior neural tube, there are probably normal anatomical variations that will identify this, for example failures of the frontal commissures. The effects that are ascribed to Waddington's canalization or buffering phenomena are not necessarily associated with gross anatomical variability.

Morriss-Kay: There is a similarity between the defects Diana Juriloff has shown and the form of exencephaly that we observed in embryos with trisomy 14: they show forebrain closure almost up to the forebrain/midbrain junction (Putz & Morriss-Kay 1981). Have you looked at the chromosomes of the SELH mice?

Juriloff: The chromosomes of the normal living mice appear to be normal. We have not looked for trisomy in the exencephalic mice. One would have to postulate a very complicated chromosomal mechanism to generate 20% affected offspring and the patterns of occurrence of exencephaly in our Mendelian crosses.

Morriss-Kay: With trisomies due to balanced Robertsonian translocation there is a range of percentages for each trisomy; we obtained just under 20% for trisomy 12 (Putz & Morriss-Kay 1981).

Juriloff: We can rule that out. We've looked at chromosomes from several mice of the SELH strain and they have the normal number of chromosomes and the normal banding pattern. There is no Robertsonian translocation present in the stock. We've also looked at meiotic as well as mitotic chromosomes.

O'Rahilly: These multiple points of closure of the rostral neuropore do not occur in the human embryo.

Hall: Can we be sure of that? We see early human embryos at only one point; we don't watch them over time while the neural tube is closing.

O'Rahilly: Quite enough human embryos have been examined to cover all the possibilities. There is simply no evidence that such a thing occurs in the human embryo.

Morriss-Kay: There is a picture of a 10-somite stage embryo in Hamilton (Boyd) and Mossman (Hamilton & Mossman 1972, Fig. 146, p 179) that is from the original collection of the embryos that you study. It has a much smaller upper hindbrain neuropore than the spindle-shaped opening observed in rodents, but it is definitely there as a closure region in the head that is separate from the rostral neuropore that closes the forebrain. I had therefore assumed that this was the normal closure process in the human embryo.

O'Rahilly: That is an isolated closure in an area different from the rostral neuropore, which shows a long, slit-like closure. The closure is advancing in only one direction, whereas the rostral neuropore closes bidirectionally (O'Rahilly & Müller 1989).

Morriss-Kay: But as well as the caudorostral direction going up the caudal hindbrain, there is a separate point apposition, which leaves, for a short time, two neuropores closing independently.

O'Rahilly: I would not call that a neuropore! It is merely a small area during the general process of closure.

Juriloff: We collected several hundred mouse embryos for a scanning electron microscopic study of this initiation site at the prosencephalon/mesencephalon boundary. We collected litters that were of exactly the correct gestational age, to maximize our chances of finding the initiation of closure 2 as it began. It appears that this initiation event is detectable during about two hours of development. In our study we could catch only about 6% actually initiating closure at that site; the other embryos were either more or less advanced. 6% is a rather low proportion to recover for a developmental event when everything has been arranged to maximize the chance of finding it. Therefore, given the difficulty of finding human embryos in sufficient quantity at exactly the right stage of development, if closure 2 is as short lived as the mouse data suggest, you might not have found it.

Jacobson: Schreckenberg and I (1975) prepared a staged series of axolotl. We kept sibling embryos from a single batch, about 50 at a time, in dishes in the same incubator at the same temperature. We followed their stages through time. They keep in phase very closely through cleavage and gastrulation. If they do get out of phase, it's especially at neurulation. Then the relationship between morphology and age broadens out.

O'Rahilly: I take the view that nothing occurs in the human embryo until it has been seen there. There is no evidence that this mechanism operates in the human embryo.

Shurtleff: Clinically, it is extremely important to relate morphological stage and gestational age, particularly in relieving parental guilt with regard to events that took place after the developmental event. I would hope that investigators studying human embryology will try to approximate the gestational age of morphological stages. Not all centres can identify the morphological stage in embryos by ultrasound or other techniques *in utero*.

Hall: These mothers will feel guilty if they have a child with a birth defect. If you can say that the defect must have arisen before a particular event, you can at least say that something they did at a later time was not the cause.

Czeizel: Last year we had two interesting probands in our genetic counselling clinic, who had agenesis of the corpus callosum with no neural tube defect. The first case had a father with spina bifida aperta; the other had a brother with anencephaly.

Opitz: Most cases of sporadic, non-syndromal agenesis of the corpus callosum in the population are asymptomatic. They are detected purely accidentally, for example at autopsy. They don't have epilepsy and they don't have mental retardation. If agenesis is seen with other defects of the neuraxis or in patients with relatives with defects of the neuraxis, the patients normally are symptomatic—they are either mentally retarded and/or have seizures. So you should take account of the mode of ascertainment, if you are going to make a connection between agenesis of the corpus callosum and neural tube defects for counselling purposes, rather than saying it is sheer coincidence.

Cuckle: With the growing use of routine ultrasound anomaly scanning, many more cases of agenesis are being discovered. We have to take account of this when assessing the clinical significance of this malformation.

Hall: It seems that the old hypothesis, before magnetic resonance imaging and ultrasound screening, was that the presence of agenesis of the corpus callosum was not necessarily a bad sign. Now that we can detect this more easily and are gaining more experience, we are finding that agenesis and hypoplasia of the corpus callosum are quite common and they are often associated with functional problems in humans.

Cuckle: It is more likely that the incidentally diagnosed cases have little clinical significance.

Opitz: We need to follow more of them prospectively.

Cuckle: I am concerned about the apparent precision in the timing of closure events. Presumably, this varies in different individuals; everything is variable in biology, but Professor O'Rahilly was talking as though there was no variability. You said that closure was complete at 28 days gestation (O'Rahilly & Müller, this volume): how many cases were examined?

O'Rahilly: I was referring to two embryos.

Cuckle: Presumably, you are studying carefully timed pregnancies, but they may be special in some way. For example, if the women are being investigated for poor fertility, the development of their babies may be different from those of fertile women.

Dolk: Are there mouse models for occipital encephalocele as opposed to anencephaly? I get the impression from human data that occipital encephalocele may itself be heterogeneous (Dolk et al 1991). Perhaps only some are closely related to other neural tube defects.

Copp: I am not aware of any animal models for encephalocele.

Juriloff: We see, rarely, some animals in our SELH mice that have what I would call an encephalocele; but it is in the rostral mesencephalon area.

Trasler: We see it very rarely.

Hall: So they are sporadic and not genetic. There are several human syndromes where encephalocele is seen.

Lindhout: It may be something like Dandy–Walker syndrome. We have seen two cases with this syndrome in our clinical unit. One was an isolated malformation, the other had spina bifida. These occurred after the mother received 5 mg folic acid supplementation periconceptionally. The mother of the first case had had a previous child with spina bifida. In the second case, the mother was on valproate. The question is: did the folic acid somehow reduce the risk, prevent the occurrence of a neural tube defect? Or could it modulate the expression into a different type of neural tube defect? What is known about the influence of folic acid or retinoic acid on the expression of the neural tube defect?

Hall: There was a concern before the MRC study that folic acid would keep affected embryos continuing in a pregnancy that would otherwise have miscarried. It doesn't look like that occurred (MRC Vitamin Study Research Group 1991).

Lindhout: Diana, have you looked at the effects of folic acid in your mouse model?

Juriloff: We've done one pilot study. It is too early to draw any conclusions. We plan to do a full study of folic acid supplementation in SELH mice.

References

Dencker L 1977 Trypan blue accumulation in the embryonic gut of rats and mice during the teratogenic phase. Teratology 15:179–184

Dolk H, De Wals P, Gillerot Y et al 1991 Heterogenicity of neural tube defects in Europe: the significance of site of defect and presence of other major anomalies in relation to geographic differences in prevalence. Teratology 44:547–559

Hamilton WJ, Mossman HW 1972 Hamilton, Boyd and Mossman's human embryology. W Heffer, Cambridge

Heid MK, Bills ND, Hinrichs SH, Clifford AJ 1992 Folate deficiency alone does not produce neural tube defects in mice. J Nutr 122:888–894

Johnson EM, Nelson MN, Monie EW 1963 Effects of transitory pteroylglutamic acid (PGA) deficiency on embryonic and placental development in the rat. Anat Rec 146:215–224

Juriloff DM, Macdonald KB, Harris MJ 1989 Genetic analysis of the cause of exencephaly in the SELH/Bc mouse stock. Teratology 40:395–405

MRC Vitamin Study Research Group 1991 Prevention of neural tube defects: results of the MRC vitamin study. Lancet 338:132–137

Moffa AM, White JA 1983 The effect of periconceptional supplementation of folic acid on the incidence of open neural tube defects in golden hamster embryos. Teratology 27:64A(abstr)

Mooij PNM, Wouters MGAJ, Thomas CMG, Doesburg WH, Eskes TKAB 1992 Disturbed reproductive performance in extreme folic acid deficient golden hamsters. Eur J Obstet Gynecol Reprod Biol 43:71–75

Nelson MN 1960 Teratogenic effects of pteroylglutamic acid deficiency in the rat. In: Congenital malformations. Churchill, London (Ciba Found Symp 59) p 134–151

O'Rahilly R, Müller F 1989 Bidirectional closure of the rostral neuropore in the human embryo. Am J Anat 184:259–268

O'Rahilly R, Müller F 1994 Neurulation in the normal human embryo. In: Neural tube defects. Wiley, Chichester (Ciba Found Symp 181) p 70–89

Putz B, Morriss-Kay GM 1981 Abnormal neural fold development in trisomy 12 and trisomy 14 mouse embryos. I. Scanning electron microscopy. J Embryol Exp Morphol 66:141–158

Schreckenberg GM, Jacobson AG 1975 Normal stages of development of the axolotl, *Ambystoma mexicanum*. Dev Biol 42:381–400

Walzem RL, Clifford CK, Clifford AJ 1983 Folate deficiency in rats fed amino acid diets. J Nutr 113:421–429

Williams KE, Roberts G, Kidston EM, Beck F, Lloyd JB 1976 Inhibition of pinocytosis in the yolk sac by trypan blue. Teratology 14:343–354

Valproic acid-induced neural tube defects

Heinz Nau

Institute of Toxicology and Embryopharmacology, Free University Berlin, Garystraße 5, D-1000 Berlin 33, Germany

Abstract. Antiepileptic drug therapy with valproic acid (VPA) during early pregnancy can result in a 1–2% incidence of spina bifida aperta, a closure defect of the posterior neural tube in the human. The predominant defect produced by VPA in the mouse is exencephaly, a closure defect of the anterior neural tube. An appropriate dosing regimen (consecutive doses of VPA on Day 9 of gestation) can also result in a low incidence of spina bifida aperta and a high incidence of spina bifida occulta in the mouse. It is likely that the parent drug and not a metabolite is the proximate teratogen. Structure–activity relationships show a strict structural requirement for high teratogenic potency: the molecule must contain an α-hydrogen atom, a carboxyl function and branching on C-2 with two chains containing three carbon atoms each for maximum activity. If these two carbon chains are different, then enantiomers are present. Pairs of enantiomers were synthesized and shown to be significantly different in regard to teratogenic potency. Both enantiomers of each compound reach the embryo to the same degree, therefore, the intrinsic teratogenic activity of the enantiomers differs. This suggests that stereoselective interaction occurs between the drugs and a chiral structure within the embryo. The molecular mechanism of the teratogenicity of VPA is not known; one hypothesis is that VPA interacts with embryonic folate metabolism.

1994 Neural tube defects. Wiley, Chichester (Ciba Foundation Symposium 181) p 144–160

Antiepileptic drug therapy with valproic acid (VPA) (2-n-propylpentanoic acid) has been related to teratogenicity, particularly the induction of neural tube defects in humans. Retrospective (Robert 1988) and prospective studies (Nau et al 1981, Jäger-Roman et al 1986, Lindhout & Schmidt 1986) have established that exposure to VPA during early pregnancy (organogenesis) can result in a significant incidence of spina bifida aperta (myeloschisis). The incidence of this extremely severe open defect of the spinal cord was estimated at 1–2% of exposed cases (Table 1). A few further cases have been reported where the lesion was covered by skin (meningocele, myelomeningocele).

Species differences

Spina bifida aperta was extremely difficult to produce with VPA in experimental animals (Nau et al 1991). A single case of a 'spina bifida-like lesion' was reported in a monkey fetus after treatment of the mother with three consecutive doses of 170 mg VPA/kg during early organogenesis (Michejda & McCollough 1987). Administration of VPA on Day 8 of gestation in the mouse resulted in high rates of exencephaly, a malformation of the anterior neural tube (Nau et al 1981b, Turner et al 1990) (Table 1). Analogous malformations of the anterior portion of the neural tube (anencephaly) were apparently not induced by VPA in the human, except in one case. This specificity of action is supported by the spina bifida:anencephaly ratio of 5:1 for cases exposed to VPA, compared to 1:1 for the control population (Lindhout & Schmidt 1986).

Thus, there is an apparent species difference in VPA-induced neural tube defects: spina bifida aperta, a defect of the posterior neural tube, is the dominant malformation in the human, while exencephaly, a defect of the anterior portion of the neural tube, is the dominant malformation in the mouse. The reason for this species difference is not known. It may be due to differences in spatial or temporal developmental processes or their sensitivities to the action of VPA.

VPA appears to induce specifically spina bifida in the human, not neural tube defects in general. We investigated whether posterior neural tube defects can also be induced by VPA in the mouse to develop a practical and relevant animal model. We found that administration of three consecutive high doses of VPA on Day 9 of gestation (one day after the peak sensitivity to the induction of exencephaly in this species) produced a significant rate of lumbosacral spina bifida aperta (4–6% of live fetuses after doses of 450 or 500 mg/kg) (Table 1). Spina bifida occulta, as demonstrated in double-stained fetal skeletons by measurement of the distance between the cartilaginous ends of the vertebral

TABLE 1 The effects of valproic acid on neural tube formation in mouse and human embryos

Teratogenic regimen		
Mouse	*Human*	*Site of neural tube defect*
190 mg/kg per day s.c. from gestational Day 6 to 9	No effect	Anterior[a]
3 × 300 mg/kg i.p. on gestational Day 9[b]	Unknown	Posterior: spina bifida occulta
3 × 450 mg/kg i.p. on gestational Day 9[b]	20–30 mg/kg per day orally	Posterior: spina bifida aperta

[a]Exencephaly in the mouse, anencephaly in the human.
[b]Injected at 8 a.m., 2 p.m. and 8 p.m.
Data taken from Nau et al (1991) with permission of the publisher.

arches, was produced with high frequency at lower doses (Table 1) (Ehlers et al 1991).

In addition to species differences, there are also pronounced strain differences in regard to VPA-induced formation of neural tube defects in the mouse. Finnell et al (1988) found that different inbred strains of mice exposed to the same VPA doses at identical developmental periods show greatly differing response frequencies; similar results were obtained with 4-en-VPA (2-n-propyl-4-pentenoic acid). Furthermore, physical agents such as hyperthermia showed the same hierarchy of susceptibility. Whether a common mechanism for pathogenesis exists for such varied agents as heat, VPA and 4-en-VPA remains to be investigated. The high stereoselectivity (see below) speaks against these agents having a common molecular mechanism. It may just be that in some mouse strains particular features of neural tube closure are more vulnerable to interference than in other strains. The precise molecular mechanism of such interference need not be the same for different agents. It is clear, however, that susceptibility to VPA-induced neural tube defects is under genetic regulation in the mouse (Finnell 1987, Finnell et al 1988).

Structure–teratogenicity relationships

Because the parent drug molecule and not a metabolite is responsible for formation of neural tube defects, we have investigated the teratogenic potency of several compounds structurally related to VPA. The aim of these studies was to determine the structural elements responsible for the teratogenic potency of this class of compounds. This should help obtain information about the mechanism of teratogenic action of VPA and help develop alternative antiepileptic agents with low teratogenicity (cf. Nau et al 1991).

A strict structural requirement for expression of teratogenic effects was found (Nau 1986, Nau & Löscher 1986) and it proved difficult to synthesize novel structures with high teratogenic potency. Most of the compounds synthesized had low or undetectable teratogenic activity in our mouse model. The results suggest that the following structural features are necessary to induce significant exencephaly in the mouse: (a) the α-carbon atom must be tetrahedral (sp^3 hybridization). It must be connected (b) to a free carboxyl function (valpromide [2-propylpentanamide] is not teratogenic); (c) to a hydrogen atom (substitution by a methyl or ethyl group abolishes the teratogenic activity); and (d) to two alkyl groups (branching on C-2 is required, unbranched acids are not active). (e) Maximal teratogenic potency is found if the two alkyl chains contain three carbon atoms each: shorter or longer carbon chains reduce the activity. (f) Introduction of a double bond between C-2 and C-3 (E-2-en-VPA) or between C-3 and C-4 (3-en-VPA) abolishes teratogenic activity. (g) Introduction of a double bond (*R*,*S*-4-en-VPA) or a triple bond (*R*,*S*-4-yn-VPA) in the 4-position results in molecules with high teratogenic activity. These are the only compounds with teratogenic activity comparable to or even higher than the activity of VPA (Nau & Löscher 1986, Hauck & Nau 1989a,b, Hauck et al 1990).

Influence of chirality

If the two chains on C-2 differ, C-1 becomes asymmetric (binding to four different substituents), resulting in enantiomer formation. We have synthesized several of these enantiomeric pairs to investigate whether the teratogenic potency is dependent on the three-dimensional structure of these compounds. The exencephaly response was found to be highly enantioselective. The compounds with higher teratogenicity (*R*,*S*-4-yn-VPA > *R*,*S*-4-en-VPA) showed greater enantioselectivity (Hauck & Nau 1989a, Hauck et al 1990). This suggests that the chiral centre on C-2 is directly involved in the process leading to neural tube defects. The nature of such a stereoselective interaction between the drug and the chiral structure within the embryo is unknown.

The enantioselective teratogenicity could be the result of differing transplacental pharmacokinetics of the two enantiomers or of different intrinsic activities. We therefore determined the maternal plasma kinetics as well as the placental transfer of two enantiomeric pairs, 2-ethylhexanoic acid (2-EHXA) and 2-n-propyl-4-pentynoic acid (4-yn-VPA), in the mouse during the sensitive stage of early organogenesis. For each compound, the enantiomer with the lower teratogenic potency (*R*-4-yn-VPA, *S*-EHXA) reached similar or even slightly higher plasma concentrations than did the more potent enantiomer. These results indicate that the intrinsic teratogenic activities differ between the enantiomers and that the pharmacokinetics are not stereoselective in these cases (Nau et al 1991).

A possible mechanism of teratogenic action: interaction with embryonic folate metabolism

Several hypotheses regarding VPA teratogenesis are being investigated: (a) interference with embryonic lipid metabolism (Clarke & Brown 1987); (b) alteration of intracellular pH as an important parameter for numerous cellular functions (Scott et al 1990); (c) interference with Zn metabolism (Vormann et al 1986, Wegner et al 1990); (d) interference with neurotransmitter metabolism; (e) interference with embryonic folate metabolism (Fig. 1).

Hypothesis (b) appears promising because the early mammalian embryo has a relatively high intracellular pH and can thus accumulate acidic compounds such as VPA via ion trapping (Nau & Scott 1986, 1987). Whether the accumulation of VPA alters the intracellular pH and whether such changes are directly related to the observed teratogenic effects remain to be determined, particularly with regard to the strict structural specificity of the teratogenic action of carboxylic acids.

We have investigated hypothesis (e) during the past few years because periconceptional administration of multivitamins, particular folate, was shown to reduce the recurrence risk of neural tube defects (Smithells 1984, Laurence et al 1981, MRC Vitamin Study Research Group 1991, Czeizel & Dudás 1992). Antiepileptic agents such as VPA (Carl 1986) and phenytoin (5,5-diphenyl-2,4-imidazolidinedione) were shown to interfere with the folate metabolic pathways in the embryo (Hansen & Billings 1985, Will et al 1985). Furthermore, folate concentrations decrease during pregnancy and treatment with antiepileptic drugs, including VPA, intensifies this effect (Hendel et al 1984). Several reports suggest that epileptic patients with malformed infants had particularly low levels of folate in their plasma (Ogawa et al 1991). Recently, a defect of homocysteine metabolism—which is interrelated with the folate pathway—was found in patients with a history of offspring with neural tube defects (Steegers-Theunissen et al 1991). Taken collectively, these studies suggest that a genetically altered folate metabolism and/or interference with folate metabolism by antiepileptic agents may be in part responsible for the malformations observed.

We first tried to reduce the occurrence of VPA-induced exencephaly in the mouse by coadministration of folinic acid (5-formyl-tetrahydrofolic acid,

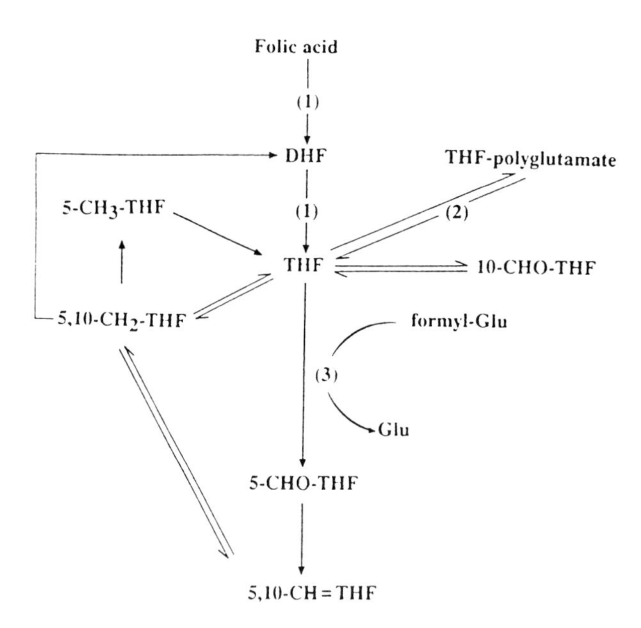

FIG. 1. Aspects of folic acid metabolism discussed in this paper. DHF, dihydrofolic acid; THF, tetrahydrofolic acid; 10-CHO-THF, 10-formyl-THF; 5-CHO-THF, 5-formyl-THF; 5,10-CH = THF, 5,10-methenyl-THF; 5,10-CH_2-THF, 5,10-methylene-THF; 5-CH_3-THF, 5-methyl-THF. (1) Dihydrofolate reductase (EC 1.5.1.3); (2) γ-glutamyl-hydrolase (EC 3.4.22.12); (3) glutamate formyltransferase. Taken from Nau et al (1991), with permission from the publisher.

TABLE 2 **Effect of valproic acid on mouse embryonic folate metabolism *in vivo* and the activity of glutamate formyltransferase *in vitro***

Dose of valproic acid (mg/kg)	*Rate of exencephaly (% live fetuses)*	*Maximum plasma concentration*[a] *(μg/ml)*	*Concentration of folate metabolites in the embryo (% of control value)*			*Glutamate formyltransferase activity (% of control activity)*
			5-CHO-THF	*10-CHO-THF*	*THF*	
356	10	445 ± 54[b]	65	70	115	45
500	42	490 ± 45[b]	28	52	173	32

THF, tetrahydrofolic acid; 5-CHO-THF, 5-formyl-THF; 10-CHO-THF, 10-formyl-THF.
[a]Mice were given a single dose s.c. on Day 8 of gestation.
[b]Mean ± SD.
[c]Samples were taken 2–4 h after administration of valproic acid.
Data taken from Wegner & Nau (1991, 1992) with permission from the publisher.

5-CHO-THF). In many cases the rate of exencephaly was significantly reduced (to 30–50% of initial rates) (Trotz et al 1987); in some cases, however, little protection was observed. These difficulties may be related to the large diurnal fluctuations of folate metabolite concentrations in the embryo (Wegner & Nau 1991, 1992).

Teratogenic doses of VPA induced a characteristic dose-dependent alteration of embryonic folate metabolism (Wegner & Nau 1992). The amounts of formylated folates, particularly 5-CHO-THF, decreased, while the amount of tetrahydrofolic acid (THF) increased (Table 2). Comparable doses of the non-teratogenic analogue 2-en-VPA did not affect embryonic folate metabolism. The relative change in folate metabolite concentrations suggests that the interconversion between THF and 5-CHO-THF may have been inhibited by VPA. The relevant enzyme, glutamate formyltransferase (EC 2.1.2.5), was inhibited by VPA *in vitro*; at teratogenic concentrations of VPA the activity of this enzyme was only one-third of its initial value (Table 2). These results indicate that interference with embryonic folate metabolism may be involved in some aspects of VPA teratogenesis (Fig. 1).

Acknowledgement

The work in our laboratory was supported by grants from the Deutsche Forschungsgemeinschaft (C-6, Sfb 174) and from the Bundesgesundheitsamt, Berlin (ZEBET).

References

Carl GF 1986 Effect of chronic valproate treatment on folate-dependent methyl biosynthesis in the rat. Neurochem Res 11:671–685

Clarke DO, Brown NA 1987 Altered cholesterol levels and metabolic response of rat conceptuses exposed to valproic acid (VPA) in vitro. Teratology 35:17A

Czeizel AE, Dudás I 1992 Prevention of the first occurrence of neural-tube defects by periconceptional vitamin supplementation. N Engl J Med 327:1832–1835

Ehlers K, Stürje H, Merker HJ, Nau H 1991 Valproic acid-induced spina bifida: a mouse model. Teratology 45:145–154

Finnell R 1987 Gene–teratogen interactions: an approach to understanding the metabolic basis of birth defects. In: Nau H, Scott WJ (eds) Pharmacokinetics in teratogenesis: experimental aspects in vivo and in vitro. CRC Press, Boca Raton, FL, vol 2:97–109

Finnell RH, Bennett GD, Karras SB, Mohl VK 1988 Common hierarchies of susceptibility to the induction of neural tube defects in mouse embryos by valproic acid and its 2-propyl-4-pentenoic acid metabolite. Teratology 38:313–320

Hansen DK, Billings RE 1985 Phenytoin teratogenicity and effects on embryonic and maternal folate metabolism. Teratology 31:363–371

Hauck RS, Nau H 1989a Asymmetric synthesis and enantioselective teratogenicity of 2-n-propyl-4-pentenoic acid (4-en VPA), an active metabolite of the anticonvulsant drug, valproic acid. Toxicol Lett 49:41–48

Hauck RS, Nau H 1989b Zu den strukturellen Grundlagen der teratogenen Wirkung des Antiepileptikums Valproinsäure (VPA): 2-n-Propyl-4-pentinsäure, das erste Strukturanalogon mit signifikant höherer teratogener Aktivität als VPA. Naturwissenschaften 76:528–529

Hauck RS, Wegner C, Blumtritt P, Fuhrhop JH, Nau H 1990 Asymmetric synthesis and teratogenic activity of (*R*)- and (*S*)-2-ethylhexanoic acid, a metabolite of the plasticizer di-(2-ethylhexyl)phthalate. Life Sci 46:513–518

Hendel J, Dam M, Gram L, Winkel P, Jorgensen H 1984 The effects of carbamazepine and valproate on folate metabolism in man. Acta Neurol Scand 69:226–231

Jäger-Roman E, Deichl A, Jakob S et al 1986 Fetal growth: major malformations, and minor anomalies in infants born to women receiving valproic acid. J Pediatr 108:997–1004

Laurence KM, James N, Miller MH, Tennant GB, Campbell H 1981 Double-blind randomised controlled trial of folate treatment before conception to prevent recurrence of neural-tube defects. Br Med J 282:1509–1511

Lindhout D, Schmidt D 1986 In-utero exposure to valproate and neural tube defects. Lancet 1:1392–1393

Michejda M, McCollough D 1987 New animal model for the study of neural tube defects. Z Kinderchir 42(suppl 1):32–35

MRC Vitamin Study Research Group 1991 Prevention of neural tube defects: results of the Medical Research Council vitamin study. Lancet 338:131–137

Nau H 1986 Species differences in pharmacokinetics and drug teratogenesis. Environ Health Perspect 70:113–129

Nau H, Löscher W 1986 Pharmacologic evaluation of various metabolites and analogs of valproic acid: teratogenic potencies in mice. Fundam Appl Toxicol 6:669–676

Nau H, Scott WJ 1986 Weak acids may act as teratogens by accumulating in the basic milieu of the early mammalian embryo. Nature 323:276–278

Nau H, Scott WJ 1987 Teratogenicity of valproic acid and related substances in the mouse: drug accumulation and pH_i in the embryo during organogenesis and structure–activity considerations. Arch Toxicol Suppl 11:128–139

Nau H, Rating D, Koch S, Häuser I, Helge H 1981a Valproic acid and its metabolites: placentral transfer, neonatal pharmacokinetics, transfer via mother's milk and clinical status in neonates of epileptic mothers. J Pharmacol Exp Ther 219:768–777

Nau H, Zierer R, Spielmann H, Neubert D, Gansau C 1981b A new model for embryotoxicity testing: teratogenicity and pharmacokinetics of valproic acid following constant-rate administration in the mouse using human therapeutic drug and metabolite concentrations. Life Sci 29:2803–2814

Nau H, Hauck RS, Ehlers K 1991 Valproic acid-induced neural tube defects in mouse and human: aspects of chirality, alternative drug development, pharmacokinetics and possible mechanisms. Pharmacol Toxicol 69:310–321

Ogawa Y, Kaneko S, Otani K, Fukushima Y 1991 Serum folic acid levels in epileptic mothers and their relationship to congenital malformations. Epilepsy Res 8:75–78

Robert E 1988 Valproic acid as a human teratogen. Congenital Anom 28:S71–S80

Scott WJ, Duggan CA, Schreiner CM, Collins MD 1990 Reduction of embryonic intracellular pH: a potential mechanism of acetazolamide-induced limb malformations. Toxicol Appl Pharmacol 103:238–254

Smithells RW 1984 Can vitamins prevent neural tube defects? Can Med Assoc J 131:273–274, 276

Steegers-Theunissen RPM, Boers GHJ, Trijbels FJM, Eskes TKAB 1991 Neural tube defects and derangement of homocysteine metabolism. N Engl J Med 324:199

Trotz M, Wegner C, Nau H 1987 Valproic acid-induced neural tube defects: reduction by folinic acid in the mouse. Life Sci 41:103–110

Turner S, Sucheston ME, de Philip RM, Paulson RB 1990 Teratogenic effects on the neuroepithelium of the CD-1 mouse embryo exposed in utero to sodium valproate. Teratology 41:421–442

Vormann J, Höllriegel V, Merker HJ, Günther T 1986 Effect of valproate on zinc metabolism in fetal and maternal rats fed normal and zinc-deficient diets. Biol Trace Elem Res 10:25–35

Wegner C, Nau H 1991 Diurnal variation of folate concentrations in mouse embryo and plasma: the protective effect of folinic acid on valproic acid-induced teratogenicity is time dependent. Reprod Toxicol 5:465–471

Wegner C, Drews E, Nau H 1990 Zinc concentrations in mouse embryo and maternal plasma. Effect of valproic acid and a nonteratogenic metabolite. Biol Trace Elem Res 25:211–217

Wegner C, Nau H 1992 Alteration of embryonic folate metabolism by valproic acid during organogenesis: implications for mechanism of teratogenesis. Neurology 42(suppl 5):17–24

Will M, Barnard JA, Said M, Ghishan FK 1985 Fetal hydantoin syndrome: inhibition of placental folic acid transport as a potential mechanism for fetal growth retardation in the rat. Res Commun Chem Pathol Pharmacol 48:91–98

DISCUSSION

Seller: What happens if you inject VPA at other times during the organogenetic period in rodents? Do you produce other abnormalities?

Nau: Yes. We can produce cleft palate, rib abnormalities, vertebral abnormalities, heart and kidney defects—all types of defects. In the human, we also see several other abnormalities, as well as neural tube defects. Neural tube defects are probably the most severe. The other defects depend on the dose you use and on the time the drug is administered.

Scott: In mammalian cells, 5-formyl-tetrahydrofolate is probably a storage form of formyl groups (Stover & Schirch 1992). Clinically, it's used in methotrexate rescue, when it is given in large amounts. Does VPA inhibit any other folate-dependent enzymes?

Nau: We checked various folate co-factors and enzymes. We were also concerned about the significance of the described changes. Glutamate formyltransferase is not, as you indicated, in the main folate metabolic pathway. However, the concentration of 10-formyl-tetrahydrofolate is also decreased by treatment with VPA. This may be very important.

We have tested only a few other enzymes. VPA had no effect on γ-glutamyl hydrolase or polyglutamate synthase. N_2O inhibits the activity of vitamin B_{12}: VPA had no effect on the N_2O-sensitive metabolic reaction. Two other anticonvulsants, trimethoprim and methotrexate, did not induce neural tube defects when given alone, but increased the frequency of VPA-induced neural tube defects when given in combination. This further demonstrates the involvement of the folate metabolic pathway in the mechanism of neural tube defects.

Wald: Does giving folic acid to someone on VPA who has epilepsy reduce the efficacy of the drug in controlling the epilepsy?

Nau: Some early studies showed that high doses of folates can change the susceptibility of these patients to seizures, but very little is known on this. For low doses, most people doubt that there could be an effect. There could be an indirect effect in the way that folate can interact with anticonvulsant metabolism. Folate administration lowers levels of phenytoin in the plasma.

Wald: How would you advise a woman with epilepsy who was planning a pregnancy who was on VPA?

Nau: I am not a clinician so I can't give you practical advice, but I see no reason not to administer low dose folate to these patients.

Oakley: When we published our guidelines on administration of folic acid to prevent neural tube defects, we explicitly excluded treatment for people who were exposed to VPA. We didn't have any data, so we didn't speak to it. Some people asked if we were suggesting that there are reasons for not doing it. The answer is: No. We just don't know whether or not it will prevent these defects.

Wald: Would you keep the woman on VPA and give low dose folic acid? And what constitutes a low dose?

Nau: We can only speculate because there are no good studies on a possible interaction between VPA and folate. Dr M. Dam's group showed some time ago that VPA can lower resorption of folate. There may be some interaction in the intestinal tract where the folate absorption is lowered by administration of VPA. Before a definitive study has been done, one could not make a conclusion. I see no reason why there should be a problem administering folate of 0.4 or 0.8 mg to patients on VPA.

Copp: The animal studies on the interaction between folate and VPA-induced defects are controversial, not all groups find the same interaction. Hansen & Grafton (1991) published rather different results.

Nau: Those results were from *in vitro* studies. Deborah Hansen cultured mouse and rat embryos and found no protective effect of folate on VPA-induced effects in the cultures.

Copp: She now has new data (Hansen 1993) that show a very clear teratogenic effect of VPA *in vitro*. She added at least three different metabolites of folic acid in the way you have done *in vivo*, but saw no protective effect on the development of VPA-induced neural tube defects.

Lindhout: I think giving a physiological dose of folic acid to women taking VPA is alright. There is no rational basis for giving a pharmacological dose. First, no good clinical trials have been done to prove their effectiveness (which is difficult because of the low denominators). Secondly, the most rational choice is to supplement with folinic acid and not with folic acid. This comes from Heinz Nau's own experimental data (Trotz et al 1987).

Wald: Would you switch the patient from VPA to another anticonvulsant?

Nau: These are very difficult questions. There is the risk of worsening the seizure control. Other anticonvulsant agents also carry some risks. Carbamazepine has a risk of spina bifida. Several groups have shown that the risk is lower than that of VPA but still higher than that of a control population: it is below 1%.

Lindhout: Most commonly used antiepileptic drugs are teratogenic. Usually, the advice to clinicians is to reconsider medication before a pregnancy. This is difficult because many pregnancies are not planned. Secondly, apply the medication that's most effective, because all the drugs are teratogenic. Then if you have equally effective drugs, let's say VPA and phenobarbital, it's a matter of discussion with the parents to decide which drug should be given in preparation for pregnancy. If prenatal diagnosis and termination of pregnancy are unacceptable to the parents, you might even choose a drug that is a little less effective in that particular woman, but prevents her bearing a child with a severe handicap. It is always an individual decision. It is very difficult to make general rules from a public health point of view.

Cuckle: There was a recent MRC randomized trial of withdrawal of antiepileptic drugs in adults who had not had seizures for several years (MRC Antiepileptic Drug Withdrawal Study Group 1991). One might be able to use the trial results to predict the chance of a woman having a fit if treatment is stopped and use this in preconceptional counselling.

Lindhout: We did a prospective study of patients who attended our prenatal clinic at a very early stage of pregnancy (Omtzigt et al 1992a). We detected 92 pregnancies that had been exposed to VPA. Five of these (six fetuses because there were monozygotic twins) had a spina bifida aperta. Five affected fetuses had been exposed to VPA only, one to VPA and carbamazepine. The mean daily dose was 1640 mg $\pm$ 136 mg* with a range of 200–2000 mg. The daily dose was 25 $\pm$ 2 mg/kg and the peak dose was 650 $\pm$ 124 mg. The doses to which other fetuses were exposed were almost half: for three fetuses with other defects the mean daily dose was 983 $\pm$ 130 mg; for 84 normal fetuses it was 941 $\pm$ 4 mg* (*P = 0.001 versus normal).

The maternal serum levels of VPA were significantly higher in pregnancies with spina bifida (73 $\pm$ 25 μg ml^{-1}) than in those with normal outcome (44 $\pm$ 22 μg ml^{-1}) (Omtzigt et al 1992b). These data are relevant, because Heinz Nau's experimental data show that it is not the total dose but the dose per administration and the peak levels that are associated with teratogenic effects. We cannot discriminate now between the total daily dose or the dose per administration. But, on the basis of these data, our advice is that if VPA medication cannot be stopped because of the neurological condition of the woman, then one should give the lowest possible dose and divide it into at least three or more doses.

If you divide the same data into three different dose groups, then you see the denominators; 0 spina bifida in 54 pregnancies exposed to doses lower than 1000 mg/per day, two affected pregnancies among 30 exposed to daily doses between 1000 and 1500 mg and three affected pregnancies among eight exposed to daily doses over 1500 mg. If you take the fetuses and not the pregnancies

Ihr nächster Arzttermin

am ____________ um ____________ Uhr

Denken Sie bitte an:

- ○ Bitte kommen Sie nüchtern
- ○ Krankenschein
- ○ Überweisungsschein
- ○ Krebsvorsorgeschein
- ○ Mutterschaftsvorsorgeschein
- ○ Sonstiges

Hinweis: Zeitverschiebungen können eintreten, wenn die Wahrnehmung medizinischer Interessen gegenüber allen Patienten dieses erforderlich macht.

Arztstempel

Brief Susanne Böhme
7788

Bitte ausdrucken u. an
Pat. schicken
(hat bisher keinen bekommen)

as the denominator, there were four affected fetuses among nine fetuses exposed to more than 1500 mg VPA/day.

Hall: Are you suggesting that there isn't a genetic predisposition to this effect: rather that it's really dose related?

Cuckle: There may be an underlying association between epilepsy and neural tube defects. Epileptic women may have a greater risk of bearing a child with a neural tube defect, even when they are not taking antiepileptic drugs. It may not necessarily be just a teratogenic effect.

Lindhout: Genetic predisposition certainly may play a role, but it is questionable whether this is related to the maternal epilepsy. Of 34 cases of spina bifida and neural tube defects in The Netherlands, known to me through multiple sources, there are six cases with a family history of neural tube defects (Lindhout et al 1992). The paternal and the maternal family contribute equally to these positive family histories. Therefore, genetic factors may play a role in the phenotype of a neural tube defect, but these are not related to the maternal epilepsy. In addition, there are now two pairs of monozygous twins concordantly affected. Monozygous twins with spina bifida are usually not concordant. We are now aware of at least three families in The Netherlands where two siblings have been exposed to the same dose of VPA and both were affected. However, these cases do not allow for discrimination between genetic and teratogenic aetiology.

Cuckle: That may be true for VPA and neural tube defects, but do we know that other anticonvulsant agents are not teratogenic?

Lindhout: The classical antiepileptic drugs (barbiturates and phenytoin), although prescribed to similar patient populations, are more associated with heart defects and clefts, indicating teratogenic rather than genetic differences (Lindhout & Omtzigt 1992). For these other birth defects, there may be more of a relationship with the phenotype of maternal epilepsy.

Czeizel: We evaluated the database of the Hungarian Case-Control Surveillance of Congenital Anomalies, 1980–1987 (Czeizel et al 1992). We found ten women who previously had epilepsy. The use of anticonvulsant was stopped several years before the study pregnancy, thus these women were not treated for epilepsy during pregnancy. Of ten children born from these epileptic women, seven had congenital anomalies (two hypospadias, one cleft lip with cleft palate, one cleft palate, one spina bifida aperta, one diaphragm defect and one syndactyly). Thus, the proportion of epileptic women without treatment is 4.6 times higher in the malformed case group (0.065%) than in the control group (0.014%). Shapiro et al (1976) have published a similar finding. So I believe the maternal epilepsy may be an independent factor in the origin of congenital abnormalities. These babies had mainly oral clefts.

Holmes: Our study on anticonvulsants does not show that a woman with epilepsy, but not on drugs, has an increased risk for having a child with microcephaly, growth retardation or major malformations compared with a woman who takes no medication and has never had seizures (Holmes et al 1990). The possibility that the genes which 'cause' epilepsy will produce fetal defects has not been established in any study in which the study infants were examined carefully as part of a protocol by study personnel.

Stanley: We need large cohorts to address this, because epilepsy is not a single disease. There will be heterogeneity in parents with epilepsy. One will probably have to find out whether the father also has epilepsy. Those studies need to be on large cohorts and we need to try and classify what types of epilepsy the women have.

Cuckle: The women with the most severe epilepsy, in some sense, are often on polytherapy, which these days may include VPA. If severity of the epilepsy was related to the teratogenic risk, this would increase the complexity in interpreting the data.

Lindhout: In The Netherlands we saw a change from polytherapy to monotherapy. At the same time there was a shift from the old medications—phenytoin, phenobarbital and primidone—to VPA and carbamazepine. All the data showed that there is an association, especially with VPA monotherapy. So polytherapy is not an issue; it is VPA itself. Experimental data from animal studies, as well as several epidemiological studies, even suggest that by adding phenobarbital you lower the risk of spina bifida with VPA, probably because you lower the levels of VPA via induced metabolism.

Secondly, in our cohort, we demonstrated that the spina bifida is associated with almost double the levels of VPA and not any of the 13 metabolites that were measured in the sera.

Wald: Were the dose data that you presented monotherapy, polytherapy or a mixture?

Lindhout: It was a mixture, but even if you corrected for that, the results were the same. Of the five affected pregnancies (six fetuses), four (five fetuses) received monotherapy and one received VPA with carbamazepine. One woman, who had been treated with 2000 mg of VPA per day, in a subsequent pregnancy again had a fetus with spina bifida.

Holmes: In Boston, US, the rate of polytherapy has fallen from about 50% of pregnant epileptic women in the 1970s to 20% of pregnant epileptic women in the late 1980s. Pregnant women in the US are treated with VPA much less often than in Europe.

We need a careful and large cohort study of women on VPA alone. Dick Lindhout showed six infants (five pregnancies) with neural tube defects and three with other major malformations. If the 5:3 ratio of neural tube defects to other major malformations is confirmed, focusing on the incidence of spina bifida will not be good enough. The other three drugs that are prescribed for pregnant women in the Boston area, i.e. phenytoin, carbamazepine and phenobarbital, each have a risk of fetal effects that is clearly less than that of VPA.

Hall: Dick, in your cohort, what were the defects in the three that didn't have spina bifida?

Lindhout: In our prenatal diagnosis clinic cohorts of 92 pregnancies exposed to VPA (96 fetuses), other defects were cleft soft palate (also exposed to carbamazepine), unilateral clubfoot and inguinal hernia requiring surgical treatment. Other studies showed that the second most frequent defect in pregnancies exposed to VPA is hypospadia (see Lindhout et al 1992).

Stanley: There have been two Australian studies, one by Les Sheffield (1988) and one by us (Stanley et al 1985). The first examined every child born to a cohort of women who had been exposed to various antiepileptic drugs during pregnancy. We did a record linkage between all women identified in our database (total Western Australian births) as being epileptic. We didn't know what therapy they were on. We found a low rate of major defects in epileptic women compared with all women and the defects tended to be minor. There were no neural tube defects. We then went back to the medical records to ascertain what drugs the women who had an affected child had used. They were mostly on Dilantin, Tegretol and/or phenobarbital. Only one woman was on VPA.

Dolk: There was a similar study in the Marseille area (Dravet et al 1992), following a cohort of epileptic women identified during pregnancy and comparing their babies with a population-based birth defect registry. They found various specific associations between anticonvulsants and malformations, but also a less specific relationship with microcephaly. It would be interesting to determine the functional significance of the microcephaly in the later development of the children, and also the extent to which it is related to the epilepsy or the anticonvulsant.

Holmes: In addition to determining the frequency of major malformations among infants exposed to VPA, we need to determine the frequency of developmental delay and cognitive dysfunction.

Opitz: What is the risk associated with phenobarbital on its own?

Holmes: Roughly, a twofold increase in the frequency of growth retardation, microcephaly and major malformations.

Opitz: Is the microcephaly associated with mental retardation or not?

Holmes: The phenobarbital-exposed children we have enrolled in our study are too young for cognitive testing. One assumes that if a child continues to have a small head, he/she will exhibit developmental delay and/or cognitive dysfunction. However, we have found in a follow-up study that some infants show catch-up in their head size and body size and this group could do much better.

Oakley: It is sad that we don't have an effective anticonvulsant that is not teratogenic. This is a gaping hole in our pharmacopoeia.

Holmes: Why is the non-teratogenic metabolite of VPA, 2-en-VPA, not available?

Nau: We are not directly involved in drug development but we were hoping that the pharmaceutical industry would pick up on some of these analogues.

They have been very slow. One reason could be that the 2-en-VPA has been known for some time as a metabolite of VPA, so it cannot be patented. There are several other compounds and several additional lead structures. I'm sure that in the future some of these structures will be followed up.

We have recently produced a new type of analogue with a methyl group in the penultimate position of one side chain of VPA—the last position from the end of the side chain. This modification completely abolishes teratogenicity, although the anticonvulsive activity is maintained or is even better than that of VPA. The enantiomers are also a good lead for drug development, although they are more costly to synthesize and purify on a large scale.

Oakley: This issue of teratogenicity is very serious. VPA causes human babies to have spina bifida. There are considerable leads that other variants of the drug would be equally effective as anticonvulsants and the drug companies have refused to do randomized trials to get one of those drugs on the market so that we could prevent these birth defects. These data have been available for at least five years. There are no randomized clinical trials, because VPA is alleged to be very profitable for some of the major pharmaceutical companies.

Nau: Drug development is very slow. One has to do a lot of animal experiments before the first human volunteer studies can be performed. Then you have to test the drug in epileptic patients. To develop one drug costs US $150 million on the average. But I fully agree with Dr Oakley that we should talk to people in the pharmaceutical industry and try to push development of one of these novel drugs.

Lindhout: The problem is not the lack of new drugs to be developed. It is the huge number of newly developed antiepileptic drugs that will be marketed in the next five years, e.g. vigabatrin, topiramate and gabapentin. The first two of these have already been shown to be weak teratogens in animal experiments. Nevertheless, they are marketed. Even if the drugs were proven safe in animal experiments, there is still the possibility of human teratogenicity. There should be some obligation on the drug companies to fund post-marketing surveillance in which all exposed pregnancies are evaluated!

Stanley: We put a proposal to our federal government in Australia for initiating post-marketing drug surveillance by doing a simple record linkage of doctors prescription data, for example for Epilim (VPA) and retinoic acid, with several Australian birth defects registries. This would be easy and quite cheap to do. The drug companies were very supportive but the federal government was reluctant to allow access to named prescription data and rejected the scheme. Why do they keep the data if no one is allowed to use them? This is very frustrating for us and potentially damaging.

Hall: There seem to be three principles. 1) Less teratogenic drugs need to be developed. 2) There has to be post-marketing surveillance. 3) We need better data on whether women who suffer seizures are at risk for having children with birth defects, with or without drugs.

Stanley: We possibly have the potential, through international collaborative studies, either to pool existing data or to set up some really large cohorts. It wouldn't take long to answer these questions.

Hall: Are the pooled data good enough? Have the kids been tested carefully enough?

Holmes: Generally, you need a careful and systematic evaluation of the drug-exposed child at birth; you need to standardize the definitions of epilepsy; you need to consider all of the obvious confounders; you need a comparison group of unexposed 'control' children who have been evaluated in the same way as those exposed to the drug; and you need a re-evaluation of both groups of children when they are old enough for cognitive testing. That takes money. But if we don't have a cohort study of sufficient size, this issue will not be resolved.

Mills: The National Collaborative Perinatal Project (NCPP) in the US did examine the children at birth. Information on the use of anticonvulsants during pregnancy was collected and the children were followed for seven years to look at developmental progress. This satisfies several of your criteria, but for older drugs because the study births occurred between 1959 and 1966.

Holmes: The different analyses of those results have led to very different findings. Hanson et al (1976) reviewed the findings in each of the 104 drug-exposed infants and a comparison group. They concluded that the drug-exposed infants showed a distinctive pattern of physical abnormalities. By contrast, a more general analysis by epidemiologists of the same group of infants for malformations within the context of the large data set of the NCPP led to the conclusion that parental epilepsy, not exposure *in utero* to anticonvulsants, was associated with an increased risk of birth defects (Shapiro et al 1976).

References

Czeizel AE, Bod M, Halász P 1992 Evaluation of anticonvulsant drugs during pregnancy in a population-based Hungarian study. Eur J Epidemiol 8:122–127

Dravet C, Julian C, Legras C et al 1992 Epilepsy, antiepileptic drugs, and malformations in children of women with epilepsy: a French prospective cohort study. Neurology 42(suppl 5):75–82

Hansen D 1993 In vitro effects of folate derivatives on valproate-induced neural tube defects in mouse and rat embryos. Toxicol In Vitro 7:735–742

Hansen DK, Grafton TF 1991 Lack of attenuation of valproic acid-induced effects by folinic acid in rat embryos in vitro. Teratology 43:575–582

Hanson JW, Myrianthopoulos NC, Harvey MA et al 1976 Risks to the offspring of women treated with hydantoin anticonvulsants, with emphasis on the fetal hydantoin syndrome. J Pediatr 89:662–668

Holmes LB, Harvey EA, Hayes AM, Brown KS, Schoenfeld DA, Khoshbim S 1990 The teratogenic effects of anticonvulsant monotherapy: phenobarbital (Pb), carbamazepime (CBZ) and phenytoin (PHT). Teratology 41:565

Lindhout D, Omtzigt JGC 1992 Pregnancy and the risk of teratogenicity. Epilepsia 33:S41–S48

Lindhout D, Omtzigt JGC, Cornel MC 1992 Spectrum of neural-tube defects in 34 infants prenatally exposed to antiepileptic drugs. Neurology 42 (suppl 5):111–118
MRC Antiepileptic Drug Withdrawal Study Group 1991 Randomised study of antiepileptic drug withdrawal in patients with remission. Lancet 337:1175–1180
Omtzigt JGC, Los FJ, Grobbee DE et al 1992a The risk of spina bifida aperta after first-trimester exposure to valproate in a prenatal cohort. Neurology 42 (suppl 5): 119–125
Omtzigt JGC, Nau H, Los FJ, Pijpers L, Lindhout D 1992b The disposition of valproate and its metabolites in the late first trimester and early second trimester of pregnancy in maternal serum, urine, and amniotic fluid: effect of dose, co-medication, and the presence of spina bifida. Eur J Clin Pharmacol 43:381–388
Shapiro S, Hartz SC, Siskind V et al 1976 Anticonvulsants and parental epilepsy in the development of birth defects. Lancet 1:272–275
Sheffield L 1988 A multicentre prospective study of the effects of anti epileptic drugs taken during pregnancy: dysmorphic features. In: Vajda FJ, Donnan GA, Berkovic SF (eds) Focus on epilepsy. University of Melbourne, Melbourne, NSW, p 141–146
Stanley FJ, Priscott PK, Johnston R, Brooks B, Bower C 1985 Congenital malformations in infants of mothers with diabetes and epilepsy in Western Australia, 1980–1982. Med J Aust 143:440–442
Stover P, Schirch V 1992 Enzymatic mechanism for the hydrolysis of 5,10 methenyl-tetrahydropteroylglutamate to form 5-formyltetrahydropteroylglutamate by serine hydroxymethyltransferase. Biochemistry 31:2155–2164
Trotz M, Wegner C, Nau H 1987 Valproic acid-induced neural tube defects: reduction by folinic acid in the mouse. Life Sci 41:103–110

Vitamins, folic acid and the cause and prevention of neural tube defects

Mary J. Seller

Division of Medical & Molecular Genetics, United Medical & Dental Schools of Guy's & St Thomas's Hospitals, 7/8th Floors, Guy's Tower, Guy's Hospital, London Bridge, London SE1 9RT, UK

Abstract. Primary prevention of neural tube defects has been demonstrated in humans by maternal therapy with multivitamins and folic acid or folic acid alone. It has also been shown in several animal models of neural tube defects. One of these, the *curly tail* mouse, has been used extensively to study which agents will prevent neural tube defects in embryos when administered to the mother in early pregnancy. Prevention is achieved with retinoic acid, inositol and the DNA inhibitors hydroxyurea, mitomycin C, 5-fluorouracil and cytosine arabinoside. In no case were neural tube defects prevented in every embryo. A possible preventive effect was seen with riboflavin, vitamin C and vitamin D_2. Despite the use of a variety of dose levels, no prevention was achieved with folic acid, folinic acid, Pregnavite Forte F® tablets, pyridoxine or vitamin B_{12}, or triamcinolone and cycloheximide (inhibitors of mRNA and protein synthesis, respectively), zinc, homocysteine, methionine and thymidine. Various studies have investigated whether there is a biochemical lesion in folate metabolism in women who have had children with neural tube defects. While there is no difference in their dietary intake of folate compared with control patients, the correlation between their dietary folate and the level of folate in both serum and red blood cells is distorted. Also they are less efficient at raising their folate levels after a folate load. The effects are minor but may hint at a lesion which, if identified, could act as a marker for 'at risk' women.

1994 Neural tube defects. Wiley, Chichester (Ciba Foundation Symposium 181) p 161–179

In 1980, my colleagues and I presented the results of a clinical trial which suggested for the first time that the primary prevention of human neural tube defects was possible (Smithells et al 1980). Women who had previously had one or more children with a neural tube defect, and so were at a relatively high risk of doing so again, were given periconceptional supplementation with a multivitamin and folic acid preparation (Pregnavite Forte F, Bencard). They had a markedly reduced recurrence of neural tube defects in their offspring compared with contemporaneous untreated women at the same risk. This apparent success was repeated in another cohort of women (Smithells et al 1983) and by other people (Holmes-Siedle et al 1982, summarized by Seller 1985, Super et al 1991). Although prevention was not total, and despite criticism of the imposed design of the original study (Wald & Polani 1984), it cannot be disputed that this work raised the possibility that the long-sought aim of the primary prevention of a major congenital malformation in humans had actually been achieved. It also stimulated a variety of related studies, most of which have supported the findings (Mulinare et al 1988, Mills et al 1989, Milunsky et al 1989). Pregnavite Forte F® was selected in our study because it was the only multivitamin preparation readily available which also contained folic acid, and folic acid had long been implicated in the cause of neural tube defects (Hibbard & Smithells 1965, Laurence et al 1981). The evidence is now strong that folic acid is the crucial agent in prevention, thanks to the case-control study of Bower & Stanley (1989) and the randomized double-blind prevention trials of the MRC Vitamin Study Research Group (1991). The latter, as did the Smithells work, focused on preventing a recurrence of neural tube defects. An intervention study by Czeizel & Dudás (1992), where folate and other vitamins and minerals were used, has prevented first occurrences.

While neural tube defects are known to be heterogeneous and to have multiple aetiologies, the majority of these defects are regarded as being multifactorial in origin, with both a genetic component and environmental factors believed to play a part. Primary prevention by maternal dietary supplementation would seem to be manipulating the environmental arm in the cause, reducing the cumulative liability so that most embryos are not pushed beyond the threshold to develop the malformation. However, the actual pathway, or mechanism, by which this is achieved remains unknown.

This principle of primary prevention had previously been demonstrated in an animal model of neural tube defects, the *curly tail* mouse. These mice are homozygous for the recessive mutation *curly tail*, but there is a penetrance of only around 60%. Roughly 10–20% of the mice have a sacral spina bifida cystica, 40–50% have a curly tail and 1% or fewer have exencephaly (Embury et al 1979, Copp et al 1982). My colleague, Kathy Cole, and I have used the *curly tail* mice to investigate the extent to which maternally administered exogenous agents will influence the genesis of the neural tube defects in the embryos. From experimental teratology, it was not unexpected that a variety of agents would increase the number of neural tube defects in the *curly tail* mice. What was unexpected was that certain of the agents could decrease the incidence of neural tube defects too. The timing of administration was often important. Vitamin A as retinoic acid (5 mg/kg), given on Day 8 of pregnancy (Day 0 = day of plug), increased the total incidence of neural tube defects to 72%. When retinoic acid was given on Day 9, this fell to around 30%: thus there was prevention, particularly of the posterior abnormalities (Seller et al 1979).

Vitamin A was the sole vitamin that could influence the number of neural tube defects in this way, probably because, as we now know, the retinoids are

TABLE 1 The effect of cytosine arabinoside administered intraperitoneally to *curly tail* mice on Day 8 or Day 9 of pregnancy and assessed on Day 16

	Dose of cytosine arabinoside (mg/kg body weight)				
	0	*5*	*7.5*	*10*	*15*
Day 8					
No. litters	8	11	ND	10	13
No. resorptions	7	4	ND	1	25
No. live embryos	46	80[a]	ND	63[a]	67[a]
Mean litter size	5.6	7.3	ND	6.3	5.2
No. with neural tube defects	20 (44%)	22 (28%)	ND	17 (27%)	14 (21%)
No. with exencephaly	1	2	ND	4	6
No. with spina bifida	4	4	ND	1	0
No. with curly tail	15	16	ND	12	8
No. with straight tail	26 (56%)	58 (72%)	ND	46 (73%)	53 (79%)
Day 9					
No. litters	6	9	10	8	10
No. resorptions	3	3	7	11	36
No. live embryos	33	47	60[b]	41[c]	26[d]
Mean litter size	5.5	5.2	6	5.1	2.6
No. with neural tube defects	14 (42%)	11 (23%)	5 (8%)	5 (12%)	11 (42%)
No. with exencephaly	1	0	0	0	0
No. with spina bifida	6	0	0	1	1
No. with curly tail	7	11	5	4	10
No. with straight tail	19 (58%)	36 (77%)	55 (92%)	36 (88%)	15 (58%)

N.D., Not done. Other abnormalities observed included anotia, micrognathia, exomphalos, microcephaly, exophthalmia, microphthalmia, anophthalmia, cleft lip and cleft palate.
[a]Other abnormalities in 1% of embryos.
[b]Other abnormalities in 3% of embryos.
[c]Other abnormalities in 15% of embryos.
[d]Other abnormalities in 69% of embryos.

fundamental to certain morphogenetic processes (reviewed for example by Morriss-Kay 1993). More effective at preventing neural tube defects were certain inhibitors of DNA synthesis—hydroxyurea (Seller & Perkins 1983), 5-fluorouracil (Seller 1983), mitomycin C (Seller & Perkins 1986) and cytosine arabinoside (Table 1). An inhibitor of mRNA synthesis, triamcinolone, and cycloheximide, which inhibits protein synthesis, have no effect. With cytosine arabinoside the preventive effect is also seen to some extent when it is given on Day 8, affecting especially the posterior lesions, but is more marked when the drug is given on Day 9. Here, as with other agents used, there is an optimal preventive dose, 7.5 mg/kg on Day 9. It is also possible to produce harmful effects with too large a dose of the same substance. For example, 15 mg/kg of cytosine arabinoside on Day 8 or 9 is lethal to a proportion of the embryos and other congenital

TABLE 2 Incidence of neural tube defects in *curly tail* mice on Day 16 of pregnancy after the intraperitoneal injection of triamcinolone or cycloheximide to the mother on Day 9

Substance	*Dose (mg/kg body weight)*	*No. of mothers treated*	*Litter size*	*% Offspring with neural tube defects*
Saline (pooled controls)	0	27	6.4	63
Triamcinolone	5	10	5.2	63
	7.5	6	5.0	50
	10[a]	6	4.8	66
Cycloheximide	5	6	6.0	64
	15	8	7.9	67
	30	6	6.8	59

All doses were given in 0.1 ml saline.
[a]Toxic to the mother.

malformations may appear in the survivors, even if they do not have a neural tube defect. This may have implications for prevention in humans.

Primary prevention of neural tube defects has also been effected in two other mouse mutants—*splotch*, using retinoic acid (Kapron-Bras & Trasler 1985) and *Axial deformity* (*Axd*) mice using methionine (Essien & Wannberg 1993).

Apart from demonstrating the principle of primary prevention, the exact relevance of the *curly tail* mice to the human cases is uncertain. We have injected them intraperitoneally with an aqueous solution of folic acid on Day 9 of pregnancy, with no preventive effect (Table 3). Because mice may differ from humans in the degree to which they rely on intestinal floral synthesis of folic acid, we also injected *curly tail* mice with the active form of folic acid, folinic acid as calcium folinate, in various doses on Day 9. Again, there was no change in the number of neural tube defects. We have administered crushed Pregnavite tablets in water by stomach tube at roughly twice the human dose to *curly tail* mice for one to two weeks prior to conception and then until Day 9 of pregnancy. We have also given folic acid and folinic acid intraperitoneally pre- and post-conceptionally. None of these treatments reduced the number of neural tube defects (Table 3). Earlier we had extensively tried folic and folinic acid on Days 7 and 8 with no effect (Seller & Adinolfi 1981).

In view of the suggestion that a zinc deficiency may be implicated in the aetiology of some human neural tube defects (Cavdar et al 1980, Soltan & Jenkins 1982), we injected *curly tail* mice with zinc. Long before it was suggested that a derangement in homocysteine metabolism might be involved in human neural tube defects (Steegers-Theunissen et al 1991), we had treated *curly tail* mice with homocysteine. Further, before methionine had been shown to be essential for closure of the rat neural tube (Coelho & Klein 1990) or to prevent neural tube defects in *Axd* mice (Essien & Wannberg 1993) and in sodium

TABLE 3 Incidence of neural tube defects in *curly tail* mice after administration of Pregnavite Forte F®, folic acid or folinic acid to the mother before and/or during pregnancy, assessed on Day 16

Substance	*Dose*[a] *(mg/kg body weight)*	*When given*	*No. mothers treated*	*Litter size*	*% Offspring with neural tube defects*
Saline (pooled controls)	0	Day 9	22	6.5	56
Folic acid	0.5	Day 9	3	5.3	69
Folinic acid	0.06	Day 9	12	5.2	55
Folinic acid	0.6	Day 9	6	6.5	62
Folinic acid	6	Day 9	6	5.8	46
Folinic acid	60	Day 9	3	7.0	67
Water (pooled controls)	0	Pre & post conception	3	5.0	54
Pregnavite Forte F®	2[b]	Pre & post conception	8	6.0	54
Folic acid	0.05	Pre & post conception	7	6.6	52
Folinic acid	0.6	Pre & post conception	3	5.3	56

[a]Folic acid and folinic acid were given intraperitoneally in 0.1 ml saline. Pregnavite Forte F® was given orally in 0.1 ml water.
[b]Dose was approximately twice the human daily dose.

TABLE 4 Incidence of neural tube defects in *curly tail* mice on gestational Day 16 after maternal intraperitoneal injection of various agents around the time of embryonic neurulation

Substance	*Dose (per kg body weight)*	*Gestational day treated*	*No. mothers treated*	*Litter size*	*% Offspring with neural tube defects*
Saline (pooled controls)	0	9	76	6.2	66
Zinc (as zinc chloride)	20 μg	9	11	5.8	47
	40 μg	9	11	6.5	52
	150 μg	9	4	7.8	71
	20 μg	7, 8 & 9	7	6.0	43
Homocysteine	50 mg	9	11	5.6	69
	100 mg	9	11	6.5	63
	200 mg	9	8	5.9	64
	400 mg	9	7	7.0	73
Methionine	200 mg	9	9	6.1	67
	400 mg	9	6	5.8	63
	800 mg	9	6	6.8	51
	1600 mg	9	7	6.2	86
Thymidine	25 mg	9	6	7.8	59
	50 mg	9	13	6.1	56
	100 mg	9	6	6.0	61
	200 mg	9	6	6.3	63
	400 mg	9	7	7.1	58

All doses were given in 0.1 ml saline.

valproate-treated rats (Nosel & Klein 1992), we had tried methionine and also thymidine because of its involvement in *de novo* DNA synthesis. All of these treatments failed to alter the number of neural tube defects (Table 4).

We also tested most of the other components of the Pregnavite Forte F® tablet: riboflavin (vitamin B_2), pyridoxine (vitamin B_6), vitamin C, vitamin D_2 (calciferol) and also vitamin B_{12} (Table 5). There was some evidence of prevention of neural tube defects with specific doses of riboflavin on Day 9 and vitamin D on Day 8, but the results were not as consistent as they were with the DNA inhibitors. Also, these experiments were performed during a summer when, as can be seen from the controls, the incidence of neural tube defects temporarily declined somewhat.

Inositol was another agent investigated. Cockcroft (1979) had cultured rat embryos through neurulation, with individual vitamins systematically removed from the culture medium. Growth retardation or dysmorphism occurred in the absence of several of the vitamins, but neural tube defects occurred only in the absence of inositol. In the *curly tail* mice, administration of inositol to the mother on Day 8 reduced the number of neural tube defects (Table 5). This could tie in with the fact that Cockcroft et al (1992) have subsequently found that

TABLE 5 Incidence of neural tube defects in *curly tail* mice on gestational Day 16 after maternal intraperitoneal injection of various vitamins around the time of embryonic neurulation

Substance	*Dose (per kg body weight)*	*Gestational day treated*	*No. mothers treated*	*Mean litter size*	*% Offspring with neural tube defects*
Saline (pooled	0	8	34	6.1	42
controls)	0	9	28	7.1	52
Riboflavin	75 mg	8	7	6.0	69
	75 mg	9	10	5.5	42
	100 mg	9	10	5.2	27[a]
	75 mg	10	7	6.4	51
Pyridoxine	1 mg	9	5	4.6	43
	2 mg	9	10	7.0	41
	6 mg	9	10	6.1	56
	12 mg	9	12	6.3	51
Vitamin C	10 mg	8	5	7.4	51
	100 mg	8	8	5.6	58
	200 mg	8	5	6.8	56
	1 mg	9	7	6.7	45
	400 mg	9	5	7.6	45
	500 mg	9	11	6.5	37
Vitamin D_2	8×10^6 IU	8	7	5.6	31[a]
	8×10^6 IU	9	5	6.6	48
Vitamin B_{12}	3 μg	8 & 9	3	4.7	64
	30 μg	8 & 9	10	6.3	49
	300 μg	8 & 9	4	5.3	57
Inositol	250 mg	8	6	6.5	31[a]
	400 mg	8	5	7.0	37
	250 mg	9	4	4.8	53
	400 mg	9	2	7.5	73

All doses were given in 0.1 ml saline.
[a]Possibly some preventive effect of the agent tested.

in vitro inositol deficiency increases the susceptibility of *curly tail* embryos to neural tube defects.

Although folic acid does not seem to be able to prevent neural tube defects in *curly tail* mice as it does in humans, these mice may well be relevant to human neural tube defects in another respect. Around 80% of the cases of human neural tube defects involve an isolated and often quite local defect (Khoury et al 1982). Apart from secondarily induced abnormalities like talipes (club foot), the rest of the body develops normally. In only a minority of babies with neural tube defects are there multiple malformations. A detailed analysis shows a clustering of specific defects related to the site of the lesion in the neural tube, rather than a random occurrence of malformations. This suggests that these defects have also arisen in conjunction with or influenced by the actual lesion in the neural tube (Seller & Kalousek 1986). What we learn from animal teratology experiments is that even if a teratogen is given as a single pulse on one day of pregnancy, it rarely produces an isolated malformation, several defects usually occur. Further, in animal experiments concerning nutritional deficiency, the same phenomenon is observed. While a chronic general dietary restriction in early pregnancy in rats results in intrauterine death and small, immature young, rather than malformations (Saitoh & Takahashi 1973), total fasting for 24 hours of pregnant mice during the organogenetic period does produce malformations in the offspring, but they are not restricted to neural tube defects, they include cleft palate and skeletal abnormalities too (Kalter 1960, Runner & Miller 1956). A lack of individual vitamins—vitamins A, B_{12}, D, nicotinic acid or riboflavin—produces a variety of defects, but not neural tube defects. With a lack of vitamin B_6, vitamin E, pantothenic acid or folic acid, neural tube defects do occur, but not alone: they are among a spectrum of defects produced (reviewed by Seller 1989). All these findings must be reconciled with the fact that human neural tube defects, in which we believe a deficiency of folic acid is somehow implicated in the cause, are, more often than not, isolated malformations. The neural tube defects of *curly tail* mice are similarly localized lesions with no other abnormalities. Copp and his colleagues have shown that this is caused by an underlying localized defect (Copp et al 1988, Copp, this volume). It is possible that the seat of human neural tube defects lies likewise in a local defect(s) in the embryo (Seller 1983, 1989), and this is the model we should be following rather than a generalized micronutrient deficiency model.

Nevertheless, we continue to search women for evidence as to how a folate deficiency might be involved in a cause of human neural tube defects. It has long been known that there is an association between low maternal folate levels and neural tube defects. Hibbard & Smithells (1965) found a folate deficiency after delivery in 69% of mothers of babies with abnormalities of the central nervous system, but in only 17% of women with normal babies. In the first trimester of pregnancy, much closer to the time when the neural tube defect lesion arises, women bearing fetuses with such defects were found to have significantly lower levels of folate in their red blood cells and of vitamin C in their white blood cells (Smithells et al 1976). Yates et al (1987) found low levels of folate in red blood cells of women who had had two or more pregnancies with neural tube defects, when they were not pregnant. There was a positive linear relationship of the levels with the number of offspring with neural tube defects. These associations continue to be found, sometimes from unexpected sources, such as in the aftermath of hurricane Gilbert in Jamaica (Duff et al 1991), when a folate deficiency due to nutritional deprivation was associated with a statistically significant increase in births of babies with neural tube defects.

Despite these multifarious associations between low levels of maternal folate and neural tube defects, there is no evidence that the folate deficiency actually causes the defects, and it is obvious that a low maternal folate level *per se* is not sufficient to produce neural tube defects. For instance, the women with the lowest folate levels in their red blood cells in the 1976 study of Smithells et al did not have children with neural tube defects (Schorah & Smithells 1991). A line of evidence frequently quoted is, on close scrutiny, either much better, or much worse, than commonly thought. This is the effect of aminopterin, a folic acid antagonist, used as an abortifacient and said to produce neural tube defects. Of all the cases documented in the literature (Shaw & Steinbach 1968, Thiersch 1952, Thiersch and Warkany in Nelson 1960; reviews by Milunsky et al 1968, Warkany 1986), there are a variety of malformations reported, but only two neural tube defects (Thiersch 1952, in Nelson 1960), and only one patient received the aminopterin early enough in pregnancy to influence closure of the neural tube. She was treated on Days 17, 18 and 19 after fertilization and she did produce an anencephalic child. The second neural tube defect, said to be a meningoencephalocele with facial anomalies (Thiersch 1952), is illustrated in a later paper (Thiersch in Nelson 1960) and could be a case of cervical cystic hygromata and hydrops, possibly fetal Turner's syndrome, rather than a neural tube defect. Anyway, the mother had been treated with aminopterin, as were all other patients, around Day 50. So, in the single valid case where a folic acid antagonist was used, a neural tube defect occurred.

Efforts are now being made to identify some form of metabolic perturbation in folate metabolism in 'at risk' women. This, of course, would be a most important advance in terms of prevention, because it might form the basis of a test by which such women could be identified before their reproductive phase, so that only they, and not the whole population of women, would receive periconceptional folate supplementation.

My colleagues and I (Wild et al 1994) have studied women who have had two children or fetuses with neural tube defects when they are neither pregnant nor on vitamin supplements. We assessed their current dietary intake of folate using a food questionnaire and measured their levels of folate in serum and red blood cells in the morning, after a minimal breakfast. There was no difference in folate intake or blood folate levels between them and matched controls who had not had children with neural tube defects. But they did differ in the following way. In the controls there was a significant correlation between the dietary and serum folate levels, between dietary and red blood cell levels, and between serum and red blood cell folate levels. In women who had had affected pregnancies, there was no correlation between dietary and serum folate or dietary and red blood cell folate, although there was between serum and red blood cell folate. Yates et al (1987) also found that the relationship between dietary and red blood cell folate was significantly different in women who had had children with neural tube defects compared with controls.

Some of the women in our study were given a folate load after the initial blood samples, through a drink of 500 ml of orange juice. One hour later, both groups had increased levels of serum folate, but the increase was less marked in the study group than in the controls. This work suggests that some women at risk for producing children with neural tube defects may have both a long-term and a short-term problem in getting dietary folate into the bloodstream.

Another approach we have taken is to culture cells from mid-trimester placentae from normal fetuses and those with neural tube defects. After extensive subculture and selection, pure cytotrophoblastic cells can be obtained. If these are then grown in a medium lacking thymidine and other DNA precursors and given DL-5-[^{14}C]methyltetrahydromonoglutamate, there is a significant lag in the rate of incorporation of the [^{14}C] label into the nucleic acid pool in the trophoblasts from the affected fetuses compared with those from normal fetuses. However, there is a gradual catch-up after about 60 hours, which is achieved totally by 90 hours (Habibzadeh et al 1993).

Vitamin B_{12} is intimately bound up in folate metabolism. I have found that in mid-trimester amniotic fluid, vitamin B_{12} levels are lower than normal where the fetuses have neural tube defects, although the folate levels are normal and the levels of transcobalamins, the vitamin B_{12} carrier proteins, are raised (Gardiki-Kouidou & Seller 1988). Further, and possibly more importantly, both the low level of vitamin B_{12} and the higher levels of transcobalamins were found in amniotic fluids of normal fetuses of women who had previously had a fetus with a neural tube defect. This implies a defect which is inherited. Low levels of vitamin B_{12} in the amniotic fluid of pregnancies with neural tube defects have also been found by Economides et al (1992). Magnus et al (1991) have reported raised levels of transcobalamins in the amniotic fluid of mothers carrying a normal fetus who had previously produced one with a neural tube defect. Depressed vitamin B_{12} levels in the presence of high levels of carrier proteins argue for a derangement in the production, transport or metabolism of vitamin B_{12}. It is not inconsistent with the observations on folates, because there could be a primary folate abnormality that secondarily affects vitamin B_{12}.

Overall, there are now several strands of evidence which support the idea that neural tube defects arise where there is a defect of folate handling at some location. The work I have presented suggests it is manifest in both the maternal and the fetal compartments. This may result in a lowered availability of folate for intracellular needs such as *de novo* DNA synthesis. This can be disastrous for the embryo, but is not otherwise detrimental to the mother, who shows no symptoms of folate deficiency. At risk women do not necessarily have a lower than normal folate intake, but because of faulty handling of folate, additional folate is needed to fulfill unusual demands, such as those of the rapidly growing embryo.

Acknowledgements

I am indebted to Mrs Kathy Cole for technical assistance and for financial support from Action Research, the Buttle Trust and the Spastics Society.

References

Bower C, Stanley FJ 1989 Dietary folate as a risk factor for neural-tube defects: evidence from a case-control study in Western Australia. Med J Aust 150: 613–619

Cavdar AO, Arcasoy A, Baycu T, Himmetoglu O 1980 Zinc deficiency and anencephaly in Turkey. Teratology 23:141

Cockcroft DL 1979 Nutrient requirements of rat embryos undergoing organogenesis in vitro. J Reprod Fertil 57:505–510

Cockcroft DL, Brook FA, Copp AJ 1992 Inositol deficiency increases the susceptibility to neural tube defects of genetically predisposed (curly tail) mouse embryos in vitro. Teratology 45:223–232

Coelho CND, Klein NW 1990 Methionine and neural tube closure in cultured rat embryos: morphological and biochemical analyses. Teratology 43:437–451

Copp AJ 1994 Genetic models of mammalian neural tube defects. In: Neural tube defects. Wiley, Chichester (Ciba Found Symp 181) p 118–143

Copp AJ, Seller MJ, Polani PE 1982 Neural tube development in mutant (curly tail) and normal mouse embryos: the timing of posterior neuropore closure in vivo and in vitro. J Embryol Exp Morphol 69:151–167

Copp AJ, Brook FA, Roberts HJ 1988 A cell type specific abnormality of cell proliferation in mutant (curly tail) mouse embryos developing spinal neural tube defects. Development 104:285–295

Czeizel AE, Dudás I 1992 Prevention of the first occurrence of neural-tube defects by periconceptional vitamin supplementation. N Engl J Med 327:1832–1835

Duff EMW, Cooper ES, Danbury CM, Johnson BE, Serjeant GR 1991 Neural tube defects in hurricane aftermath. Lancet 337:120–121

Economides DL, Ferguson J, Mackenzie IZ, Darley J, Ware II, Holmes-Siedle M 1992 Folate and vitamin B_{12} concentrations in maternal and fetal blood and amniotic fluid in second trimester pregnancies complicated by neural tube defects. Br J Obstet Gynaecol 99:23–25

Embury S, Seller MJ, Adinolfi M, Polani PE 1979 Neural tube defects in curly-tail mice. I. Incidence, expression and similarity to the human condition. Proc R Soc Lond Ser B Biol Sci 206:85–94

Essien FB, Wannberg SL 1993 Methionine but not folic acid or vitamin B_{12} alters the frequency of neural tube defects in Axd mutant mice. J Nutr 123:27–34

Gardiki-Kouidou P, Seller MJ 1988 Amniotic fluid folate, vitamin B_{12} and transcobalamins in neural tube defects. Clin Genet 33:441– 448

Habibzadeh N, Schorah CJ, Seller MJ, Smithells RW, Levene MI 1993 Uptake and utilisation of DL-5-[methyl-C-14]tetrahydropteroylmonoglutamate by cultured cytotrophoblasts associated with neural tube defects. Proc Soc Exp Biol Med 203: 45–54

Hibbard ED, Smithells RW 1965 Folic acid metabolism and human embryopathy. Lancet 1:1254

Holmes-Siedle M, Lindenbaum RH, Galliard A, Bobrow M 1982 Vitamins and neural tube defects. Lancet 1:276

Kalter H 1960 Teratogenic action of a hypocaloric diet and small doses of cortisone. Proc Soc Exp Biol Med 104:518–520

Kapron-Bras CM, Trasler DG 1985 Reduction in the frequency of neural tube defects in splotch mice by retinoic acid. Teratology 32:87–92

Khoury MJ, Erickson JD, James LM 1982 Etiologic heterogeneity of neural tube defects: clues from epidemiology. Am J Epidemiol 115:538–548

Laurence KM, James N, Miller MH, Tennant GB, Campbell H 1981 Double-blind randomised controlled trial of folate treatment before conception to prevent recurrence of neural-tube defects. Br Med J 282:1509–1511

Magnus P, Magnus EM, Berg K 1991 Transcobalamins in the etiology of neural tube defects. Clin Genet 39:309–310

Mills JL, Rhoads GG, Simpson JL et al 1989 The absence of a relation between the periconceptional use of vitamins and neural-tube defects. N Engl J Med 321:430–435

Milunsky A, Graef JW, Gaynor MF Jr 1968 Methotrexate-induced congenital malformations. J Pediatr 72:790–795

Milunsky A, Jick H, Jick SS et al 1989 Multivitamin/folic acid supplementation in early pregnancy reduces the prevalence of neural tube defects. JAMA (J Am Med Assoc) 262:2847–2852

Morriss-Kay GM 1993 Retinoic acid and craniofacial development: molecules and morphogenesis. BioEssays 15:9–15

MRC Vitamin Study Research Group 1991 Prevention of neural tube defects: results of the Medical Research Council vitamin study. Lancet 338:131–137

Mulinare J, Cordero JF, Erickson JD, Berry RJ 1988 Periconceptional use of multivitamins and the occurrence of neural tube defects. JAMA (J Am Med Assoc) 260:3141–3145

Nelson MM 1960 Teratogenic effects of pteroylglutamic acid deficiency in the rat. In: Congenital malformations. Churchill, London (Ciba Found Symp 59) p 134–157

Nosel PH, Klein NW 1992 Methionine decreases the embryotoxicity of sodium valproate in the rat: in vivo and in vitro observations. Teratology 46:499–507

Runner MN, Miller JR 1956 Congenital deformity in the mouse as a consequence of fasting. Anat Rec 124:437–438

Saitoh M, Takahashi S 1973 Changes of embryonic wastage during pregnancy in rats fed low and high energy diets. J Nutr 103:1652–1657

Schorah CJ, Smithells RW 1991 Maternal vitamin nutrition and malformations of the neural tube. Nutr Res Rev 4:33–49

Seller MJ 1983 The cause of neural tube defects: some experiments and a hypothesis. J Med Genet 20:164–168

Seller MJ 1985 Periconceptional vitamin supplementation to prevent recurrence of neural tube defects. Lancet 1:1392–1393

Seller MJ 1989 Vitamins and neural tube defects. In: Rodeck CH (ed) Fetal medicine. Blackwell Scientific Publications, Oxford, p 1–25

Seller M, Adinolfi M 1981 The curly-tail mouse: an experimental model for human neural tube defects. Life Sci 29:1607–1615

Seller MJ, Kalousek DK 1986 Neural tube defects: heterogeneity and homogeneity. Am J Med Genet Suppl 2:77–87

Seller MJ, Perkins KJ 1983 Effect of hydroxyurea on neural tube defects in the curly-tail mouse. J Craniofac Genet Dev Biol 3:11–17

Seller MJ, Perkins KJ 1986 Effect of mitomycin C on the neural tube defects of the curly-tail mouse. Teratology 33:305–309

Seller MJ, Embury S, Polani PE, Adinolfi M 1979 Neural tube defects in curly-tail mice. II. Effects of maternal administration of vitamin A. Proc R Soc Lond Ser B Biol Sci 206:95–107

Shaw EB, Steinbach HL 1968 Aminopterin-induced fetal malformation: survival of infant after attempted abortion. Am J Dis Child 115:477–482

Smithells RW, Sheppard S, Schorah CJ 1976 Vitamin deficiencies and neural tube defects. Arch Dis Child 51:944–950

Smithells RW, Sheppard S, Schorah CJ et al 1980 Possible prevention of neural tube defects by periconceptional vitamin supplementation. Lancet 1:339–340

Smithells RW, Nevin NC, Seller MJ et al 1983 Further experience of vitamin supplementation for the prevention of neural tube defect recurrences. Lancet 1:1027–1031

Soltan MH, Jenkins DM 1982 Maternal and fetal plasma zinc concentration and fetal abnormality. Br J Obstet Gynaecol 89:56–60

Steegers-Theunissen RPM, Boers GHJ, Trijbels FJM, Eskes TKAB 1991 Neural tube defects and derangement of homocysteine metabolism. N Engl J Med 324:199–200

Super M, Summers EM, Meylan B 1991 Preventing neural tube defects. Lancet 338:755–756

Thiersch JB 1952 Therapeutic abortions with a folic acid antagonist, 4-aminopteroylglutamic acid (4-amino PGA) administered by the oral route. Am J Obstet Gynecol 63:1298–1304

Wald NJ, Polani PE 1984 Neural tube defects and vitamins: the need for a randomised clinical trial. Br J Obstet Gynaecol 91:516–523

Warkany J 1986 Aminopterin and methotrexate: folic acid deficiency. In: Sever JL, Brent RL (eds) Teratogen update: environmentally induced birth defect risks. Alan R Liss, New York p 39–43

Wild J, Seller MJ, Schorah CJ, Smithells RW 1994 Investigation of folate intake and metabolism in women who have had two neural-tube defect pregnancies. Br J Obstet Gynaecol, in press

Yates JRW, Ferguson-Smith MA, Shenkin A, Guzman-Rodriguez R, White M, Clark BJ 1987 Is disordered folate metabolism the basis for the genetic predisposition to neural tube defects? Clin Genet 31:274–287

DISCUSSION

Mills: Your studies on folate metabolism in humans were very interesting in that you did not find a correlation between the folate in the diet and in serum, or between folate in the diet and in the red blood cells. This suggests there is either an intestinal problem or an enzyme problem, such as a conjugase defect. Have you explored those possibilities?

Seller: Not yet.

Scott: If malabsorption was a significant factor, wouldn't one expect it to reflect itself in different plasma folate status? There is no obvious difference in status between cases and controls. If it was a malabsorption problem, the only way the embryo could be signalled that there was a problem presumably would be through circulating levels of plasma maternal folate.

Magnus et al (1991) and yourself (Gardiki-Kouidou & Seller 1988) have done some nice work on the transcobalamins in amniotic fluid. It is very interesting because the signals are there to be interpreted, so to speak. Transcobalamin II has an obvious function; it's the specific transport protein that takes vitamin B_{12} around from the liver to all the other cells. But transcobalamin I doesn't seem to have any function. I was the first one to identify transcobalamin III (Bloomfield & Scott 1972) 20 years ago and it still doesn't have any known function! So there are two proteins looking for a role.

Lindhout: How do you explain the higher levels of transcobalamin II in the amniotic fluid?

Scott: I don't have an explanation. I think it's an important result, but it is unclear what it means.

Lindhout: Is the transcobalamin II of fetal or maternal origin?

Scott: Transcobalamin II is made by a wide variety of cells, including the endothelial cells from human umbilical vein (Quadros et al 1989), but I don't think the site of synthesis of amniotic transcobalamin II has been established. Porck et al (1983) suggest that transcobalamin II in cord blood is fetal in origin.

Lindhout: Then if there is a feedback mechanism, it could be playing a role in the maternal organism. What you find in the amniotic fluid is secondary to maternal effects.

Scott: If that were true, in pernicious anaemia the concentration of apo-transcobalamin II in the plasma should go up. It doesn't. So there's no evidence for feedback. You can say, maybe it only happens in the embryo. This is possible, but I don't know of any evidence that suggests that transcobalamin II is responsive to either vitamin B_{12} status or folate status.

Hall: There is lots of evidence that the fetus orchestrates the mother and the mother orchestrates the fetus through humoral kinds of messages.

Stanley: The general feeling that most of us have is that the problem is not straight folate deficiency. Mary Seller showed that in case-control studies, women with lowest levels of folic acid were in the control group but not amongst the cases. Therefore, there is now an interest in looking at abnormalities of folic acid metabolism. Carol Bower and I have just done some very preliminary studies looking at absorption (Bower et al 1993). We like the hypothesis that there may be a conjugase deficiency. We felt that this could explain the familial patterns, as well as the fact that free folate given to mothers overcomes the problem. We have looked at a group of mothers who had previously had children with neural tube defects. There were two control groups: one was our own research group; the other consisted of friends of the mothers, who were much better controlled for social class. We gave a conjugated folate drink (like a Vegemite milk shake). We had taken a blood sample a week before and given 5 mg of folic acid with the aim of overcoming any differences in intake. We then gave them the Vegemite milk shake after fasting and bled them every hour after that. The women certainly deconjugated the folate and there was absorption in all groups. The research group had much higher levels of folic acid than either the control or the neural tube defect group of mothers and absorbed much more dramatically as well.

When we looked at the pattern of absorption between the cases and controls, there did seem to be some differences. This may relate to the way folate is transported across the gut or possibly taken out of the system and into the tissues. This is very preliminary but it does suggest that we should pursue studies of folic acid handling. This might identify the 50–70% of women who are at risk—the ones whose defects seem to be preventable by folic acid supplementation.

Hall: Were your subjects all Northern Europeans?

Stanley: Yes.

Nau: Folinic acid has a very short half-life in mice. In your *curly tail* mouse studies, it might be better to administer the folinic acid several times to get long-term exposure. We saw the best protective effect in our mice using infusion with a minipump.

Seller: We gave either a single injection on one day during pregnancy or for the pre- and post-conception we gave one injection daily for several weeks.

van Straaten: I recently repeated the methionine supplementation in *curly tail* mice. There was little or no change in the frequency of curly tails or open spina bifida (unpublished results), as you found.

You said that the increased demand of the fetus for folate might not be fulfilled in women with faulty handling of folate, thereby leading to a neural tube defect. I can understand how that applies to the mouse or the rat because the embryos are relatively large (3mm) at the beginning of neurulation. But in humans, a 2–3 mm embryo can, quantitatively, put hardly any demand on the mother.

Seller: I agree. This is my dilemma: I think it's a localized defect (not the same defect), like that in *curly tail.* But I have to try to make sense of all the work in humans. I'm just grasping at straws to try to explain these other observations.

Copp: You say a defect. Do you think there's a single localized defect that is responsible for neural tube defects?

Seller: I have no idea, but something local.

Holmes: I agree that children with anencephaly, spina bifida and other neural tube defects tend to have a single, isolated major malformation. But we must acknowledge the fact that people who have done a lot of autopsies on children with anencephaly, such as Ronald Lemire, report that the infant with anencephaly almost always has a non-neural malformation. This statement requires that you review the definition of a major malformation. Others have reported that 20–25% of infants with anencephaly or spina bifida have associated non-neural malformations. Regardless of the percentage, there is no reason why the abnormal genes which cause anencephaly and spina bifida cannot produce additional non-neural malformations as well.

Hall: Mary (Seller) is distinguishing things that sort of make sense as being part of the neural tube defect (such as rib anomalies) from types of defects that wouldn't be expected to be part of the neural tube closure (such as heart defects).

Seller: There will also be hypoplasia of the adrenal gland.

Holmes: I don't know of anyone who has done autopsies on an unselected population of infants and children and reported the frequency of internal structural abnormalities that will be identified by a post-mortem examination.

Stanley: The difficulty now is that many of these pregnancies are terminated in a destructive way, so we can't determine the level of defect or even the sex.

Dolk: There is a well known difference in prevalence of neural tube defects between the United Kingdom and Ireland versus continental Europe, the former having a prevalence two to three times higher than the latter in the 1980s. We looked at whether the United Kingdom and Ireland have a higher prevalence of both isolated neural tube defects and neural tube defects in association with other major anomalies (multiply malformed cases) and found that they do (Dolk et al 1991). We don't know whether the difference in prevalence in the first place is folate related, either genetically or environmentally. However, this suggests that whatever aetiological factor underlies the geographical difference in prevalence, it is important for both isolated neural tube defects and at least some forms of multiple malformations including a neural tube defect. These are 'real' multiple malformations, such as omphalocele with a neural tube defect. In this registry-based study, overall 31% of all cases of neural tube defect had another anomaly, but this was reduced to 18% on exclusion of minor anomalies or anomalies which were secondary to the neural tube defect, like club foot or adrenal hypoplasia.

Czeizel: I disagree with Lewis Holmes. The term sequence represents a pattern of multiple anomalies derived from a single known or presumed prior anomaly with a cascade of secondary anomalies. Neural tube defects are a typical sequence. In the case of spina bifida aperta, there are many secondary consequences—hydrocephalus, club foot, etc. In Hungary autopsy was obligatory in infant death. Thus, all live-born anencephalic infants had a pathological evaluation. Pathology departments of children's hospitals have pathologists with appropriate expertise in this field. Anencephalic infants had a high frequency of cleft palate (54%). In my opinion, this is a secondary consequence of the anencephaly, therefore we can consider the anencephaly with cleft palate as an isolated malformation. Another important point was the dilatation of the urinary system in cases with spina bifida aperta and anencephaly. Again, this is probably connected with the primary defect. I don't believe that it is an independent malformation; it's a secondary consequence of neural tube defects.

Dolk: What about omphalocele?

Czeizel: This is difficult. I suppose that it is independent, but I'm not sure, because it is a schisis-type abnormality. Such abnormalities (i.e. anencephaly-spina bifida aperta, oral clefts, omphalocele, diaphragm defect) have a non-random tendency for association (Czeizel 1981).

Seller: 70–80% of neural tube defects in a population of fetuses that I have looked at were isolated lesions in the way I define it.

Opitz: In my autopsy work on anencephalics, I also see these other anomalies. They are not necessarily just consequences of the neural tube defect. There are other vertebral malformations of the neuraxis above or below the lesion, and also cleft palate. There is sometimes a curious cleft in the nose when you look at the face down the middle of the nose. There may be congenital heart defects, but then you never know whether to count a patent ductus arteriosus as normal or as a malformation. One of the recurrences for neural tube defects is, in fact, a risk for diaphragmatic abnormalities. When I look at these fetuses, I am always impressed that they have not just a neural tube defect but rather have a defect of gastrulation. You are looking at the end-point, some less complex, some more complex, of a much earlier defect of gastrulation. Something may have gone wrong with the inner cell mass, for example.

There are two other reasons I think this. One is this very useful concept of the schisis association Andrew Czeizel published in 1981. Among the still-born fetuses with multiple congenital anomalies that I autopsy, schisis associations are a common entity. Neural tube defects, specifically spina bifidas, are common in those infants. In infants with a pure schisis association, the defect acts during blastogenesis and gives rise to all midline anomalies.

The third item concerns the infants of diabetic mothers. Neural tube defects are common in this group, not just spina bifida, but also alobar holoprosencephaly and so forth. Biochemists and nutritionists have said there are many items that could be pathogenetically relevant: hypoglycaemia, hyperglycaemia, somatomedin inhibitors or the arachidonic acid defects that Alan Goldman (1985) has been working on. I think it is extremely helpful to think about neural tube defects as one end of a spectrum of defects of gastrulation that may involve multiple biochemical metabolic abnormalities. Folic acid handling could be one of these. For example, in the infants of diabetic mothers, just correcting the haemoglobin A_1 levels with a very strict diet is a wonderful way of preventing neural tube defects.

Hall: Wouldn't you separate the whole bundle of neural tube defects into one group of gastrulation problems and one of pure multifactorial neural tube defects? In our experience, those with congenital anomalies, which could be said to have gastrulation problems, have a very low recurrence rate.

Opitz: Andrew Czeizel has said that the schisis association has a low recurrence rate for the individual component anomalies of the association (Czeizel 1981).

Hall: Is this for the whole groups of midline problems or just the anomalies which have been called a schisis association?

Opitz: This is for one of the component anomalies in siblings and schisis associations. I have tried hard to separate them, but I can't. This is simply from doing lots of autopsies. One doesn't routinely do vertebral X-rays on every anencephalic still-born to count the number of vertebrae and to find all spina bifida occulta. I think if fetuses with neural tube defects were studied more carefully, we might find more of these associated midline anomalies that probably represent a defect of gastrulation.

Hall: But that's a subgroup. If you knew this was a gastrulation defect, you would worry about recurrence of anything to do with gastrulation. There is a subgroup that really does have neural tube defects and that's all they are at risk for. There are probably also other subgroups. The more closely we define those, the more likely we are to discover the environmental risks and what the real recurrence for that group is.

Opitz: The provisional and arbitrary working distinction that I have made is that the group that appears to have only a neural tube recurrence risk may, in fact, represent a relatively late, pure neural tube defect without a general defect of gastrulation. However, I am uneasy with that concept and would prefer to operate with the null hypothesis that *all* apparently non-syndromal, 'multifactorial' neural tube defects are defects not just of neurulation but of gastrulation and, that if looked for carefully, more midline defects will be found in those infants. I have often wondered if many schisis association cases with a neural tube defect are not simply 'multifactorial' neural tube defects with onset earlier in blastogenesis. If Hungary continues to supplement all pregnant women with 0.8 mg folic acid daily and schisis associations subsequently disappear, then I will feel justified in holding this opinion. I would then recommend that mothers who have had such a child, including diabetics, also receive folic acid prophylaxis.

Hall: When we see the effect of folic acid on populations, will it be specifically on one of these subgroups? That to me is a really exciting question.

Opitz: I would bet that folic acid would prevent all defects of blastogenesis.

Copp: We have to be very careful when we talk about gastrulation defects. To some of us gastrulation has a particular meaning: it's the event in embryogenesis that Gary Schoenwolf described (this volume). There is no evidence to my mind that any of these malformations are related to the gastrulation process. You are considering malformations that involve different systems and somehow you imagine that they all arose from a common event, namely gastrulation. This misses the point that many of the midline closure events may use the same gene products. The schisis association (Czeizel 1981) might reflect the use of the same gene products in different closure events, therefore those events don't have to be related in terms of cell lineage. They don't have to go back to gastrulation to be related to each other aetiologically. I would like to see some evidence that the so-called gastrulation defects do involve gastrulation.

References

Bloomfield FJ, Scott JM 1972 Identification of a new B_{12} binder (transcobalamin III) in normal human sera. Br J Haematol 22:33–42

Bower CI, Stanley FJ, Nicol DJ 1993 Maternal folate status and the risk for neural tube defects: the role of dietary folate. Ann NY Acad Sci 678:146–155

Czeizel AE 1981 Schisis-association. Am J Med Genet 10:25–35

Dolk H, De Wals P, Gillerot Y et al 1991 Heterogenicity of neural tube defects in Europe: the significance of site of defect and presence of other major anomalies in relation to geographic differences in prevalence. Teratology 44:547–559

Gardiki-Kouidou P, Seller MJ 1988 Amniotic fluid folate, vitamin B_{12}, and transcobalamins in neural tube defects. Clin Genet 33:441–448

Goldman AS, Baker L, Piddington R, Marx B, Herold R, Egler J 1985 Hyperglycemia-induced teratogenesis is mediated by a functional deficiency of arachidonic acid. Proc Natl Acad Sci USA 82:8227–8231

Magnus P, Magnus EM, Berg K 1991 Transcobalamins in the etiology of neural tube defects. Clin Genet 39:309–310
Quadros EV, Rothenberg SP, Jaffe EA 1989 Endothelial cells from human umbilical vein secrete functional transcobalamin II. Am J Physiol 256:C296–C303
Porck HJ, Frater-Schroder M, Frants RR, Kierat L, Eriksson AW 1983 Genetic evidence for the origin of transcobalamin II in human cord blood. Blood 62:234–237
Schoenwolf GC 1994 Formation and patterning of the avian neuraxis: one dozen hypotheses. In: Neural tube defects. Wiley, Chichester (Ciba Found Symp 181) p 25–50

Folic acid metabolism and mechanisms of neural tube defects

J. M. Scott,* D. G. Weir†, A. Molloy†, J. McPartlin†, L. Daly‡ and P. Kirke°

Department of Biochemistry, Department of Clinical Medicine†, Trinity College, Dublin 2, Department of Community Medicine and Epidemiology‡, University College, Earlsfort Terrace, Dublin 2 and Health Research Board°, 73 Lower Baggot Street, Dublin 2, Ireland*

Abstract. Folate acts as a cofactor for enzymes involved in DNA and RNA biosynthesis. Folate is also involved in the supply of methyl groups to the so-called methylation cycle, which uses methionine and makes homocysteine. The folate cofactor, N^5-methyltetrahydrofolate, donates its methyl group to a vitamin B_{12}-dependent enzyme, methionine synthase, which recycles homocysteine back to methionine. The cell's ability to methylate important compounds such as proteins, lipids and myelin will be compromised by deficiency of folate or vitamin B_{12}, resulting in impaired cellular function. Methionine synthase plays another role: it converts circulating N^5-methyltetrahydrofolate into tetrahydrofolate. The latter but not the former can act as a substrate for polyglutamate synthase, thereby becoming retained in the cell as polyglutamate. Interruption of DNA biosynthesis or methylation reactions could prevent the proper closure of the neural tube. Such inhibition could be caused by simple deficiency of either folic acid or vitamin B_{12}. Studies comparing serum folate and vitamin B_{12} status in women who have had an affected pregnancy to those in control women indicate no difference between the two groups and show that most cases are not clinically deficient in either vitamin. A small number of studies using the level of folate in red blood cells, which is a better reflection of tissue stores, confirm this, suggesting instead a metabolic impairment in the biochemical functions of one of these vitamins. The trials using folic acid to prevent neural tube defects thus seem to be effectively overcoming a metabolic block rather than treating folate deficiency.

1994 Neural tube defects. Wiley, Chichester (Ciba Foundation Symposium 181) p 180–191

In mammalian cells the central form of folate, namely tetrahydrofolate, accepts single carbon units from a variety of donors to form a pool of interchangeable cofactors, including N^5,N^{10}-methylenetetrahydrofolate, N^{10}-formyltetrahydrofolate and N^5,N^{10}-methenyltetrahydrofolate (Blakley & Benkovic 1984). The principle single carbon donor is the C-3 of serine (Fig. 1).

The cofactor N^{10}-formyltetrahydrofolate donates its formyl group to two of the enzymes in *de novo* purine biosynthesis. Both reactions produce tetrahydrofolate which can then be re-used for further single carbon transfer reactions. The cofactor N^5,N^{10}-methylenetetrahydrofolate donates its C-1 group to deoxyuridine monophosphate to produce deoxythymidine monophosphate in a reaction catalysed by the enzyme thymidylate synthase (EC 2.1.1.45). Similarly, folates are also necessary for the *de novo* biosynthesis of pyrimidines and thus the synthesis of DNA.

Another enzyme, 5,10-methylenetetrahydrofolate reductase (EC 1.7.99.5), can convert this cofactor to N^5-methyltetrahydrofolate. At first sight this reaction might not seem important, because it produces a form of folate that has no role in DNA or RNA biosynthesis and thus no role in cell replication. However, this reaction is the channel whereby single carbon units are fed from amino acids such as serine, histidine, glycine, and also from formate, to participate in what is sometimes called the methylation cycle.

Figure 1 shows that the methyl group of N^5-methyltetrahydrofolate is used by the enzyme methionine synthase (5-methyltetrahydrofolate–homocysteine methyltransferase, EC 2.1.1.13) to methylate homocysteine to methionine. The methionine so produced can be activated by ATP to make S-adenosylmethionine (SAM). There are 115 different methyltransferase enzymes in nature that transfer the methyl group of SAM to a wide range of compounds, including proteins (e.g. myelin basic protein), lipids, dihydroxyphenylalanine, DNA and RNA. As well as the methylated product, these enzymes also produce S-adenosylhomocysteine (SAH), i.e. SAM minus its methyl group. In cells the SAH is enzymically hydrolysed to homocysteine. In most cells, this homocysteine is methylated to methionine by the action of vitamin B_{12}-dependent methionine synthase, thereby completing the methylation cycle. The cycle can thus be seen as self-regenerating with methyl groups being added from N^5-methyltetrahydrofolate to homocysteine to make methionine and SAM, then this SAM being used to methylate a wide range of substrates. In this way the carbon sulphur skeleton of methionine is conserved.

All cells have a second fate for homocysteine, namely its irreversible degradation via the vitamin B_6-dependent enzyme, cystathionine β-synthase (EC 4.2.1.22), to cystathionine and then to cysteine. In most cells this pathway is probably not used extensively, but in some, such as hepatocytes, it is very important. This is because the liver uses homocysteine degradation as a method of catabolizing methionine, ultimately supplying the liver and the body with their principle source of the amino acid cysteine, which in turn is used for protein biosynthesis and to make other important cysteine-containing compounds, such as glutathione. A non-folate, non-vitamin B_{12}-dependent enzyme, betaine–homocysteine methyltransferase (EC 2.1.1.5), is also able to recycle homocysteine back to methionine. In humans, this enzyme is found only in the liver and the kidney and it depends for its activity upon a supply of betaine which comes from the degradation of dietary or cellular choline (McKeever et al 1991).

Apart from providing methyl groups for the methylation cycle, vitamin B_{12}-dependent methionine synthase also plays a role in the retention of folate

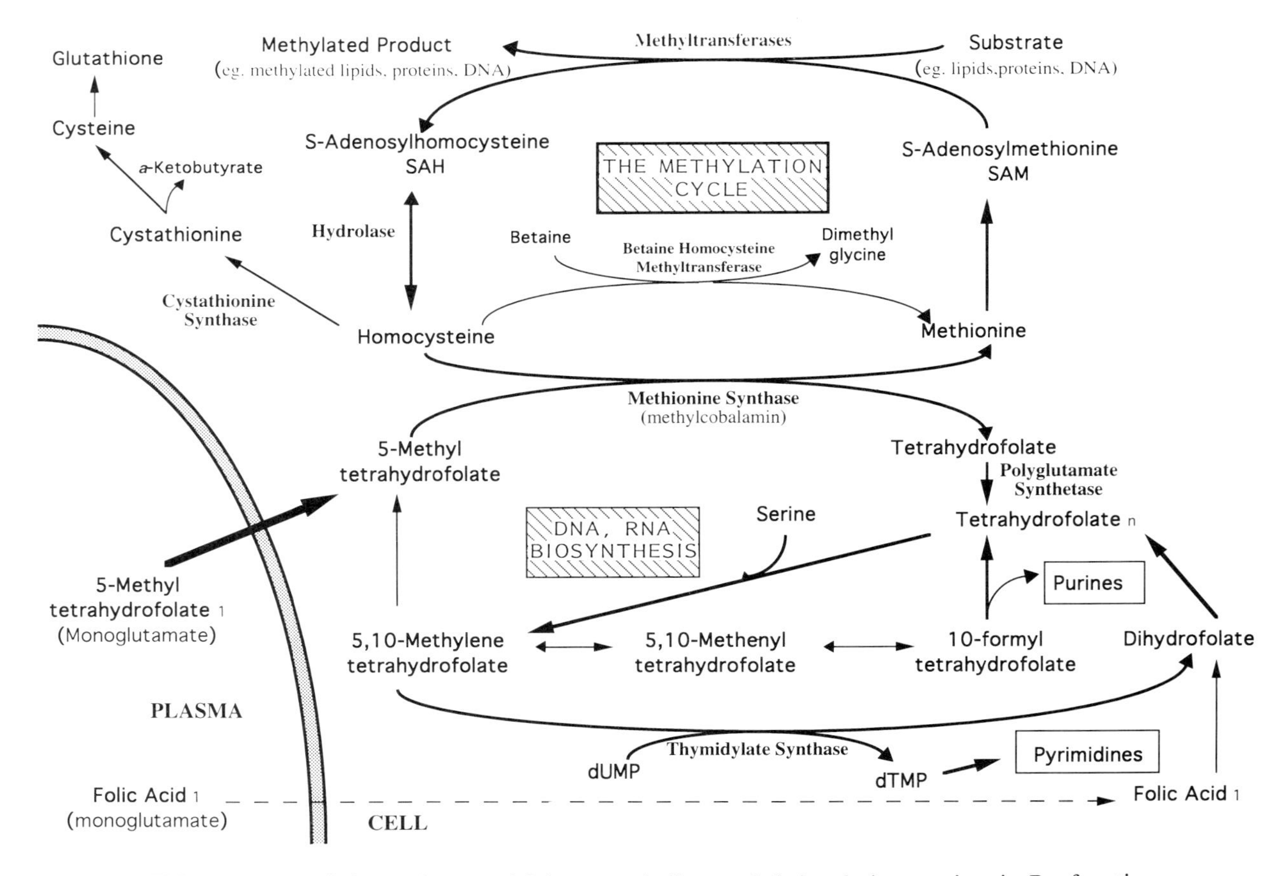

FIG. 1. Intracellular pathways of folate metabolism and their relation to vitamin B_{12} function.

by cells. Almost all of the folate in the circulation is N^5-methyltetrahydrofolate, which contains a single glutamic acid residue (Fig. 1). This cofactor can pass into the cell, but it does not act as a substrate for folylpolyglutamate synthase (EC 6.3.2.17) until the methyl group is removed by the action of methionine synthase (Cichowicz et al 1981). The product of this reaction, tetrahydrofolate, has four or five glutamate residues added to it, ensuring that it is retained by the cell. During vitamin B_{12} deficiency, where methionine synthase activity is impaired, there is decreased ability to demethylate circulating folate with subsequent lack of its retention by cells. This is most easily seen in the lower levels of folate in red blood cells that are associated with vitamin B_{12} deficiency.

How does folic acid prevent neural tube defects?

Folic acid supplements at different levels have been found to prevent most neural tube defects (Smithells et al 1983, MRC Vitamin Study Group 1991, Czeizel & Dudás 1992). From our knowledge of folate biochemistry outlined above, can we predict how folic acid achieves this protection? There are two main possibilities:

(1) Folic acid corrects folate deficiency;
(2) Folic acid overcomes a metabolic block, for which there are several candidates.

Is the present state of knowledge, based on the results of ourselves and others, sufficient to distinguish between these two broad alternatives or to identify the nature of a possible metabolic block?

Even before the advent of intervention trials, investigators suggested that deficiency of or increased requirements for folate might have a role in the aetiology of neural tube defects. Perhaps the pioneer in this was Dick Smithells. Work that he did with Hibbard found higher concentrations of urinary formiminoglutamic acid, which occur in folate deficiency and reflect abnormal folate metabolism, in women with an affected pregnancy when compared to those with normal pregnancies (Hibbard & Smithells 1965). Their results were obtained either after delivery or very late in pregnancy and, as the authors themselves conceded, may not have reflected folate status some 8–9 months earlier at the time of the closure of the neural tube. Similar studies by others on women in late pregnancy did not help resolve this dilemma (Fraser & Watt 1964, Giles 1966, Emery et al 1969, Pritchard et al 1970).

The first attempt to overcome the problem of the time of sampling was by Hall (1972), who obtained serum from nearly 3000 women at their first antenatal clinic. Of these, nine had outcomes affected by neural tube defects. The folate status of these nine, as determined by their serum folate level, showed no evidence of deficiency and was similar to that found in controls. A similar study by our own group was able to expand the number of cases by taking advantage of the fact that serum samples are routinely collected in Dublin at the first booking clinic to test for Rubella antibody (Molloy et al 1985). From a bank of 18 000 such sera, we were able to obtain samples from 32 women who delivered a fetus with a neural tube defect. The concentrations of folate and vitamin B_{12} in serum from these women were distributed in a manner similar to those in 395 randomly selected controls. In addition, the range of values found in the controls, as expected, showed no evidence of folate deficiency and this was also true of the cases.

A similar and more extensive study was undertaken in Finland by Mills et al (1992). This study compared the folate and vitamin B_{12} levels of sera from 89 women with affected pregnancies with those of 178 women with a normal pregnancy. They concluded that there was no apparent difference between the two groups and no evidence of deficiency of either vitamin in either group. The above studies seem to indicate that women who have affected pregnancies do not have greater deficiency than controls. The problem, however, is that serum folate levels depend upon recent dietary intake. They can change markedly in any one subject if folate intake is changed. Thus, they have limitations in determining true tissue status, particularly when this has to be extrapolated back to the event of the closure of the neural tube some weeks or even months earlier. For determining tissue stores, measurement of folate levels in red blood cells is far superior (Hoffbrand et al 1966). Unlike serum folate levels, they are not subject to daily fluctuations, reflecting instead the average folate status over the previous 120 days when the circulating populations of red cells were synthesized. Furthermore, Leck (1977) found that concentrations of folates in red blood cells taken from 98 women in early pregnancy showed a significant correlation with those from the same women sampled one year later.

Yates et al (1987) compared folate concentrations in the red blood cells of 20 women with a previous history of bearing fetuses with neural tube defects at a time when they were not pregnant with those of a similar number of controls matched for age, parity and social class. They found significantly lower values ($P<0.005$) in the former. Almost all of the red cell folate values for the cases, as well as the controls, were in the normal range. This suggests that the cause of the defect is a metabolic block or increased requirement for one of the many folate-dependent enzymes rather than simple folate deficiency. The main criticism of this study was that it did not involve the index pregnancy, i.e. the actual pregnancy where the fetus had a neural tube defect. Nevertheless, it did yield positive results.

While many studies referred to above have retrospectively used sera collected in different ways to assess folate status in the index pregnancy, red cells need to be prepared specially for folate analysis. Studies such as those of Yates et al (1987) looked at women who had had a previous pregnancy affected by a neural tube defect but who were not pregnant at the time. A different study was done by Smithells and his colleagues (1976) who measured folate concentrations in the red blood cells and serum of six women in the index pregnancy. The mean values $\pm$ SD were 141 ± 25.5 μg/l for the red blood cells and 4.9 ± 1.7 μg/l for the serum. Mean values of 228 and 6.3 μg/l, respectively, were found in nearly 1000 controls sampled at the same time. The number of cases was too small to indicate a distribution but they gave a variance which differed significantly from that of the control population. Thus a 2-sample Student's *t*-test could not be used. However, because the control group was large and the standard errors of its mean were low, the authors considered a 1-sample *t*-test to be valid. With this test they found that the concentrations of folate in the red blood cells were significantly lower in the six cases than in the 1000 controls ($P < 0.001$). The serum folate values were not different in the two groups. Thus, while small numbers make significance hard to gauge, there did appear to be differences. Subsequently, a further analysis by the author found that three of these six women had vitamin B_{12} levels suggestive of deficiency (Schorah et al 1980). During a double-blind controlled trial on the effect of folic acid on the prevalence of pregnancies affected by neural tube defects, the folate concentrations of six women carrying an affected fetus were determined (Laurence et al 1981). These samples were taken in the 7th and 8th week of gestation. The red blood cell and serum folate values (μg/l), respectively, were 70, 2.9; 228, 3.7; 275, 11.0; 380, 5.0; 155, 4.8; 259, 212.0. A group of 51 women who did not receive a folate supplement and who had normal pregnancies had a red blood cell folate concentration of 278 ± 16 μg/l. Of the six women listed above who had affected pregnancies, the first four received no folic acid supplement during their pregnancy. The last two had been in the treatment group but were deemed not to have complied with the regime—the last on the basis of an admission, the second last because of her low level of folate in red cells. It is difficult to make any conclusions other than that five of the six who had affected pregnancies did so in the presence of fairly normal levels of folate in red blood cells with reference to the control population and from general clinical experience.

More recently, Economides et al (1992) measured folate concentrations in 32 women undergoing termination of pregnancy between 14 and 21 weeks gestation. Of these, eight had been diagnosed as carrying a fetus with a neural tube defect and 24 controls were having their pregnancies terminated for social reasons. No difference with respect to folate concentrations in red cells or serum vitamin B_{12} was found between the two groups. In addition, the red cell folate values for the eight cases showed a mean of 435 μg/l with a range of 270–1055, indicating that none of them was deficient with reference to the controls in the study or the normal range of folate concentrations in red cells used by most laboratories.

In conclusion, studies on folate status, particularly those using the more probing measurement of folate concentration in red cells, are very unsatisfactory and limited. They appear, however, to suggest that while folate status plays some role in the generation of neural tube defects, the values seen are not those associated with overt clinical deficiency.

Acknowledgements

This research was supported by the Health Research Board, Ireland, and by BioResearch Ireland.

References

Blakley RL, Benkovic SJ 1984 Folates and pterins, vol 1: Chemistry and biochemistry of folates. Wiley, New York

Cichowicz DJ, Foo SK, Shane B 1981 Folypoly-γ-glutamate synthesis by bacteria and mammalian cells. Mol Cell Biochem 39:209–228

Czeizel AE, Dudás I 1992 Prevention of the first occurrence of neural-tube defects by periconceptual vitamin supplementation. N Engl J Med 227:1832–1835

Economides DL, Ferguson J, Mackenzie IZ, Darley J, Ware II, Holmes-Siedle M 1992 Folate and vitamin B_{12} concentrations in maternal and fetal blood and amniotic fluid in second trimester pregnancies complicated by neural tube defects. Br J Obstet Gynaecol 99:23–25

Emery AEH, Timson J, Watson-Williams EJ 1969 Pathogenesis of spina bifida. Lancet 2:909–910

Fraser JL, Watt HJ 1964 Megaloblastic anemia in pregnancy and the puerperium. Am J Obstet Gynecol 89:532–534

Giles C 1966 An account of 335 cases of megaloblastic anaemia of pregnancy and the puerperium. J Clin Pathol 19:1–11

Hall MH 1972 Folic acid deficiency and congenital malformation. J Obstet Gynaecol 79:159–161

Hibbard ED, Smithells RW 1965 Folic acid metabolism and human embryopathy. Lancet 1:1254

Hoffbrand AV, Newcombe BF, Mollin DL 1966 Method of assay of red cell folate activity and the value of the assay as a test for folate deficiency. J Clin Pathol 19:17–28

Laurence KM, James N, Miller MH, Tennant GB, Campbell H 1981 Double-blind randomised controlled trial of folate treatment before conception to prevent recurrence of neural-tube defects. Br Med J 282:1509–1511

Leck I 1977 Folates and the fetus. Lancet 1:1099–1100

McKeever MP, Weir DG, Molloy A, Scott JM 1991 Betaine-homocysteine methyltransferase: organ distribution in man, pig and rat and subcellular distribution in the rat. Clin Sci 81:551–556

Mills JL, Toumilehto J, Yu KF et al 1992 Maternal vitamin levels during pregnancies producing infants with neural tube defects. J Pediatr 120:863–871

Molloy AM, Kirke P, Hillary I, Weir DG, Scott JM 1985 Maternal serum folate and vitamin B_{12} concentrations in pregnancies associated with neural tube defects. Arch Dis Child 60:660–665

MRC Vitamin Study Research Group 1991 Prevention of neural tube defects: results of the Medical Research Council vitamin study. Lancet 338:131–137

Pritchard JA, Scott DE, Whalley PJ, Haling RF 1970 Infants of mothers with megaloblastic anemia due to folate deficiency. JAMA (J Am Med Assoc) 211:1982–1984

Schorah CJ, Smithells RW, Scott JM 1980 Vitamin B_{12} and anencephaly. Lancet 1:880

Smithells RW, Sheppard S, Schorah CJ 1976 Vitamin deficiencies and neural tube defects. Arch Dis Child 51:944–950
Smithells RW, Nevin MC, Seller MJ et al 1983 Further experience of vitamin supplementation for prevention of neural tube defect recurrences. Lancet 1:1027–1031
Yates JRW, Ferguson-Smith MA, Shenkin A, Guzman-Rodriguez R, White M, Clarke BJ 1987 Is disordered folate metabolism the basis for genetic predisposition to neural tube defects? Clin Genet 31:279–287

DISCUSSION

Stanley: Is there any other pathway whereby these methyl groups can be incorporated? Does incorporation depend totally on folate?

Scott: The enzyme betaine–homocysteine methyltransferase (EC 2.1.1.5) is folate independent, but it occurs only in the liver and the kidney in humans (McKeever et al 1991).

Another way for cells to get methyl groups would be to take up methionine, make it into S-adenosylmethionine and use it for a methylation reaction. However, each time a cell takes methionine in and uses its methyl group, it makes S-adenosylhomocysteine and homocysteine, which it then has to dispose of. This is not a sensible thing to do. The sensible thing is to keep recycling the carbon sulphur skeleton of the methionine and just keep adding methyl groups from N^5-methyltetrahydrofolate. In neural tissue, you can make serine from the glycolytic pathway in three enzymic steps: the C-3 of serine gets converted to N^5,N^{10}-methylenetetrahydrofolate, then to N^5-methyltetrahydrofolate and then to methionine.

Copp: In *curly tail* (*ct*) mice, the basic defect underlying the failure to close the neural tube is one of cell proliferation. Would you expect a defect in folate metabolism to underlie such a localized tissue-specific abnormality in cell proliferation? It's possible that just adding folate to such a system might not correct the defect. It would affect all tissues, therefore the differential between the genetically normal and genetically affected tissues would not be corrected. This might explain Mary Seller's (this volume) inability to correct neural tube defects in *ct* mice with folate. But it hasn't yet been shown that the cell proliferation abnormality is related to folate.

Scott: I think it is more likely that the folate-correctable event is some specific metabolic problem associated with certain people than that it is simple deficiency. However, it could be folate deficiency with the closure of the neural tube being at risk at levels of folate which we don't normally associate with clinical deficiency symptoms such as anaemia.

Copp: You view it as a problem in the mother, not a problem in the embryo, where we see the defect in *ct* mice?

Scott: It could be either. It could be that the mother has the defect and the embryo inherits it. Otherwise, it has to be something that the mother is transmitting or not transmitting to the embryo. The mother may not be giving the embryo enough plasma folate or not enough of a folate metabolite.

Oakley: The mother's cells obviously aren't dividing as fast as the embryo's. Is there something going on at the time that neural tube closure occurs? Is the rate of cell division faster then than later? It is curious that only neural tube defects can be prevented with folic acid. There are some reports of non-randomized prevention of cleft palate and so on, but the epidemiology doesn't fit.

Copp: In mammalian embryos, the cell proliferation rate is very rapid in all tissues at the stage of neurulation. But 24 h earlier it's even more rapid. From the moment of implantation cell proliferation is extremely rapid; as you go through gastrulation, it begins to slow; as you go through neurulation, it slows more, and it slows more after that. Then there are some tissue-specific differences. That is quite a general statement, but there is by no means a peak of cell proliferation during neurulation.

Scott: We find that the rate of folate catabolism is fairly directly tied to the rate of cell division (McPartlin et al 1993). In the later stages of pregnancy, folate catabolism starts to outstrip the amount of folate that's available to the fetus. This is why there is folate deficiency in late pregnancy. But in early pregnancy the fetal mass is so small that it is very unlikely to be a significant drain on the mother.

Oakley: Not so much a drain. But there might be something about the demand right at the point where the embryo is; somehow the embryo isn't getting quite enough, although the mother is.

Scott: Then you have to explain why women have high recurrence risks. If the woman continues on the same diet, that would explain it. But these women seem to be marked in some way.

Oakley: That only tells you the lesion is familial, it doesn't tell you it's genetic.

Hall: It might tell you that the yolk sac is not functioning in a normal way.

Wald: One could make a simple model that looked at the rate of cell division multiplied by the number of cells dividing. This would give a measure of the use of folate. If you plotted the product according to gestation, it may reach a maximum at the time of neural tube formation, which would explain the specificity of the folate effect on neural tube defects.

van Straaten: It depends what cell type you are talking about. There are major tissue-specific differences, starting even before the time of gastrulation.

Oakley: You need a third factor, which would be when fetal circulation is established.

Scott: The business of halving the cellular content at each division is important. After the fetal cells divide in the neural tube, they have to double their folate content through uptake just to restore the folate level they had before division. This might give rise to cells that are vulnerable to low levels of circulating folate.

van Straaten: As well as cell proliferation, another important mechanism in neurulation is apical contraction, in which actin plays a role. There is controversy over its precise role, but it might contribute to the elevation and convergence of the neural folds or at least stabilize the neural walls during elevation. Coelho & Klein (1990) demonstrated that L-methionine is essential for normal neurulation in the rat embryo. They suggested that lack of this amino acid might interfere with proliferation, but also with contraction of actin filaments which need methylation of lysine and histidine residues for proper functioning.

Scott: Again, this is an area about which very little is known. Many proteins, for example myelin basic protein, are methylated post-transcriptionally. The functions of some of these methylations are known, some are not. Proteins are only methylated to a relatively small extent, but look at the effect of the lack of methylation of myelin in vitamin B_{12} deficiency—it causes drastic changes in myelin generation and in nerve conduction leading to neuropathy (Weir et al 1988).

Copp: Is there any way that folate interacts with inositol? We have seen two separate pieces of evidence that inositol (Seller, this volume, Cockroft et al 1992) does have an interaction in the *ct* system.

Scott: I doubt it.

Hall: What is your feeling about absorption? Is the theory that folate-preventable neural tube defects are due to a problem of absorption by the mothers just one of many options?

Scott: I don't favour the malabsorption hypothesis. The only way I could see that work would be through the plasma folate. In that case, one would have expected differences in maternal plasma folate levels of such a magnitude that they would affect supply of folate to the embryo. If such differences exist, present studies would say that they are small. If the hypothesis was that a difference in intestinal absorption were a marker for some similar defect in tissues, that would be different.

Wald: If you had to say, where do you think the block would be?

Scott: That's a very difficult question. Folate is involved in two cycles: the one involving DNA biosynthesis is clearly essential for cell division and thus for closure of the neural tube. The methylation cycle is also essential in that it provides methyl groups for a wide range of cellular methyltransferases, most of which have important functions. A reduction in the level of an enzyme or a damaged enzyme in either cycle could cause neural tube defects.

Wald: How would just having a large amount of folate overcome the block?

Scott: Having lots of folate would drive both cycles and could overcome a metabolic block in one of them. There could be some abnormality in the folate-binding site for one of these key enzymes, which means that it just requires more folate to make it work properly.

If you don't accept a theory like that, either you have to propose a dietary theory, a total lack of folate, which I don't think the data support, or you have to propose a transport problem. It is possible that folate is not transported as well into the growing embryo in cases with neural tube defects.

Morriss-Kay: What are the principal dietary sources of folic acid?

Scott: Green vegetables are a good source. There are also foods that, although they are low in folate, are an important source in some countries because they are eaten in such vast amounts. For example, in Ireland 15% of the folate in the diet comes from potatoes.

Hall: Does cooking destroy the folic acid?

Scott: Prolonged cooking certainly destroys some food folate. Folic acid itself is quite stable. This is important because cereals fortified with folic acid are now a major source. One third of folic acid in the British diet and certainly in the Irish diet would be from fortified cereals. Such folic acid is not just a lot more stable than natural folates but it has twice the bioavailability (Gregory et al 1991).

Mills: John Scott described the data from our study in Finland (Mills et al 1992). We had 89 women bearing a fetus with neural tube defects, from whom blood was taken fairly early in pregnancy. We saw no difference at all in serum folate between that group and the control group. This might suggest that conjugase is not the relevant issue.

There is an alternative explanation. In the MRC trial, 28% of the expected recurrence was not prevented despite enormous amounts of folate being available (MRC Vitamin Study Research Group 1991). In Finland, where the rates of neural tube defects are extremely low, the proportion of cases not related to folate may be much higher. Thus, we see equal levels of serum folate in the case and control groups because folate is not part of the aetiology in an area with such a low risk.

Opitz: How does folate get into the fetus from the mother?

Scott: In other species, there are several different forms of folate in the circulation, but in humans it's almost exclusively N^5-methyltetrahydrofolate monoglutamate. This is what the mother is giving the fetus.

Opitz: Does it require active transport or does it just diffuse passively across the placenta?

Scott: I think it requires an active transport system. There is a receptor with a high specificity for folate present in the human placenta that is almost certainly involved in its transport (Antony et al 1981). The end result is that folate is concentrated in the fetal blood at term by a factor of five. This could be achieved by binding proteins, but I do believe there is a transport system. When that transport system is established, I don't know. Whether it is of fetal or maternal origin is obviously important. One could see how increasing the concentration of N^5-methyltetrahydrofolate in the circulation, as you do even with 0.4 mg of folic acid, could overcome a transport problem.

Stanley: We discussed fetal circulation yesterday (p 86–87): basically, at the time of neurulation there isn't a placenta.

Wald: Steegers-Theunissen et al (1991) in Nijmegen found high levels of serum homocysteine after oral methionine loading in mothers who had babies with neural tube defects. Would you like to comment on that?

Scott: They administered a methionine load to 16 women with a previous history of a pregnancy with a neural tube defect and to 15 control women. Five of the former seemed to have an impaired rate of metabolism of the homocysteine that was produced from the methionine load, although there was no difference in baseline homocysteine levels between cases and controls. They suggest that cases may have lower levels of cystathionine β-synthase, the enzyme responsible for homocysteine breakdown.

Wald: Isn't it a remarkable coincidence that folate supplementation prevents neural tube defects and women who have an affected pregnancy also exhibit an abnormality in a folate metabolic pathway? Does it not suggest that this is the biochemical abnormality involved?

Scott: I think it's a flag. It's disappointing that the effect is seen in only a few of the cases and there were no differences in baseline plasma homocysteine levels between cases and controls.

References

Antony AC, Utley C, Van Horne KC, Kolhouse JF 1981 Isolation and characterization of a folate receptor from human placenta. J Biol Chem 256:9684–9692

Cockroft DL, Brook FA, Copp AJ 1992 Inositol deficiency increases the susceptibility to neural tube defects of genetically predisposed (curly tail) mouse embryos in vitro. Teratology 45:223–232

Coelho CND, Klein NW 1990 Methionine and neural tube closure in cultured rat embryos: morphological and biochemical analyses. Teratology 42:437–451

Gregory JF III, Bhandari SD, Bailey LB, Toth JP, Baumgartner TG, Cerda JJ 1991 Relative bioavailability of deuterium-labeled monoglutamyl and hexaglutamyl folates in human subjects. Am J Clin Nutr 53:736–740

McKeever MP, Weir DG, Molloy A, Scott JM 1991 Betaine-homocysteine methyltransferase: organ distribution in man, pig and rat and subcellular distributions in the rat. Clin Sci 81:551–556

McPartlin J, Halligan A, Cott JM, Darling M, Weir DG 1993 Accelerated folate breakdown in pregnancy. Lancet 341:148–149

Mills JL, Toumilehto J, Yu KF et al 1992 Maternal vitamin levels during pregnancies producing infants with neural tube defects. J Pediatr 120:863–871

MRC Vitamin Study Research Group 1991 Prevention of neural tube defects: results of the Medical Research Council vitamin study. Lancet 338:131–137

Seller MJ 1994 Vitamins, folic acid and the cause and prevention of neural tube defects. In: Neural tube defects. Wiley, Chichester (Ciba Found Symp 181) p 161–179

Steegers-Theunissen RPM, Boers GHJ, Trijbels FJM, Eskes TKAB 1991 Neural-tube defects and derangement of homocysteine metabolism. N Engl J Med 324:199–200

Weir DG, Keating S, Molloy A et al 1988 Methylation deficiency causes vitamin B_{12}-associated neuropathy in the pig. J Neurochem 51:1949–1952

Folic acid and neural tube defects: the current evidence and implications for prevention

N. J. Wald

Department of Environmental and Preventive Medicine, Wolfson Institute of Preventive Medicine, St. Bartholomew's Hospital Medical College, Charterhouse Square, London EC1M 6BQ, UK

Abstract. The results of the MRC Vitamin Study have established the specific role of folic acid in the prevention of neural tube defects. Folic acid supplementation (4 mg/day) at the time of conception reduced the recurrence rate by about 70%. Evidence from observational studies and the Hungarian randomized trial (that used 0.4–0.8 mg/day supplemental folic acid) indicates that a similar level of prevention can be achieved among women who have not already had an affected pregnancy and that this may be achieved with a lower dose of folic acid. The medical implications of these results are important. (1) Women planning a pregnancy should take folic acid supplements. (2) Because a large proportion of pregnancies are unplanned and many of those that are planned will occur in women who will not have taken folic acid supplements, a complementary general public health strategy is needed, based on dietary advice and food fortification designed to achieve the benefit with adequate safety. (3) Any public health prevention strategy needs to be monitored for efficacy and possible harm. (4) Research is needed on the dose–response relationship between folic acid and neural tube defect prevention and the mechanism of action.

1994 Neural tube defects. Wiley, Chichester (Ciba Foundation Symposium 181) p 192–211

It is now known that lack of folic acid in certain individuals is an important cause of neural tube defects. The evidence needs to be disseminated and appropriate public health preventive strategies introduced. A recent review summarizes the scientific background to the work on folic acid and the prevention of neural tube defects (Wald 1993). This paper draws on the earlier review with the aim of focusing on the opportunities for prevention.

The evidence

Three observations provided important clues to the environmental aetiology of neural tube defects. Firstly, women in low socioeconomic groups have had a higher risk of neural tube defect pregnancies. Secondly, women who have eaten relatively poor quality diets have had a higher risk of neural tube defect pregnancies. Thirdly, migrants from an area with a high prevalence of neural tube defects to an area with a low prevalence acquire the new lower risk. Hibbard & Smithells (1965) showed that the formiminoglutamic acid excretion test was more often positive in women carrying a fetus with a neural tube defect than in controls. (This test is a measure of a person's ability to metabolize histidine: if there is a relative lack of folic acid or if folic acid metabolism is disturbed, an excessive amount of formiminoglutamic acid is excreted in the urine.)

The strongest evidence that lack of folic acid is a cause of neural tube defects comes from the results of randomized prevention trials. Four have been carried out; these are summarized in Table 1. One investigated folic acid alone (4.0 mg/day), but had too few participants to be definitive (Laurence et al 1981). The MRC Vitamin Study Research Group (1991) used a factorial design to investigate both folic acid (4.0 mg/day) and a mixture of other vitamins; this trial provided the strongest evidence that folic acid was protective. Kirke et al (1992) used a three-way randomization (folic acid [0.36 mg/day] alone or folic acid with other vitamins or other vitamins without folic acid) but observed only one affected pregnancy. Czeizel & Dudás (1992) used folic acid (0.8 mg/day) with other vitamins. Collectively, there were 40 pregnancies associated with neural tube defects in the four randomized trials. The rate of neural tube defects in the supplemented women was 76% lower than in the women receiving no supplements (Table 1). The MRC Vitamin Study found no evidence that the vitamins other than folic acid had any effect in the prevention of neural tube defects; these vitamins were (with their daily dose) vitamin A (4000 U), B_1 (thiamine 1.5 mg), B_2 (riboflavine 1.5 mg), B_6 (pyridoxine 1.0 mg), C (40 mg), D (400 U) and nicotinamide (15 mg). The rate of neural tube defect pregnancies in the women receiving this mixture of vitamins was similar to that in the controls (relative risk 0.8, 95% confidence interval 0.38–1.70, $P=0.7$).

Table 2 shows the results from the two non-randomized intervention studies (Smithells et al 1983, Vergel et al 1990). The more extreme results suggest that part of the effect in these studies may have been due to bias, but the numbers of cases in the randomized and the non-randomized studies were too small to exclude chance as an explanation for the difference in the relative risk estimates. Table 3 is a summary of the results of observational studies of neural tube defects and estimated *dietary* folate intake; all show a reduced risk in women considered to have a higher intake of folic acid.

There have been six observational studies of neural tube defects and folic acid supplementation (that is, taking folic acid in the form of tablets or capsules) (Winship et al 1984, Mulinare et al 1988, Mills et al 1989, Milunsky et al 1989, Werler et al 1993, Bower & Stanley 1992). All but one involved asking women who had pregnancies associated with a neural tube defect whether they took extra vitamins, then comparing the proportion that did so with the corresponding

TABLE 1 Use of vitamin supplements in randomized prevention trials

Trial	*Recurrence (R) or occurrence (O)*	*Folic acid (FA) or multivitamin (M)*	*Daily dose of folic acid supplement (mg)*	*Number of pregnancies*	*Number of pregnancies with neural tube defects*	*Relative risk*	*95% Confidence interval*
MRC Vitamin Study Research Group (1991)	R	FA	4	1031	27	0.29	0.10–0.74
Laurence et al (1981)	R	FA	4	111	6	0.42	0.04–2.97
Czeizel & Dudás (1992)	O	M	0.8	4156	6	0.00	0.00–0.85
Kirke et al (1992)	R	FA	0.36	261	1	0.00	Insufficient data
Combined	—	—	—	5559	40	0.24	0.11–0.52

TABLE 2 Use of vitamin supplements in non-randomized prevention studies

Study	*Recurrence (R) or occurrence (O)*	*Folic acid (FA) or multivitamin (M)*	*Daily dose of folic acid supplement (mg)*	*Number of pregnancies*	*Number of pregnancies with neural tube defects*	*Relative risk*	*95% Confidence interval*
Smithells et al (1983)	R	M	0.36	973	27	0.14	0.03–0.47
Vergel et al (1990)	R	FA	5	195	4	0.00	0.00–2.13
Combined	—	—	—	1168	31	0.12	0.04–0.41

TABLE 3 Observational studies on dietary folate and neural tube defects

Study	*Number of neural tube defects*	*Relative risk (high vs low folate diet)*	*95% Confidence interval*
Neural tube defect recurrence:			
Laurence et al (1980)	8	0	0–0.33
Yates et al (1987)	20	0.33[a]	0.09–1.27
Neural tube defect occurrence:			
Bower & Stanley (1989)[b]	77	0.31[c]	0.10–0.97
		0.16[d]	0.06–0.49
Milunsky et al (1989)	39	0.42	0.16–1.15

[a]Estimated from figure in cited reference.
[b]Study includes use of vitamin supplements in assessing intake.
[c]Using controls with congenital abnormalities other than neural tube defects.
[d]Using normal controls.

proportion among women who had unaffected pregnancies. One study was prospective in design (Milunsky et al 1989); women who were being screened for neural tube defects in early pregnancy were asked whether they had taken extra vitamins immediately before and in the early weeks of pregnancy. The pregnancies were followed up to determine the outcome and the proportion of women who took vitamin supplements was compared in women with and without affected pregnancies. The other studies were retrospective, a design that has the advantage of economy and statistical power but is open to the possibility of recall bias. Neither type of study can determine conclusively whether a reduced risk of neural tube defects in women taking vitamin supplements is a causal association. Such an association may be due to some protective factor that is itself associated with characteristics of individuals who choose to take vitamin supplements. Selection bias is likely to operate more strongly in an intervention study in which the selective pressures on gaining access to a specialist centre that offers a possible remedy are likely to be greater than the selective pressures that influence women buying vitamin supplements over the counter.

All six observational studies show a relative risk of less than one; four were individually statistically significant (Table 4). Taken together, the studies show conclusively an increased risk of neural tube defects among women who did not take supplements. The NIH study (Mills et al 1989) did not show an effect. The reason for this is uncertain, but it may be the fact that many of the women recruited into the study had a relatively high dietary folate intake (most of the women came from California, where folate intake is likely to be higher than average) so the expected effect of extra folate would have been less than that found in other studies.

TABLE 4 Observational studies of the use of vitamins containing folic acid and the occurrence of neural tube defects in the general population

Study	*Relative risk*[a] *(95% confidence interval)*
Winship et al (1984)	0.14 (0.003–1.11)
Mulinare et al (1988)	0.41 (0.26–0.66)
Mills et al (1989)[b]	0.94 (0.80–1.10)
Milunsky et al (1989)	0.29 (0.15–0.55)
Werler et al (1993)	0.6 (0.4–0.8)
Bower & Stanley (1992)[c]	0.11 (0.01–1.33)
Combined with Mills et al	0.47 (0.29–0.76)
without Mills et al	0.41 (0.27–0.61)

Dose of folic acid was usually between 0.4 and 0.8 mg/day.
[a]Relative risk estimate given is that cited by the authors and may be adjusted for possible confounding factors.
[b]The relative risk cited was based on the use of normal controls; it was 0.87 (0.73–1.02) using controls with congenital abnormalities other than neural tube defects.
[c]The relative risk was 0.69 (0.06–8.53) for controls with congenital abnormalities other than neural tube defects.

A subset analysis from the study reported by Milunsky et al (1989), based on the women who said that they took vitamin capsules before pregnancy, showed that the relative risk of a neural tube defect in a pregnancy among women who took vitamin capsules that contained folic acid compared to women who took vitamin capsules that did not contain folic acid was 0.29. This is identical to the point estimate observed in the MRC Vitamin Study. Though this was not statistically significant (the confidence interval was 0.07–1.63), taken with the other evidence it weighs in favour of the specific effect of folic acid. The result is unlikely to have been influenced by bias because it is unlikely that the women who took the vitamin supplements before pregnancy would have segregated themselves into a high and low risk group for neural tube defects on the basis of whether the vitamins contained folic acid.

Scientific conclusions

Preventive effect

Folic acid supplementation prevents about three out of four cases of neural tube defects. The mechanism of action is not known; while this should encourage further research, it is not a reason for delaying the introduction of preventive measures.

Dose of folic acid: efficacy and safety

Two issues with regard to dose are the quantification of efficacy and safety. The ideal would be to determine the minimum fully effective dose of folic acid, an aim that may never be realized satisfactorily because of the scale and costs of the studies involved. The results of the observational studies and the Hungarian randomized trial (Czeizel & Dudás 1992) suggest that supplements between 0.4 and 0.8 mg a day confer a substantial protective effect. Whether increasing the intake of folic acid will further increase the protective effect is not known. Most of the folic acid given in the 4 mg per day supplement is excreted in the urine and it is likely that the minimum fully effective dose may be less than a half or a quarter of this amount.

The main concern over safety may be more academic than practical. None of the randomized trials suggested any harm to the fetus or mother, though the statistical power of the trials in this respect is limited. Two concerns have been raised, namely the exacerbation of pernicious anaemia and grand mal epilepsy associated with taking several mg of folic acid a day.

Exacerbation of pernicious anaemia. Before the aetiology of pernicious anaemia (which is due to a lack of vitamin B_{12}) was recognized, and before there were simple and reliable assays for serum B_{12}, pernicious anaemia was treated with high doses of folic acid. This resolved the megaloblastic anaemia, but tended to mask, or occasionally precipitate, the neurological signs of pernicious anaemia. If the problem was recognized, it was always reversed by treatment with vitamin B_{12}, but if folic acid treatment persisted, the changes could be irreversible. Pernicious anaemia is extremely rare in women of child-bearing age. Also, in current medical practice the neurological features of pernicious anaemia will invariably lead to measurement of serum vitamin B_{12} and, if necessary, vitamin B_{12} supplementation, in which case the problem will be effectively treated. It is likely that the presentation of the neurological signs of vitamin B_{12} deficiency manifested by folic acid supplementation is an extremely rare clinical event. It is important to recognize that the cause of the problem (or potential problem) is B_{12} deficiency, not folic acid excess; withholding the prevention of one disorder because of failure to prevent another is illogical. Each needs attention in its own right. Some people require supplementation with B_{12}, while others require extra folic acid and some may require both extra folic acid and extra B_{12}.

Current opinion is that a dose of 4 mg folic acid per day is acceptably safe when given under medical supervision for women planning a pregnancy, but not as a widely available general supplement which could readily be obtained by elderly people at risk of pernicious anaemia. A dose of 0.4 or 0.8 mg per day is considered acceptably safe for general use.

Grand mal epilepsy. A practical management issue arises among women with epilepsy receiving anticonvulsants. Anticonvulsants have antifolate activity and there is a concern that supplemental folic acid may antagonize the anticonvulsant activity. The issue is further complicated by the fact that some anticonvulsants (notably valproic acid) are themselves associated with an increased risk of neural tube defects (see Nau, this volume, Holmes, this volume). A simple, practical approach would be to advise women on anticonvulsants to take extra folic acid to reduce their risk of a neural tube defect pregnancy, but warn them that there is a risk that their epilepsy will be less well controlled. Whether monitoring blood levels of anticonvulsants would be helpful in avoiding the problem is uncertain, but such monitoring may be of value to ensure that anticonvulsant levels are neither lower nor higher than intended.

Other hazards. Other than concerns over women with pernicious anaemia and women receiving anticonvulsants, there is no suggestion that folic acid supplementation poses any hazard.

Prevention of recurrence and occurrence

Over 95% of infants with neural tube defects are born to women who have not previously had an affected pregnancy. The observational studies that showed a reduced neural tube defect birth prevalence associated with taking vitamin supplements were all 'occurrence' rather than 'recurrence' studies. These studies, taken on their own, would be inconclusive, because of the possibility of selective bias. Taken with the evidence from the MRC Vitamin Study showing a specific protective effect of folic acid, it is reasonable to conclude that folic acid explained at least a substantial part of the results of the observational occurrence studies. This would be sufficient to draw a firm conclusion that folic acid prevents neural tube defects in general, but the evidence from the Hungarian trial (Czeizel & Dudás 1992) leaves no doubt—it provided direct evidence of a protective effect of folic acid in women who had not already had an affected pregnancy using a randomized design that excluded bias.

Mechanism of action

The mechanism of action of folic acid in the prevention of neural tube defects is not known. The rates of miscarriage in the MRC Vitamin Study were similar in women in the folic acid and control groups of the trial. Folic acid is therefore unlikely to act by encouraging a miscarriage of a fetus with a neural tube defect. It is more likely to prevent the inception of the defect and so decrease its incidence. Not all neural tube defects can be prevented by folic acid. Women in the folic acid groups of the MRC Vitamin Study who had affected pregnancies had high serum folic acid levels, so the lack of a protective effect was not due to failure to take the capsules or to absorb the folic acid. Neural tube defects, like many other medical disorders, have more than one cause. The important conclusion is that most of them can be prevented by consuming sufficient folic acid.

The narrow window effect

A paradox that requires explanation is the observation that although neural tube defects can be prevented by consuming more folic acid, the serum levels of women with pregnancies with neural tube defects are surprisingly similar to those in women with unaffected pregnancies (Table 5); the median estimate of the relative odds of having an affected pregnancy at the 90th centile of serum folic acid compared to the 10th was 0.92, an 8% decrease. The red cell folic acid levels (Table 5) suggest a small difference between women who have affected and unaffected pregnancies, but this is not consistent across studies. One would have expected women with affected pregnancies to have unequivocally lower blood folic acid levels than those with unaffected babies.

An explanation for the paradox is offered by the narrow window effect which has recently been described (Wald 1993). Different women in the population will have different genetic predisposition to producing a child with a neural tube defect (Fig. 1). In addition, folate intake, and hence serum folate (since this is directly related to folate intake), will be related to risk, so the combined risk can be represented as a series of parallel lines declining with increasing serum folate and decreasing genetic predisposition. There is a relatively narrow range of serum folate in the population, represented by the two vertical lines (Fig. 1). Any study comparing serum folic acid levels between women with affected and unaffected pregnancies will necessarily reveal only a small difference in serum folate levels because it will be constrained within the two vertical lines. The actual difference might be only about half an arbitrary unit, indicated by the short horizontal line. Such a small difference would be unlikely to reach statistical significance except in very large studies. The narrow window of serum folate and folic acid intake constrains the scope for identifying differences in serum folate between women with affected and unaffected pregnancies. If serum folate levels were increased outside the narrow window, the position would be different. Supplementation effectively does this, raising individual serum folate levels above the range in the general population, thereby substantially reducing the risk independently of the genetic predisposition.

Results from several sources show that serum folic acid levels are approximately linearly related to dietary folate intake expressed on a log scale. (1) Background levels of dietary folate (about 0.2 mg) are associated with serum folate levels about a quarter of this, namely 6 ng/ml (Office of Population Censuses and Surveys 1990). (2) The intervention study of Smithells et al (1976) was based upon total dietary folate intake of about 0.6 mg a day (0.2 mg actual intake and about 0.36 mg from a supplement) and this was associated with a serum folic acid level of about 22 ng/ml (Schorah et al 1983). (3) The MRC Vitamin Study Research Group (1991) found that women with an intake of about 4.2 mg (0.2 mg + 4 mg supplement) had serum levels of about 50 ng/ml. On present evidence it is reasonable to conclude that a public health prevention

TABLE 5 Folate levels in serum and red blood cells in women with pregnancies affected by neural tube defects

			Mean (SD) level of folate (ng/ml) in	
Publication	*Number of pregnancies*	*Blood taken*	*serum*	*red blood cells*
Smithells et al (1976)[a]				
affected	5[b] (6[b])	1st trimester	4.9 (1.7)	141[c] (25)
unaffected	943 (959)		6.3 (3.5)	228[c] (58)
Mills et al (1992)				
affected	89	1st trimester	4.1 (2.4)	ND
unaffected	172		4.3 (2.5)	ND
Hall (1977)				
affected	11	Antenatal booking	6.3 (2.8)	ND
unaffected	>1000		6.6 (2.9)	ND
Molloy et al (1987)				
affected	32	Antenatal booking	3.4[d] (3.2)	ND
unaffected	384		3.4[d] (3.2)	ND
Laurence et al (1981)				
affected	4	Pregnancy	ND	238 (106)
unaffected	47		ND	281 (119)
Economides et al (1992)				
affected	8	2nd trimester	9.8[d] (4.5)	435 (275)
unaffected	24		7.4[d] (4.0)	400 (194)
Emery et al (1969)				
affected	19	Post partum	4.9 (2.5)	ND
unaffected	37		4.6 (1.9)	ND
Yates et al (1987)				
affected	20	Post partum	2.8 (0.7)	178[c] (64)
unaffected	20		3.3 (1.4)	268[c] (110)
Bower & Stanley (1989)				
affected	61	Post partum	5.6 (3.7)	301 (478)
unaffected	140		5.7 (3.6)	308 (133)

ND, not determined.
[a]The number of pregnancies assessed in each part of this study differs; the number of women for whom levels of folate in red blood cells were measured is given in parentheses.
[b]One case was microcephaly, not a neural tube defect.
[c]Statistically significant.
[d]Median.

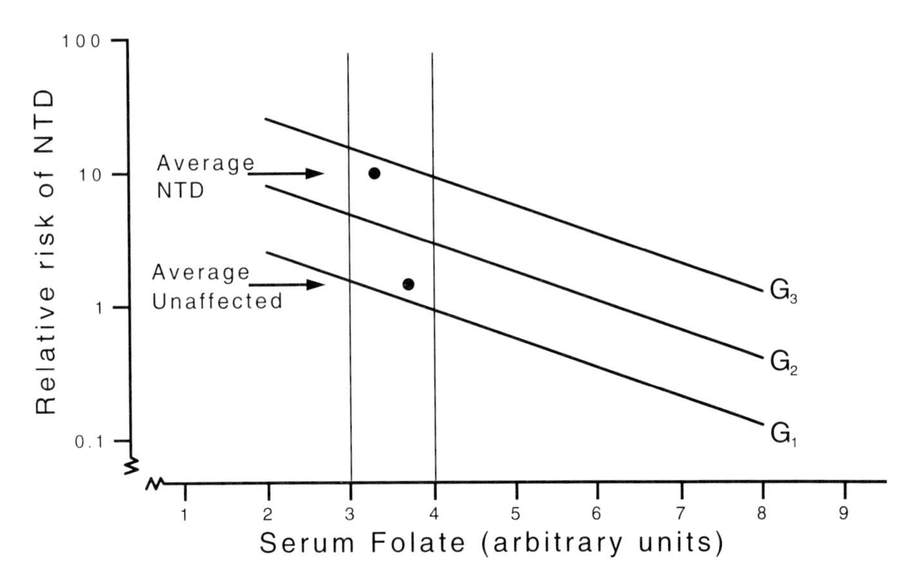

FIG. 1. The narrow window effect. G1, G2 and G3 represent women with a successively greater genetic predisposition to having a child with a neural tube defect (NTD). The vertical lines depict the range of serum folate concentrations observed in the general population.

strategy will need to increase serum folic acid levels in the population to at least 15–20 ng/ml. The dietary studies of folic acid intake and neural tube defects suggest that lower levels of intake may be sufficient to confer some protection (see Table 3). These data showed a reduced risk in women with a relatively high dietary folic acid intake but not taking a supplement (Laurence et al 1980, Yates et al 1987, Milunsky et al 1989, Bower & Stanley 1989). However, the direct evidence from the trials in which higher intakes were used and the evidence on serum folate suggesting that there is a 'narrow window effect' should take precedence over the dietary studies until better evidence on the relationship between the amount of folate intake and the incidence of neural tube defects is available.

Implications for prevention

The consumption of extra folic acid can prevent most neural tube defects. Three complementary preventive approaches can be adopted: (1) women can take folic acid supplements when planning a pregnancy; (2) women can be encouraged to eat folate-rich foods, and (3) staple foods can be fortified with folic acid.

Taking a folic acid supplement is a practical method of prevention among women who have already had an affected pregnancy, since all such women will be receiving medical attention and will be counselled about the increased risk of recurrence. They can be given folic acid supplements as soon as they plan to have another baby. Given the evidence from the MRC Vitamin Study, it would be sensible to offer such women 4 mg a day (or 5 mg until a 4 mg capsule becomes available for this purpose). In these circumstances, supplementation is provided on prescription and therefore limited to the women concerned and limited in duration.

The position is different with regard to women in the general population. Here a supplement should be easy to obtain and available without prescription. A daily dose of 0.4 mg, unlike the 4 or 5 mg supplement, does not require a prescription. It should be readily available in all chemists, but unfortunately this is currently not the case. A licensed 'over-the-counter' 0.4 mg folic acid capsule is urgently needed. There is a further difficulty. Embryonic closure of the neural tube is completed by about four weeks after conception, which is about six weeks from the first day of the last menstrual period. It is important that the consumption of extra folic acid is started before a woman knows that she is pregnant. Intervention later may be too late. Many pregnancies, perhaps the majority, are unplanned, so that advising women who are planning a pregnancy to take folic acid supplements will in many instances be ineffective. Neural tube defects are more common among the socioeconomically disadvantaged groups in society. These groups are even less likely to be aware of the need to take folic acid before pregnancy. Vitamin supplementation through taking special capsules or tablets is not likely to be a practical method of reaching the majority of the population. The only way of doing so is to increase the folic acid intake by dietary means. A survey of British women showed the median intake of folate was 0.2 mg per day, with a range of about between 0.1 and 0.4 mg a day (Office of Population Censuses and Surveys 1990). Consumption of an extra 0.4 mg of folic acid would mean tripling the folic acid intake. Pure folic acid is approximately 100% bioavailable, but dietary folate is only about 50% bioavailable (Bailey 1992). It may therefore be necessary to increase the intake of dietary folate by even more than three times the median level. To do this by changing the foods one eats would be extremely difficult. The only practical public health strategy is fortification of foods with folic acid, in the same way that table salt is fortified with iodine to prevent goitre and cretinism (Markel 1987).

Folic acid is stable and cheap. It is not, in its simple molecular form, present naturally in food. In food, folic acid exists as a more complex molecule, which is unstable. The degree of instability increases as the molecule is hydrogenated and as the number of glutamic acid residues increases. There are no practical difficulties in fortifying food with folic acid and many breakfast cereals are already fortified in this way. The range of foods fortified needs to be extended to achieve an adequate intake among the population, while avoiding an excessive exposure that would pose a hazard to some groups in the population.

Cereals are a convenient vehicle for folic acid fortification. It would be a simple step to extend the fortification of breakfast cereals to a wider range of cereals, including bread. Fortifying flour may be better, since flour is already supplemented with certain minerals and vitamins in many parts of the world (e.g. calcium, iron, thiamine and niacin which have, in the UK, been added for more than 50 years at the mill) and it would probably be simpler and more economical to fortify flour in the mill than for each bakery to add folic acid to dough. Fortified flour will also reach groups who rarely eat bread or breakfast cereal. The general aim of fortification is to choose a food that is widely consumed over a relatively limited range of intake, so that it is possible to set a fortification level that will be effective for nearly all without providing too little to some or too much to others. Such a food may vary from country to country.

The sociopolitical perspective in prevention

The barrier to introducing a fortification policy is more sociopolitical than it is medical or scientific. There is a concern that individuals will lose control over what they eat. Currently, individual autonomy is rated highly in social values. Taken too far, the cost of such individual autonomy will be that the public health measures will be less effective and families will needlessly continue to have babies with spina bifida. A balance needs to be achieved; my own view is that the balance, in this case, weighs in favour of a collective approach, since many will benefit and few, if any, be harmed. For individuals who wish to eat unfortified bread or flour, there would be specialty breads and other cereals without added folic acid, just as many 'natural' food products are available. Many such 'niches' already exist in the food manufacturing market in economically developed countries. Choice can be preserved, but the default position should be fortification with folic acid while maintaining the option of buying unfortified products if one wishes to do so. It is neither reasonable nor just to deny the majority of people access to healthier foods because some may wish an alternative. Those who want an alternative should be entitled to obtain foods of their choice, but it is they who should make the special effort, not others.

The extent of fortification

Any system of food fortification should ideally operate according to certain guidelines but be sufficiently flexible to deal with changing needs and special circumstances. In a country where iodine lack is extremely rare and goitre is not a problem, salt need be fortified with only modest amounts of iodine, but elsewhere higher levels of fortification are needed. Adjusting the level of fortification of salt with iodine according to need is accepted public health practice. A similar policy could and should be adopted with folic acid. The level of fortification should be selected to ensure that nearly everyone in the population consumes at least 0.4 mg, with a median intake of about 0.6 mg

TABLE 6 Bread consumption among British women in 1987 and the maximum increase in intake of folate if bread were fortified

Bread (g/day)	*Slices per day*[a]	*Cumulative percent of population*	*Maximum increase in intake of folate if bread is fortified with 0.3 mg folic acid per 100 g bread*
⩽20	0.5	3.5	0.06
⩽40	1	12.6	0.12
⩽80	2	50.1	0.24
⩽160	4	95.3	0.48
⩽320	8	100.0	0.96

From MAFF/DH Dietary Information survey (1987), cited by Dr D. H. Buss (personal communication).
[a]Assuming 1 slice = 40 g.

(i.e. 0.4 mg greater than the natural median intake). Table 6 illustrates how, with information on the distribution of bread consumption in the population, the appropriate level of fortification can be estimated. In the example given, the median additional folic acid intake is about 0.2 mg (achieved by adding 0.3 mg of folic acid to 100 g of bread). At this level of fortification, virtually no one would consume more than 1 mg a day. In practice, perhaps, a somewhat higher level of fortification could be justified, but a level four times higher (1.2 mg of folic acid per 100 g of bread) would lead to some individuals consuming more than 3 mg a day, and this, for the considerations given above, would at present be regarded as too high. There should be a standing committee responsible for public health implementation that would guide the food industry as to the level of fortification. The level would be subject to periodical review on the basis of national nutritional surveys, as is currently performed in many countries, including the UK. Of particular importance in this respect is the use of surveys based on biochemical analysis of blood samples. The median serum folic acid concentration is about 5 ng/ml (Office of Population Censuses and Surveys 1990). In people taking a daily 0.4 mg supplement, the median serum folic acid concentration is about 20 ng/ml (Schorah et al 1983), so a population survey showing that average levels had increased from 5 to 20 ng/ml would demonstrate that the public health strategy was working. If this were not the case, the position would need to be reviewed and alternative, more effective methods introduced.

National recommendation on folic acid intake

National recommendations on folic acid intake have consistently suggested values in excess of actual median intake. The US recommended daily allowances for folic acid are 0.4 mg per day for non-pregnant adults and 0.8 mg for pregnant women (Food and Nutrition Board 1980), although a recent National Research

Council advisory committee suggested that these be reduced to 0.18 and 0.4 mg, respectively (Food and Nutrition Board 1989). In the UK, the Department of Health specified a 'reference nutrient intake' of 0.2 mg a day for non-pregnant adults and 0.3 mg for pregnant women (Committee on Medical Aspects of Food Policy 1991) as interim values until the results of the MRC Vitamin Study were available. Folic acid is perhaps unusual in being a nutrient that in developed countries has been consumed at lower levels than those recommended by national expert groups.

Recent government recommendations on folic acid and neural tube defects

At the end of 1992 an Expert Advisory Group, set up by the Chief Medical Officers of the UK, produced a report entitled *Folic acid in the prevention of neural tube defects* (Expert Advisory Group 1992). It produced 21 recommendations, the most important of which were that women who have previously had an affected pregnancy should take 4 or 5 mg a day folic acid supplements and that women in the general population should take a 0.4 mg supplement from the time they begin trying to conceive until the 12th week of pregnancy and should eat more folate-rich foods. The Committee recommended that the range of breads (including wholemeal breads) and breakfast cereals fortified with folic acid should be increased, though such fortification should be restricted to these foods and the present levels of folic acid fortification in them should not be greatly exceeded. There should be a method of monitoring the prevention policy in order to review it as necessary.

The US Public Health Service stated in September 1992 (Centers for Disease Control 1992): 'In order to reduce the frequency of neural tube defects and their resulting disability, the United States Public Health Service recommends that all women of childbearing age in the United States who are capable of becoming pregnant should consume 0.4 mg folic acid per day for the purpose of reducing their risk of having a pregnancy affected with spina bifida or other neural tube defects'. A recommendation to fortify selected foods with folic acid was also made and in November 1992 an advisory committee to the Food and Drug Administration recommended fortification.

Conclusions

The key to the prevention of neural tube defects is food fortification. Cereals are a convenient vehicle; flour is perhaps the best choice, but bread and breakfast cereals are an alternative if there are legislative difficulties involved in fortifying flour. The level of fortification should be determined and reviewed by a central authority or advisory committee, in the light of information on the distribution of folate intake in the population concerned and evidence on efficacy and safety. Initially, the aim should be to increase median intake to about 0.6 mg a day while avoiding high intake in some—ensuring that the 97th centile is no more than about 1.5–2.0 mg a day.

The fortification policy can, and should, be complemented by advice on diet and advice on folic acid supplements for women planning a pregnancy, but these measures are unlikely to have a major impact on the incidence of neural tube defects in the general community.

Acknowledgements

I thank David Smith and Hilary Watt for computing and statistical assistance.

References

Bailey LB 1992 Evaluation of a new recommended dietary allowance for folate. J Am Diet Assoc 92:463–468, 471

Bower C, Stanley FJ 1989 Dietary folate as a risk factor for neural tube defects: evidence from a case-control study in Western Australia. Med J Aust 150:613–618

Bower C, Stanley FJ 1992 Periconceptional vitamin supplementation and neural tube defects; evidence from a case-control study in Western Australia and a review of recent publications. J Epidemiol Community Health 46:157–161

Centers for Disease Control 1992 Recommendations for the use of folic acid to reduce the number of cases of spina bifida and other neural tube defects. Morbid Mortal Wkly Rep 41(RR-14)1:7

Committee on Medical Aspects of Food Policy 1991 Dietary reference values of food energy and nutrients for the United Kingdom. HMSO, London

Czeizel AE, Dudás I 1992 Prevention of the first occurrence of neural-tube defects by periconceptional vitamin supplementation. N Engl J Med 327:1832–1835

Economides DL, Ferguson J, Mackenzie IZ, Darley J, Ware II, Holmes-Siedle M 1992 Folate and vitamin B_{12} concentrations in maternal and fetal blood, and amniotic fluid in second trimester pregnancies complicated by neural tube defects. Br J Obstet Gynaecol 99:23–25

Emery AEH, Timson J, Watson-Williams EJ 1969 Pathogenesis of spina bifida. Lancet 2:909–910

Expert Advisory Group 1992 Folic acid and the prevention of neural tube defects. Department of Health, UK

Food and Nutrition Board 1980 Recommended dietary allowances, 9th edn. National Academy of Sciences, Washington, DC

Food and Nutrition Board 1989 Recommended dietary allowances, 10th edn. National Academy of Sciences, Washington, DC

Hall MH 1977 Folates and the fetus. Lancet 1:648–649

Hibbard ED, Smithells RW 1965 Folic acid metabolism and human embryopathy. Lancet 1:1254

Holmes LB 1994 Spina bifida: anticonvulsants and other maternal influences. In: Neural tube defects. Wiley, Chichester (Ciba Found Symp 181) p 232–244

Kirke PN, Daly LE, Elwood JH 1992 A randomised trial of low dose folic acid to prevent neural tube defects. Arch Dis Child 67:1442–1446

Laurence KM, James N, Miller MH, Tennant GB, Campbell H 1980 Increased risk of recurrence of pregnancies complicated by fetal neural tube defects in mothers receiving poor diets, and possible benefit of dietary counselling. Br Med J 281:1592–1594

Laurence KM, James N, Miller MH, Tennant GB, Campbell H 1981 Double-blind randomised controlled trial of folate treatment before conception to prevent recurrence of neural tube defects. Br Med J 282:1509–1511

Markel H 1987 'When it rains it pours': endemic goiter, iodized salt, and David Murray Cowie. Am J Public Health 77:219–229

Mills JL, Rhoads GG, Simpson JL et al 1989 The absence of a relation between the periconceptional use of vitamins and neural-tube defects. N Engl J Med 321:430–435

Mills JL, Tuomilehto J, Yu KF et al 1992 Maternal vitamin levels during pregnancies producing infants with neural tube defects. J Pediatr 120:863–871

Milunsky A, Jick H, Jick SS et al 1989 Multivitamin/folic acid supplementation in early pregnancy reduces the prevalence of neural tube defects. JAMA (J Am Med Assoc) 262:2847–2852

Molloy AM, Kirke P, Hillary I, Weir DG, Scott JM 1987 Maternal serum folate and vitamin B_{12} concentrations in pregnancies associated with neural tube defects. Arch Dis Child 60:660–665

MRC Vitamin Study Research Group 1991 Prevention of neural tube defects: results of the MRC vitamin study. Lancet 338:132–137

Mulinare J, Cordero JF, Erickson D, Berry RJ 1988 Periconceptional use of multivitamins and the occurrence of neural tube defects. JAMA (J Am Med Assoc) 260:3141–3145

Nau H 1994 Valproic acid-induced neural tube defects. In: Neural tube defects. Wiley, Chichester (Ciba Found Symp 181) p 144–160

Office of Population Censuses and Surveys 1990 Dietary and nutritional survey of British adults. HMSO, London

Schorah CJ, Wild J, Hartley R, Sheppard S, Smithells RW 1983 The effect of periconceptional supplementation on blood vitamin concentrations in women at recurrence risk for neural tube defect. Br J Nutr 49:203–211

Smithells RW, Sheppard S, Schorah CJ 1976 Vitamin deficiencies and neural tube defects. Arch Dis Child 51:944–949

Smithells RW, Seller MJ, Harris R et al 1983 Further experience of vitamin supplementation for prevention of neural tube defect recurrences. Lancet 1:1027–1031

Vergel RG, Sanchez LR, Heredero BL, Rodriguez PL, Martinez AJ 1990 Primary prevention of neural tube defects with folic acid supplementation: Cuban experience. Prenatal Diagn 10:149–152

Wald N 1993 Maternal nutrition and pregnancy outcome. Ann NY Acad Sci 678:112–129

Werler MM, Shapiro S, Mitchell AA 1993 Periconceptional folic acid exposure and risk of occurrent neural tube defects. JAMA (J Am Med Assoc) 269:1257–1261

Winship KA, Cahal DA, Weber JCP, Griffin JP 1984 Maternal drug histories and central nervous system anomalies. Arch Dis Child 59:1052–1060

Yates JRW, Ferguson-Smith MA, Shenkin A, Guzman-Rodriguez R, White M, Clark BJ 1987 Is disordered folate metabolism the basis for the genetic predisposition to neural tube defects? Clin Genet 31:279–287

DISCUSSION

Shum: Is there any information on the recurrence risk for neural tube defects of women who are made pregnant by different men? We need to distinguish, particularly for folic acid supplementation, whether there is something wrong with the woman in handling folic acid or a genetic predisposition of the embryo to neural tube defects.

Wald: There are not many data, but there was some information in *The epidemiology and control of neural tube defects* (Elwood et al 1992).

Cuckle: The women who had babies with neural tube defects despite the folic acid supplementation in the MRC trial—do you intend to follow them up in subsequent pregnancies?

Wald: There are no plans to do so.

Cuckle: You would expect them to have a higher recurrence risk than those for whom supplementation worked in the previous pregnancy. If that was not the case, it would be a useful piece of information.

Hall: It would be hard to tell if folic acid supplementation works in a subsequent pregnancy because the recurrence risk is only 5–10%.

Seller: Two of the women from the MRC trial came to me and were given Pregnavite and had normal babies.

Czeizel: The Hungarian Randomised Controlled Trial of Periconceptional Multivitamin/Trace Element Supplementation was launched in 1984 in the Centre of the Hungarian (Optimal) Family Planning Programme. We used multivitamin preparations, including 0.8 mg folic acid (Elevit pronatal, Roche), and our so-called placebo group was treated with trace elements. The randomization code was broken at the end of 1991 (Czeizel & Dudás 1992). Since then, we have had no new fetus or newborn infant with a neural tube defect (Table 1). We stopped the randomization at the end of April 1992; after that, all women were given multivitamins. Evaluation of all the pregnancy outcomes was complete by the end of April 1993.

The Hungarian Optimal Family Planning Programme operates through the network of periconceptional care clinics. In 1990, a periconceptional folic acid supplement (4 mg daily) was introduced into this programme. At the end of April 1993, we had 1779 informative pregnancies: one fetus had anencephaly and one had lumbosacral spina bifida aperta with very severe hydrocephalus, but this male fetus had only three toes in the right lower limb and polydactyly in the left hand. I think this is not a real neural tube defect case; it is a congenital abnormality syndrome or something else.

Finally, I would like to inform you of an unexpected finding of our Hungarian trial. The rate of all congenital abnormalities was lower in the group given multivitamins than in the group given trace elements. This difference cannot be explained completely by the significant reduction in neural tube defects. Four groups of congenital abnormalities were distinguished: neural tube defects, genetic syndromes (e.g. Down's syndrome), and major and mild congenital abnormalities. The rate of major congenital abnormalities other than neural tube defects and genetic syndromes was 9.0/1000 in informative pregnancies (those for which the outcome was known) in the multivitamin group and 16.6/1000 in the trace element group. This gives a relative risk of 1.85 (95% confidence interval 1.02–3.38). The difference is 7.6/1000 between the above two figures, which is 2.7 times higher than the total (fetal and birth) prevalence of neural tube defects in

TABLE 1 (*Czeizel*) The incidence of neural tube defects in offspring of women treated with multivitamins or trace elements in the Hungarian Randomised Controlled Trial

	Neural tube defects			
Group	*Expected total*	*Rate per 1000*	*Observed total*	*Rate per 1000*
Multivitamins	6.9	2.78	0	0.00
Trace elements	6.6	2.78	6	2.51

Hungary. Infants have been followed: they were examined after the eighth month of their life. The rate of all major congenital abnormalities without neural tube defects and genetic syndromes was 14.7/1000 in the multivitamin group and 28.3/1000 in the trace element group ($P = 0.0025$). The rate of some congenital abnormalities was lower in the multivitamin group than in the trace element group but the differences for each group of abnormalities were not significant (Czeizel 1993).

Stanley: Professor Wald, were the cases of neural tube defects in the intervention arm of the MRC trial anencephalies or spina bifidas and were there any sex-specific differences?

Wald: There was no remarkable bias in either the type of defect or the sex specificity, but it would be very hard to discern such effects with so few cases.

Copp: Is there is a preferential effect of folate for preventing defects at a particular level of the body axis? Could you be seeing a massive preventive effect on anencephaly and a relatively small effect on spina bifida?

Wald: No. The data from the vitamin study show that.

Lindhout: The narrow window effect sounds like a nice model. The 4 mg of folic acid dose effect fits this, but how does one fit the 0.4 mg dose effect? This lower dose is not far from the range in the daily intake in a general population, but its preventive effect seems to be of the same magnitude as that of the 4 mg dose.

Wald: The estimated range of folate intake in the general population is about 0.1–0.4 mg per day. So with the 0.4 mg dose folic acid supplement, total folate intake is outside the estimated range. The estimated range is wider than the true range because of the fluctuations in folate intake. It would be quite easy to find out what the true range is. One would estimate folate intake in about 200 people and rank them according to intake, then estimate their intake again. The second measure would be an unbiased measure of the original one. This needs to be done to establish how narrow that window is, but we know it has to be less than 0.1–0.4 mg.

Scott: The other thing to take into account with the natural folates and the daily intake of 0.2 mg is the bioavailability. The folate in food is only about half as available to the body as folic acid (Gregory et al 1991). It is also less stable to cooking, etc, than folic acid. So the background is really 0.1 mg, not 0.2 mg. The consensus now seems to be that a supplement of 0.4 mg per day would prevent neural tube defects, but it is difficult to construct a diet that would provide an equivalent amount if one allows for the reduced bioavailability of dietary folate. Fiona Stanley has shown that diet may have an influence on the incidence of neural tube defects and it may be important. But one cannot get anything near 0.4 mg daily by even the most superb diet, because you would always have to discount the natural folates by a factor of two. One can achieve 0.4 mg per day with a fortified diet, but not with a natural diet.

References

Czeizel AE 1993 Prevention of congenital abnormalities by periconceptional multivitamin supplementation. Br Med J 306:1645–1648

Czeizel AE, Dudás I 1992 Prevention of the first occurrence of neural-tube defects by periconceptional vitamin supplementation. N Engl J Med 327:1832–1835

Elwood JM, Little J, Elwood JH (eds) 1992 The epidemiology and control of neural tube defects. Oxford University Press, Oxford

Gregory JF III, Bhandari SD, Bailey LB, Toth JP, Baumgartner TG, Cerda JJ 1991 Relative bioavailability of deuterium-labeled monoglutamyl and hexaglutamyl folates in human subjects. Am J Clin Nutr 53:736–740

Prevention of folic acid-preventable spina bifida and anencephaly

Godfrey P. Oakley, Jr., J. David Erickson, Levy M. James, Joseph Mulinare and José F. Cordero

Division of Birth Defects and Developmental Disabilities, National Center for Environmental Health, Centers for Disease Control and Prevention, Building 101, 4770 Buford Highway, Atlanta, Georgia 30341-3724, USA

Abstract. The results of the British Medical Research Council's randomized controlled trial proved that folic acid can prevent spina bifida and anencephaly. The trial provided critical scientific data upon which to base public health policy for preventing folic acid-preventable spina bifida and anencephaly. Within weeks of publication of the results, the Centers for Disease Control and Prevention in the US developed and issued guidelines for women who had had a pregnancy affected by spina bifida or anencephaly. A year later, the US Public Health Service issued the recommendation that all women of child-bearing age who are capable of becoming pregnant should consume 0.4 mg of folic acid per day. The Public Health Service needed a year to make inferential judgements about dose, target groups, safety, timing of ingestion, and existing and proposed vitamin and drug policies and regulations. Current policy discussions concern whether to permit manufacturers of vitamins or food products to claim that folic acid will prevent folic acid-preventable spina bifida and anencephaly and whether to allow a food staple to be fortified with folic acid.

1994 Neural tube defects. Wiley, Chichester (Ciba Foundation Symposium 181) p 212–231

'One of the most exciting medical findings of the last part of the 20th century is that folic acid, a simple, widely available, water-soluble vitamin, can prevent spina bifida and anencephaly (SBA). Not since the rubella vaccine became available 30 years ago have we had a comparable opportunity for primary prevention of such common and serious birth defects' (Oakley 1993).

Prevention of folic acid-preventable spina bifida and anencephaly is now possible; we can prevent these serious birth defects in a large proportion of the 400 000 infants who would otherwise be born each year with spina bifida and anencephaly throughout the world. We are excited that we have been able to work on building public health recommendations that are the basis for a policy to prevent folic acid-preventable spina bifida and anencephaly in the United States. Like other health policies, these recommendations required informed judgements, for seldom do we have sufficient data to be absolutely certain that a policy will have its intended effect. Crucial to our approach to the development of a policy for preventing folic acid-preventable spina bifida and anencephaly was our conclusion that folic acid will prevent these birth defects. We describe highlights of how we reached that conclusion and then recount some of the activities that occurred between the conclusion being reached and publication of recommendations by official bodies of the United States Government (Centers for Disease Control 1991, 1993). We conclude by discussing clues gleaned from epidemiological studies that may help our colleagues working in more basic areas of research.

The search for an efficient treatment/cause

The work of Smithells et al (1981) thrust the issue of substantial primary prevention for spina bifida and anencephaly into the public arena. It showed that women who had previously had a pregnancy affected by spina bifida or anencephaly who agreed to consume daily preparations of multivitamins including 0.36 mg of folic acid were seven times less likely to have another affected pregnancy than were women who did not take the vitamins. This study raised the question of whether or not these and other data were sufficiently strong to warrant developing a public health policy based on the assumption that one or more vitamins would prevent spina bifida and anencephaly. Because the Smithells' study did not randomly assign women to either treatment or placebo groups, we concluded that factors associated with vitamin taking, rather than one or more vitamins, could explain Smithells' very interesting findings. We, therefore, concluded that the data were insufficient for the development of a treatment policy, but that they were sufficiently important to warrant further studies to determine whether one or more vitamins would prevent spina bifida and anencephaly.

Because we were interested in conducting a randomized controlled trial, we organized several CDC workshops and met with United States Public Health Service (PHS) colleagues from other agencies to discuss conducting such a trial. Throughout the 1980s, these colleagues agreed that the evidence, including the new observational studies that were consistent with one or more vitamins preventing a substantial proportion of the incidence of spina bifida and anencephaly in the general population, was incomplete and that it was necessary and ethical to conduct a placebo, double-blind controlled trial.

In the 1980s, the Medical Research Council (MRC) in the UK began its recurrence randomized controlled trial and Dr Andrew Czeizel began the Hungarian occurrence randomized controlled trial (MRC Vitamin Study Research Group 1991, Czeizel & Dudás 1992). (By recurrence we mean a trial involving women who have had a previous pregnancy affected by spina bifida or anencephaly; by occurrence we mean a trial involving women who have never had such a pregnancy.) As the results of three positive (Mulinare et al 1988, Bower & Stanley 1989, Milunsky et al 1989) and one negative (Mills et al 1989) observational studies were published, we worked to design and find resources for an occurrence randomized controlled trial in China. In public presentations, one of us (GPO) argued that it was justified to spend substantial research resources on the randomized controlled trial in China because, if the results were positive, the prevention impact would be similar to that of the prevention of polio by the development of a vaccine. In addition, it would be possible to prevent spina bifida by food fortification, just as rickets and iodine-deficient goitre are prevented by fortification. The final point, however, was that the available data could simply have been a red herring; although the data were consistent with the hypothesis that one or more vitamins would prevent spina bifida and anencephaly, it was entirely possible that results of a randomized study would show that vitamins would not prevent these defects.

During the early months of 1991, the CDC group and a group from Beijing Medical University were in the last phases of planning and implementing a pilot for an occurrence placebo randomized controlled trial in northern China, an area with high endemic rates of spina bifida and anencephaly. This area was selected because we decided an occurrence trial could be conducted and completed there within a reasonable time.

Efficacy established

At 5 p.m. on 24 June 1991, Professor Wald informed one of us (GPO) that the results of the MRC's study showed that 4.0 mg of folic acid per day had a striking protective effect in women who had had a prior pregnancy affected by spina bifida or anencephaly. That night, we concluded that the data were now sufficiently strong for us to believe that folic acid would prevent both the recurrence and occurrence of these defects. We began then to work toward policies that we hoped would lead to the worldwide eradication of folic acid-preventable spina bifida and anencephaly.

Recurrence prevention recommendations

As the lead agency for prevention in the PHS in the Department of Health and Human Services, CDC customarily develops and publishes prevention recommendations. We judged that the MRC data were sufficiently strong that we could reach rapid agreement with our scientific colleagues for recurrence recommendations (i.e. recommendations relating to women who had already had an affected pregnancy). Two weeks after the publication of the MRC study, CDC issued interim recurrence recommendations (Centers for Disease Control 1991) which suggested that women who had previously had a pregnancy with a neural tube defect and who were planning a pregnancy should consult their physician and should ingest 4.0 mg of folic acid daily in order to prevent such defects.

One might ask how it was possible to move so quickly to recurrence recommendations. First, there was a small, well defined high-risk group. The MRC study had shown the efficacy of 4.0 mg of folic acid daily in preventing a large proportion of these serious birth defects in this high-risk group. Second, obtaining a dose of 4.0 mg a day of folic acid requires a prescription; therefore, these women would be under the care of a physician who would be alert to any possible adverse effects from the treatment, should they occur. There is no LD_{50} for folic acid, although some patients with pernicious anaemia can develop neurological complications if they are taking 5.0 mg of folic acid daily and not receiving vitamin B_{12} injections. Although pernicious anaemia is a rare disease among women of reproductive age, it does occur and it is possible that taking 4.0 mg of folic acid a day could obscure the diagnosis of an undiagnosed case. The recommendations, therefore, called this issue to the attention of physicians.

Public health plan to prevent folic acid-preventable spina bifida and anencephaly in the general population

As important as it is to prevent spina bifida or anencephaly among the offspring of women who have had previously affected pregnancies, these women account for fewer than 5% of such pregnancies. It is of substantial public health importance to prevent spina bifida and anencephaly among offspring of women who have never had an affected pregnancy. We outlined a prevention plan that included 1) publishing occurrence recommendations, 2) commercial sector actions, 3) CDC actions to assist poor women to prevent spina bifida and anencephaly, 4) fortification of a food staple, and 5) CDC surveillance of effectiveness and safety. Here, we will focus on the process that led to issuing the occurrence recommendations in the United States.

Occurrence prevention recommendations

Although we had concluded that CDC should recommend that all women who might get pregnant should consume 0.4 mg daily of folic acid to prevent folic acid-preventable spina bifida and anencephaly, we correctly thought it likely that reaching agreement on occurrence recommendations would take longer and would require considerable public and internal discussion.

In September 1991, we met in Atlanta with experts from the private sector, representatives of professional and voluntary organizations and scientists from other agencies in the PHS to hear their individual assessments of the data and advice about occurrence recommendations. The primary message that we took from that public workshop was that, on the basis of current evidence, it would be a desirable, reasonable and safe public health policy for CDC to issue a recommendation that all women of child-bearing age should consume 0.4 mg of folic acid daily in order to prevent folic acid-preventable spina bifida and anencephaly. Although 0.4 mg was presumed to be sufficient to prevent folic acid-preventable spina bifida and anencephaly, we did not have sufficient evidence to know what was the smallest dose that would prevent these defects.

CDC occurrence briefing paper

For CDC to issue occurrence recommendations, we would need the approval of our supervisors, including the Director of CDC. In addition, peer review within CDC would help formulate a clear statement of the rationale for occurrence recommendations. We prepared a paper detailing this rationale. We noted that before publication of the MRC paper the consensus

> 'within the PHS has been that the evidence available was not sufficient to warrant a public health recommendation for the use of vitamins to prevent NTDs [neural tube defects] in the general population (i.e., a recommendation for folic acid use by all women who have not had a previous NTD-affected pregnancy). It had been felt that evidence from an RCT [randomized controlled trial] would be needed before such a recommendation could be made. Although, at this time, no published RCT data are available on women who have never had an NTD-affected pregnancy, the results of the British MRC study have suggested that there is a need to reevaluate this consensus position.
>
> The crucial questions are these: 1) Do the data derived from a study of women who have had prior NTD-affected pregnancies predict a protective effect among women who have not had a prior affected pregnancy? 2) Do these data cause us to alter our view of the results of the nonexperimental studies of low-dose folic acid? 3) If so, what would be the public health impact (prevention effectiveness, cost, and risk) of a recommendation for the use of low-dose folic acid for the prevention of NTDs in the general population?'

We addressed these issues.

Is folic acid use likely to prevent neural tube defects in the general US population?

We argued that folic acid use in the general population probably would prevent neural tube defects. We noted that support for our argument came from two lines of evidence.

One line of reasoning was based on a re-evaluation of the observational studies (Smithells et al 1981, 1983, Mulinare et al 1988, Milunsky et al 1989, Bower &

Stanley 1989) in light of the MRC findings. The positive results in these earlier studies could have been due to bias or to the vitamins. The nearly identical results of the MRC study and the Smithells' study argued strongly, however, that at least some part of the protective effect in the positive observational studies was due to vitamins rather than to bias.

The other line of reasoning was simply that although there may be many causes for a woman to have a pregnancy affected by a neural tube defect, it was a reasonable assumption that the cause in her second pregnancy, on average, was more likely to be the same cause as in her first rather than a different cause. If an agent would prevent second cases, it was highly likely that it would have prevented some of the first cases. Thus, we concluded that although we did not have randomized controlled data from a general population, it was likely that folic acid supplements could prevent some of the occurrence cases.

What dose of folic acid would prevent neural tube defects in the general population?

The positive observational studies with vitamin supplements used about 0.4 mg of folic acid a day. The nearly identical results in the MRC and Smithells' study made a strong argument that this dose would be effective. We concluded that it was very likely that 0.4 mg per day would be effective in the general population. We acknowledged that multivitamins were used with the folic acid in all the observational studies and that we depended on the MRC study to infer that multivitamins were unnecessary. Clearly, a judgment would have been easier to make if a study had been done administering only low doses of folic acid in a general population.

Although we did not use the United States Recommended Daily Allowance in making a judgment about the efficacy of 0.4 mg daily in the general population, we noted that for folic acid this is 0.4 mg per day. A recommendation to consume folic acid to prevent spina bifida and anencephaly should increase the proportion of persons receiving the US recommended allowance for folic acid.

Other issues

We estimated that about half of all neural tube defects in the United States could be prevented if all women of reproductive age ingested 0.4 mg daily of folic acid, but we noted that this proportion could be substantially higher or lower. We estimated that the monetary cost of a programme based on taking pills would save about what it would cost to implement.

What are safety concerns about the administration of low doses of folic acid?

We reviewed the evidence of possible harmful effects of 0.4 mg of folic acid per day to users of anticonvulsants, possible harmful effects to people with undiagnosed vitamin B_{12} deficiency, possible interference with zinc metabolism or absorption, and the possibility of an unknown but serious adverse effect. We concluded that 0.4 mg per day was unlikely to be harmful. Even though 20% of the US population daily consume a supplement with 0.4 mg of folic acid and there is no known serious side-effect, we noted that, on the basis of current evidence, we could not exclude the possibility, but a serious side-effect was unlikely.

Draft recommendation

We proposed that CDC issue a recommendation, the wording of which should be finalized after a public review of the following draft statement: 'All women in the United States who are capable of becoming pregnant should consume 0.4 mg of folic acid per day for the purpose of preventing spina bifida and other neural tube defects.'

Publication of draft recommendation

We reached these conclusions in March 1992 and prepared a final copy for further internal CDC review. On 9 May 1992 the CDC Associate Director for Science and a group of scientists from several units within CDC reviewed the final version of the briefing paper. After that meeting, the Associate Director sent us a memo suggesting that we transmit the briefing paper to the Director of CDC, that we propose to develop the exact words of a draft occurrence recommendation, that we publish the draft recommendation in the Federal Register, and that we invite public comment in writing and at a public workshop on this issue. The public meeting was held on 27 July 1992.

Although one of us (JDE) was a member of the review team for the Hungarian occurrence study, the remainder of the CDC staff were not aware of the findings until April 1992 when Dr Andrew Czeizel, the study director, confirmed that he was going to present the results in May and that his findings could be summarized for CDC. These data demonstrated a powerful protective effect on occurrence with the use of a preparation that contained both a low dose of folic acid (0.8 mg per day) and multiple vitamins (Czeizel & Dudás 1992). These results were reassuring and may have been the data necessary for some of our colleagues to agree that we should issue a recommendation.

At about the same time, we learned that Dr Martha Werler would present her recent case-control study at the same meeting in May. She gave us a verbal summary of her data: these were reassuring because they suggested that in recent years (1988–1991) in the north-eastern United States folic acid-preventable spina bifida and anencephaly accounted for about half of all cases of such defects (Werler et al 1993). Thus, when we transmitted the briefing paper to the Director of CDC, we provided addenda. One addendum summarized the Hungarian findings, the other summarized the results of Werler.

Public health service discussion

The Food and Drug Administration (FDA) is the part of the PHS that is responsible for regulations relating to food and drugs. On a separate track, the FDA had been given the responsibility to promulgate regulations on whether to allow health claims for folic acid reducing the risk of neural tube defects. In June 1992, FDA had prepared a draft final rule which concluded that low-dose folic acid did not reduce the risk of neural tube defects.

Because the two agencies were developing official statements whose conclusions about the efficacy of folic acid in the prevention of spina bifida were different, officials from PHS and its agencies held additional discussions. After the CDC workshop, agreement was reached that recommendations should be issued, but by PHS rather than by CDC.

14 September 1992 Announcement of PHS occurrence recommendations

The CDC had planned a scientific meeting on birth defect prevention and epidemiology for 14–16 September 1992, to honour the 25th anniversary of the Metropolitan Atlanta Congenital Defects Program. Dr William Roper, the Director of CDC, opened the conference by releasing the PHS recommendations that were published on 11 September 1992, as Recommendations and Reports by the Morbidity and Mortality Weekly Report. The text of the primary recommendation follows:

> '. . . the United States Public Health Service recommends that: All women of childbearing age in the United States who are capable of becoming pregnant should consume 0.4 mg of folic acid per day for the purpose of reducing their risk of having a pregnancy affected with spina bifida or other NTDs' (Centers for Disease Control 1993).

Implementation of PHS recommendations

Although the recommendations have been published and have attracted some public attention, they are far from being universally implemented. Because half of all pregnancies in the United States are unplanned, prevention of all folic acid-preventable spina bifida and anencephaly will require a programme which assures that most women of child-bearing age will consume 0.4 mg of folic acid a day. We judge that the most effective way to prevent folic acid-preventable spina bifida and anencephaly would be to fortify a food staple sufficiently with folic acid so that most women would consume 0.4 mg of folic acid daily. Building a prevention programme based on taking a pill daily for a number of years will be substantially less effective and more costly. This approach, however, is reasonable where fortification is not possible and until sufficient fortification is possible. Improving diets may prevent folic acid-preventable spina bifida and anencephaly, but current evidence demonstrates that it is folic acid supplements, as an addition to the diet in a developed country, that prevent folic acid-preventable spina bifida and anencephaly.

FDA Food Advisory Committee and its folic acid subcommittee

After the release of the PHS guidelines, the FDA quickly appointed a Folic Acid Subcommittee of the Food Advisory Committee. The subcommittee met in November 1992 and recommended that FDA devise a plan of food fortification that would permit most women of reproductive age to ingest 0.4 mg per day of folic acid, while minimizing the number of people who ingested more than 1.0 mg of folic acid.

In April 1993, the subcommittee met to review the options prepared by FDA for fortification of food. In summary, FDA proposed that the only workable fortification plan would be to fortify cereal grains, not fruit juices or milk. The FDA suggested that the level of fortification be either restoration (the amount of folic acid removed in milling)—0.07 mg of folic acid per 100 g of cereal grain products—or twice restoration (Table 1). FDA did not favour a fortification level that was five times the restoration level. The CDC testified that it favoured a fortification level that was five times the restoration level because such a level assured that most women of reproductive age would receive 0.4 mg of folic acid while only a very small proportion of persons 50 years of age or older would ingest more than 1.0 mg of folates. A final decision has not been made on the level of fortification.

Advice from epidemiologists to colleagues in clinical and basic research

In large measure, we think that clinicians and epidemiologists have provided the scientific basis for preventing folic acid-preventable spina bifida and anencephaly. Along the way, epidemiologists have found clues that may help scientists in other disciplines who are working to find the pathogenesis of folic acid-preventable spina bifida and anencephaly or to find other causes of neural tube defects.

Clue 1. Black infants have a lower rate of spina bifida and anencephaly than white infants. Do differences in folic acid metabolism explain this difference?

Clue 2. Known human environmental causes of spina bifida and anencephaly affect fewer than 5% of those exposed. Is there a folic acid metabolism variant that is highly susceptible to folic acid-preventable spina bifida and anencephaly?

TABLE 1 Food and Drug Administration model for folic acid fortification

	Daily intake of total folates (μg) with a fortification level of folic acid in food of									
	2X (0.14 mg/100 g cereal grain)					*5X (0.35 mg/100 g cereal grain)*				
% of subpopulation receiving less than specified amount of total folates	*25*	*50*	*75*	*90*	*95*	*25*	*50*	*75*	*90*	*95*
No supplement use reported										
Females aged 19–50	200	270	350	440	500	310	410	530	670	750
Males aged 11–18	330	410	520	630	710	500	610	750	930	1090
Males aged 51 +	270	350	450	540	610	400	530	640	770	870
Supplement use reported										
Females aged 19–50	250	390	610	740	870	380	550	750	920	1010
Males aged 11–18	350	470	650	840	960	530	700	910	1110	1220
Males aged 51 +	320	430	670	810	860	460	610	860	1030	1110

X represents 'restoration' fortification, i.e. the amount required to replace folic acid lost in milling and processing of grains.
Supplement use indicates that the folic acid that people consumed as a vitamin pill was included in the model.

Clue 3. Most, but not all, of spina bifida and anencephaly is folic acid-preventable spina bifida and anencephaly. We need to look for clues to other causes.

Advice

Study folic acid metabolism in order to improve our understanding of folic acid-preventable spina bifida and anencephaly. Look for errors in metabolizing folic acid as a cause of neural tube defects in susceptible mice strains.

Closing remarks

We do not know how long it will take to eradicate folic acid-preventable spina bifida and anencephaly in the United States, but we predict that at times the process will seem too slow. We expect that others will also be impatient. During such moments, we suggest reading Elizabeth Etheridge's *The Butterfly Caste*, which describes Goldberger's work to prevent pellagra, and Howard Markel's article that discusses Dr Cowie's work to initiate programmes to prevent iodine-deficient goitre (Etheridge 1972, Markel 1987). Even with strong evidence available to them, both Goldberger and Cowie were frustrated about the length of time it took from when they concluded the data were sufficient to prevent these diseases until effective programmes were begun. We hope that in future efforts to prevent spina bifida and anencephaly, frustration will be met with at least equal levels of perseverance.

References

Bower C, Stanley FJ 1989 Dietary folate as a risk factor for neural-tube defects: evidence from a case-control study in Western Australia. Med J Aust 150:613–619

Centers for Disease Control 1991 Use of folic acid for prevention of spina bifida and other neural tube defects—1983–1991. Morbid Mortal Wkly Rep 40:513–516

Centers for Disease Control 1993 Recommendations for the use of folic acid to reduce the number of cases of spina bifida and other neural tube defects. JAMA (J Am Med Assoc) 269:1233,1236 and 1238 (First published in 1992 in: Morbid Mortal Wkly Rep 41[RR-14]:1–7)

Czeizel AE, Dudás I 1992 Prevention of the first occurrence of neural-tube defects by periconceptional vitamin supplementation. N Engl J Med 327:1832–1835

Etheridge EW 1972 The butterfly caste. A social history of pellagra in the South. Greenwood Publishing Company, Westpoint, CT

Markel H 1987 'When it rains it pours': endemic goiter, iodized salt, and David Murray Cowie, MD. Am J Public Health 77:219–229

Mills JL, Rhoads GG, Simpson JL et al 1989 The absence of a relation between the periconceptional use of vitamins and neural-tube defects. N Engl J Med 321:430–435

Milunsky A, Jick H, Jick SS et al 1989 Multivitamin/folic acid supplementation in early pregnancy reduces the prevalence of neural tube defects. JAMA (J Am Med Assoc) 262:2847–2852

MRC Vitamin Study Research Group 1991 Prevention of neural tube defects: results of the Medical Research Council vitamin study. Lancet 338:131–137
Mulinare J, Cordero JF, Erickson JD, Berry RJ 1988 Periconceptional use of multivitamins and the occurrence of neural tube defects. JAMA (J Am Med Assoc) 260:3141–3145
Oakley GP Jr 1993 Folic acid-preventable spina bifida and anencephaly. JAMA (J Am Med Assoc) 269:1292–1293
Smithells RW, Sheppard S, Schorah CJ et al 1981 Apparent prevention of neural tube defects by periconceptional vitamin supplementation. Arch Dis Child 56:911–918
Smithells RW, Nevin NC, Seller MJ et al 1983 Further experience of vitamin supplementation for the prevention of neural tube defect recurrences. Lancet 1:1027–1031
Werler MM, Shapiro S, Mitchell AA 1993 Periconceptional folic acid exposure and risk of occurrent neural tube defects. JAMA (J Am Med Assoc) 269:1257–1261

DISCUSSION

Mills: I believe we should fortify food. The FDA did some calculations regarding how one can do this. They used FDA data on how often Americans eat folate-containing foods and in what quantities they eat them, to establish folate intake. The strategy in the US is to put folic acid into cereal grains, such as wheat, corn, rye and rice flour, cornmeal, corn grits, macaroni and spaghetti, in order to reach different ethnic groups.

The calculations included fortified cereals with daily supplements of 0.1 mg and 0.4 mg folic acid and health supplements taken by the US population. They were based on two different levels of fortification: replacement or restoration level, which is 1X in Godfrey Oakley's terms, and twice restoration. These were chosen because at these levels most people's intake would not exceed the subcommittee's maximum recommended level of 1 mg folic acid per day. The FDA believes that food supplementation should be accompanied by a massive education programme to alert people to the fact that folic acid is not like hydrochloric acid or nitric acid—it is a good thing!

A critical point is that you would not improve folate status very much in the people who have low dietary folate intake and do not eat large quantities of supplemented food. They would come nowhere near the level of 0.4 mg that we are trying to achieve.

At fortification levels twice the restoration amount, very few people will exceed the 1 mg limit. The problem is that women with low folate intake (including those who do not consume very much of the fortified foods) will not get 0.4 mg of folic acid with this level of fortification. If the level of fortification is raised to five times restoration, more women will consume 0.4 mg but at the cost of many other people consuming more than 1 mg. This is the critical message: you cannot get most women up to 0.4 mg without exceeding the 1 mg level in a substantial number of people.

Oakley: The 95th percentile of 1 mg will get you about 75% over 0.4 mg.

Mills: That gives about 12 million people in the US above 1 mg/day.

Oakley: They are primarily teenagers who are in the low-risk category.

Mills: On 15 April 1993, the FDA subcommittee voted six to four in favour of supplementation. There was quite a lot of controversy. Some of the issues raised by the group were:

1) The Netherlands is not currently recommending supplementation and Hungary has not yet made a decision.

2) It has been reported that 3–7% of the elderly are vitamin B_{12} deficient, although the data documenting this have been criticized.

3) By one estimate, 13 000 people may be at risk for peripheral neuropathy.

The subcommittee's tentative recommendation is that either 70 or 140 mg per 100 mg of flour be used. But they are not approving that yet. They are setting questions to the experts and then making a final decision later.

The subcommittee strongly recommended that we correct the lack of monitoring, either for the benefits of folate fortification or for its toxicity. It's getting harder to document rates of neural tube defects because of prenatal diagnosis and termination. There are no firm data at the moment on the rates of B_{12} deficiency or B_{12} deficiency with subacute combined degeneration. So it would be hard to tell whether they increase or not.

Personally, I think fortification is the only way we are really going to have an impact on this problem. But given that approximately 2000 children in the US will benefit each year from folate supplements and 250 million people are going to be getting those supplements in their food, we have to approach this very carefully. We have to come up with a proposal that's going to be politically and scientifically acceptable to the general public.

Wald: In the UK, an advisory committee has reported to the government and the government has accepted their recommendations. There were 21 recommendations: the important ones were:

1) Women at high risk should have a 4–5 mg folic supplement a day;

2) Women in the general population should be able readily to obtain 0.4 mg folic acid supplement a day over the counter without prescription.

3) There should be education about diet.

4) The range of fortified cereals should be widened.

5) The level of fortification could be increased above the existing one, but not greatly.

Unfortunately, in the general population the impact may be small. There are only one or two brands of bread that are fortified with folic acid. It is difficult to buy the 0.4 mg folic acid supplement in a chemist's shop. A usual initial response is that a prescription is required. After persistence it is accepted that one is not required, but the chemist often notes that there is none in stock and that it is a low-volume, low-profit item. The most practical advice the chemist usually provides is that one can buy folic acid supplements in health food shops!

The position should improve. There is a move to get a preparation of 0.4 mg folic acid licensed and available through Boots. The Committee on Safety of Medicine is to consider granting a product licence that will enable folic acid to be sold with the claim that it can help prevent neural tube defects. There are also supposed to be discussions between the Ministry of Agriculture, Food and Fisheries and bakeries to encourage supplementation of bread with folic acid, but to my knowledge nothing practical has happened yet.

Stanley: In Australia, we had a recommendation from the National Health and Medical Research Council that folic acid intake be increased in women who are capable of having children. In Western Australia, we have started a statewide programme aimed at both women and health professionals to encourage dietary supplementation for those contemplating pregnancy and general consumption of a diet rich in folic acid. This is being done through a media campaign as well as through general practitioners, pharmacists, nurses and others. They are using this issue as a health promotion activity. They have posters, pamphlets and other literature in their offices and pharmacies.

Czeizel: In Hungary, we prepared a recommendation for the authorities of the Welfare Ministry, but there has been no official statement. There is a regular television programme and people have been informed about the beneficial effect of taking multivitamins including folic acid or folic acid alone around the time of conception. I think it is relatively widely used.

Hall: Does everybody accept these data and does everybody in this room agree that there is now evidence that folic acid will prevent a significant percentage of neural tube defects?

Mills: I believe folate prevents neural tube defects, but we have to put this into a geographical perspective. If I were the Minister of Health in China, I would be a lot more excited about giving everybody folate than if I were the Minister of Health in Finland. In Finland, the basic risk is very low and the data do not show a relationship between folate and neural tube defects, so I'm not sure I would recommend folate supplementation there at all. Even in the US, folic acid supplementation would probably have a much bigger effect in New England and western Georgia than in California. Because it is high risk, the population in the United Kingdom would clearly benefit from having a lot more folate.

Hall: Is your statement based on geography or on differences among ethnic groups?

Mills: We discussed at the CDC whether the Chinese have an increased risk because they are different ethnically or because of a different diet or because of differences in the environmental risk factors. All these could be influential. The most important studies on the preventive effect of folic acid have been done in the UK and Hungary. In the UK study, roughly half the subjects were Hungarian. But there have been enough other studies to show that in Caucasian populations there is a beneficial effect. I do think, however, that there is a lot of genetic variability.

Oakley: With the data, there is an inferential leap to get to a recommended dose of 0.4 mg of folic acid per day. The MRC trial used 4 mg and prevented recurrence. Those are the data that show that the preventive effect is on both anencephaly and spina bifida. All the other studies used folic acid plus multivitamins. There are no data from studies that used 0.4 mg of folic acid per day without multivitamins. The 0.4 mg dose clearly is an inferential public health decision; it is one I agree with. It could be the wrong decision, but I think the chance of that is very low.

Scott: Using 0.4 mg folic acid as a supplement has two advantages compared with fortification. Firstly, it is targeted to women who may become pregnant and therefore doesn't influence anybody else. Secondly, women of child-bearing age are usually not at risk of pernicious anaemia, although this may not be true of all racial groups (Carmel et al 1987). However, supplementation has the huge disadvantage that it will probably reach only a fraction of women before they know they are pregnant.

In our enthusiasm to do the right thing, we must be careful to achieve a balance. 0.4 mg may be needed, it may not be needed. As we try to supply most people with 0.4 mg, some of the population will get much more folic acid than this. I don't think the upper safety limit of 1 mg has any real scientific basis. Some people could argue that's a dangerous dose. I don't think they could make that argument for 0.8 mg or lower doses. So there is a balance to be struck between being ambitious and trying to introduce what is, after all, an arbitrary amount of folic acid into the general population. The old Recommended Daily Allowance (RDA) of 0.4 mg was worked out by people like me calculating what was in food and then allowing for absorption rates and so on. Folic acid in supplements or fortified food is twice as available within the body as is folate in food. The new European RDA that I have been involved in formulating is 0.2 mg, but again these amounts are for food folate and not folic acid.

Nau: To decide how high a dose would be risky for humans, we should find out what folate metabolites are being produced. Normally, the N^5-methyltetrahydrofolate is the major transport form. With high doses, folic acid itself may be present in human blood and maybe other folates too.

Scott: We are doing studies along those lines; the level of conversion is a critical issue. If all the folate is carried as N^5-methyltetrahydrofolate, the risk of pernicious anaemia disappears (Scott & Weir 1981).

Jacobson: For folic acid supplementation, would there be a difference in efficacy in having a given dose over several meals a day compared to taking one pill a day?

Oakley: Smithells' data were for women taking a pill three times a day with 0.12 mg of folic acid in each pill. The other trials are for once a day: the results are similar.

Scott: I think it would be better to have food supplementation rather than a pill because it would sustain plasma folate levels for longer. Red cells and tissues and the fetus get their folate from the maternal plasma. However, with a dose as high as 0.4 mg, I don't think it matters too much, because the maternal plasma folate level gets really very high at this dose.

Wald: There are two main decisions to be made: whether to select flour or bread for fortification and at what level. The outcomes will depend on local custom and the extent to which action can be taken without administrative and regulatory delays. If flour were chosen, the level might be 0.2 mg/100 g; for bread, it might be 0.4 mg per 100 g—0.1 mg would be too little and 1 mg would probably be too much. The range of fortification is therefore not that large. Any debate would be over small differences of minor practical consequence. We should choose reasonable fortification levels and ensure that such fortification is implemented, as soon as possible.

Hall: On the west coast of North America, you would have to fortify rice and cornmeal.

Wald: Choose a food and then make a political decision. You may not solve everything in one go, but start and get the principle of food fortification accepted.

Lindhout: If I belonged to a minority group, I would be upset with such a recommendation.

Wald: Choice is reasonable, *if* one has the relevant knowledge and the ability to exercise appropriate choice. In most areas of public health, there is neither the knowledge nor the opportunity for making the necessary choices. Usually, collective action is needed; the individual has little practical discretion. We are in a building now; it doesn't collapse because of building codes of which we are probably largely ignorant. Even if we weren't, it would be impractical judging which building to enter and which to avoid on grounds of safety. There are substances in foods—preservatives, micronutrients in flour (including the vitamins, niacin and thiamine)—that most people don't know about. We are not breaking new ground by saying that there should be one more, we are simply applying the same policy with the advantage of better information and more scientific data than was available in the past. Unfortunately, we are bedevilled by arguments that are based on a spurious notion of liberty and choice that serves to delay the introduction of a valuable preventive public health measure.

Nobody is saying there will be no bread and no flour without added folic acid. All that has been suggested is that the default option is the one that we believe to be in the public interest (namely, that bread or flour would be fortified), acknowledging that almost any public health strategy involves the possibility of certain risks as well as benefits. If consumers want to buy natural food with nothing added, they still can. There is a thriving specialist market in such foods. Choice is preserved, but it is the group that wishes to avoid fortification of bread and flour that would need to seek out the 'natural' product; others should be able to assume that the food they buy is appropriately fortified without special enquiry.

Mills: The US position does not leave this element of choice: all cereal grains are going to have folate in them if food fortification is approved. This creates a more difficult political situation, because objectors will not be able to buy unsupplemented flour.

Scott: People are not opposing the introduction of folate into food just to be difficult. There is a well recognized risk in the elderly that treating them with folic acid will mask symptoms of vitamin B_{12} deficiency. If that risk didn't exist, we wouldn't be arguing about it, because the other risks are self-identified. If somebody is on anticonvulsants, they know about it. This risk has to be eliminated. We don't know the level or prevalence of B_{12} deficiency in the elderly in the US or anywhere else.

Wald: If people really believe that the deficiency of vitamin B_{12} is a serious problem, the population requires B_{12} supplementation. Pernicious anaemia is not caused by folate excess, it is caused by lack of B_{12}. Excess folate can bring forward the natural history.

Scott: Absorption is the problem.

Wald: Give enough, say 0.1 mg of vitamin B_{12} a day, and you could overcome the absorption problem.

Scott: That is not true. Treatment of pernicious anaemia with large doses of oral vitamin B_{12} does not work.

Wald: I would like to propose a package of monitoring measures that would complement a fortification policy. There would have to be:

1) A public health system of monitoring blood levels in the area where there was supplementation so that there was a mechanism to determine change in folate intake.

2) A system to monitor trends in neural tube defects, including terminations as well as births. This is not properly organized in many places.

3) A national register of neurological pernicious anaemia. This would record all complications and have a research resource to carry out case-control studies as necessary. I suspect that neurological complications of pernicious anaemia due to fortification of cereals with folic acid will not prove to be a problem, but the public interest in this justifies vigilance.

In this way, a start could be made that will at least in part do the job. The principle is set —implementation of fortification with monitoring activities and a degree of flexibility built in. Otherwise, no progress will be made.

Oakley: And women will continue to have babies with spina bifida.

Scott: When you are talking about a decision to fortify food for millions of people, you have to be absolutely certain that you have got it right. I'm not suggesting that CDC and the FDA aren't approaching this responsibly, but we have to speak up if we are concerned. Studies on elderly people in the US have shown that they have raised levels of methylmolonic acid, which is biochemical evidence of vitamin B_{12} deficiency, and that they also have neuropathy (Lindenbaum et al 1988). This neuropathy that occurs in vitamin B_{12} deficiency is very difficult to diagnose. The symptoms are just like those of old age: ataxia and loss of memory. If such elderly people were given high levels of folic acid, it might prevent the timely diagnosis of vitamin B_{12} deficiency through the emergence of anaemia and the neuropathy would proceed undiagnosed until the later stages, when it is largely irreversible.

Hall: Supplementation doesn't have to start in a whole country. There could be one area where this gets going and another where it doesn't, allowing comparison.

Stanley: I agree. Nick Wald's package, which I strongly support, should be applied in an area where there is a good population-based registry of birth defects and where you could easily set up a monitoring system for neurological pernicious anaemia. This is a terrific suggestion, because it provides you with the data on whether this is a significant problem or not.

Hall: The monitoring of blood levels could be done at a local blood bank.

Oakley: There is a unit at CDC that is interested in the diseases of older citizens. They are looking at setting up a national surveillance system for combined systems disease. We have socialized medicine for people over 65, and there is a diagnostic dataset that should have all cases of combined systems disease. We don't know how useful that's going to be, but I agree with Nick Wald that we should start some programme of fortification while we set up effective surveillance systems. If we have to choose between surveillance systems for neural tube defects and those for neurological issues, I would choose the latter. If there is a problem from the current proposed policy, which I doubt, we ought to be the first people to find it and fix it.

Mills: The FDA has considered the issues of monitoring. In order to get better information on folate and vitamin B_{12} status, they suggested that the National Health and Nutrition Examination Survey start measuring levels of folate in the serum and in red blood cells, levels of vitamin B_{12} in serum, also transcobalamins, methyl malonic acid and homocysteine. This would give a better idea of B_{12} status, albeit probably not a perfect one. They are also looking into working with the National Center for Health Statistics to review national data on hospital discharge and laboratory medical care data. This should identify patients with pernicious anaemia, other anaemias caused by a deficiency in vitamin B_{12}, combined subacute degeneration without anaemia, etc. There are two problems: one is how well those data are recorded. The other is that at the moment there are no funds to conduct the studies. In the US, there are a lot of very good ideas that are going to be rather hard to put into practice for financial reasons.

Hall: The way this issue is being presented in some of the lay journals is that folic acid causes neurological disorders. They are missing the fact that the neuropathology is caused by B_{12} deficiency, not by excess folic acid. This point should be clarified through public education.

Stanley: Is this issue of vitamin B_{12} deficiency and pernicious anaemia the only adverse effect of folic acid?

Oakley: Folic acid does not have an LD_{50}. You could take 100 g a day and it wouldn't hurt you.

Scott: There are some questions. What is the effect of giving folic acid to women in the first trimester of pregnancy? This hasn't created a big problem in the States, where a lot of women are on folate. There have not been any obvious signs in the pregnancy outcomes that we have already seen. It does require monitoring, because it would be a new venture.

There are some data on patients using anticonvulsants losing control of their epilepsy when they are given folic acid (Chanarin et al 1960, Reynolds et al 1968), but those are very suspect. Plus, people know they are on anticonvulsants, so they are a self-identifying group.

Mills: Those data come from a study where people with known epilepsy were given intravenous doses of folic acid while they had EEG monitoring. One of the eight people tested had abnormal EEG findings when the folic acid was given.

Dr Joseph Drage of the Neurology Institute made an interesting point in our discussions at NIH. He said there are 200 000 people in the US whose epilepsy is not well controlled. We do not know what will happen if all those people are exposed to food fortified with folate. Does a study that gave large doses of folic acid intravenously have any relevance to the level of exposure anticipated in food fortification? We do not know.

Scott: These patients are being treated by physicians, so they are under supervision.

Oakley: There are some countries where fortification just isn't an option because there is no centralized distribution of a common food. This is another public health problem. You cannot wait for people planning pregnancies. In the US, as many as half of pregnancies are not planned.

Cuckle: One approach is to target women who have had a miscarriage. About one quarter to one fifth of pregnancies will miscarry and many of the women will be pregnant again within the year. Since they will be receiving medical care immediately following the miscarriage, they form a captive group who can be given supplements straight away, and they are at a high risk of bearing children with neural tube defects.

Shurtleff: Nishimura et al's study in 1968 showed that miscarriages in the general population, including those of planned pregnancies, have a much higher rate of neural tube defects than pregnancies delivered at term. Monitoring the frequency of neural tube defects following fortification with folates would require knowledge of the background frequency amongst early terminated pregnancies as well as for full term pregnancies, because of the widespread use of prenatal diagnosis and elective termination of affected fetuses.

Hall: That is why an ideal place to start is an area that already has a monitoring system.

Oakley: Another point is that in the US, the vitamin manufacturers have spent lots of money trying to get everybody in the country to take vitamins; they have succeeded with 20% of the population. Adding a message that folic acid prevents neural tube defects may increase this, but it may not increase it much.

Hall: This confirms the importance of educating the public. Clearly, the physicians don't know what's going on. We have to make the public aware so that women go to their physician or pharmacist and ask for folic acid.

Morriss-Kay: What do others think about the possible facilitatory effects of Nick Wald's suggestion that folic acid should be called vitamin B_3?

Stanley: Carol Bower suggested it should be called vitamin B_4 because it is taken *before* pregnancy.

Czeizel: In Hungary, when we started our programme, we suggested using the name 'fetus-protecting vitamin'.

Scott: Changing the name might be a good suggestion. It used to be called vitamin BC. But we don't want to have a distraction.

References

Carmel R, Johnson CS, Weiner JM 1987 Pernicious anaemia in Latin Americans is not a disease of the elderly. Arch Intern Med 147:1995–1996

Chanarin I, Laidlaw J, Loughridge LW, Mollin DL 1960 Megaloblastic anaemia due to phenobarbitone. The convulsant action of therapeutic doses of folic acid. Br Med J 1:1099–1102

Lindenbaum J, Healton EB, Savage DG et al 1988 Neuropsychiatric disorders caused by cobalamin deficiency in the absence of anemia or macrocytosis. N Engl J Med 318:1720–1728

Nishimura H, Takano K, Tanimura T 1968 Normal and abnormal development of human embryos: first report of the analysis of 1,213 intact embryos. Teratology 1:281–290

Reynolds EH, Chanarin I, Matthews DM 1968 Neuropsychiatric aspects of anticonvulsant megaloblastic anaemia. Lancet 1:394–397

Scott JM, Weir DG 1981 The methyl folate trap. Lancet 2:337–340

Spina bifida: anticonvulsants and other maternal influences

Lewis B. Holmes

Embryology-Teratology Unit, Massachusetts General Hospital, Department of Pediatrics, Harvard Medical School, Boston, MA 02114-2696, USA

Abstract. At least two maternal conditions have been shown to be associated with an increased risk for the occurrence of anencephaly, spina bifida and encephalocele: insulin-dependent diabetes mellitus in the mother and maternal epilepsy for which anticonvulsant treatment is being used. Of 147 infants born at the Brigham and Women's Hospital with anencephaly, spina bifida and encephalocele among 123 489 live-born, still-born and elective terminations surveyed, 7.5% were infants of insulin-dependent diabetics and 0.7% had been exposed to anticonvulsants *in utero*. In addition to these proven causal associations neural tube defects have been postulated by some, but not all, studies as being more common among infants exposed to hyperthermia *in utero* and to drugs used to induce ovulation.

1994 Neural tube defects. Wiley, Chichester (Ciba Foundation Symposium 181) p 232–244

Spina bifida in the mouse has been shown to be caused by mutations in any one of several genes, as well as the interaction of mutant genes and non-genetic or 'environmental' factors (Copp et al 1990). Progress has been slower in identifying the responsible mutant genes and environmental factors in humans. I describe here the experience with several environmental factors that affect the occurrence of spina bifida and the related neural tube defects in humans, focusing on maternal conditions, such as insulin-dependent diabetes mellitus, maternal epilepsy and anticonvulsant medication, hyperthermia, drugs administered to induce ovulation, and other more speculative possibilities.

Materials and methods

To determine the frequency of anencephaly, spina bifida and encephalocele among infants of diabetic mothers and infants of mothers with epilepsy, we (Nelson & Holmes 1989) have used the Malformations Surveillance Program at Brigham and Women's Hospital. This surveillance programme has been carried out among all infants of at least 20 weeks gestational age born at this large university maternity hospital since 16 February 1972, except for a four year period (15 February 1975 to 31 December 1978). Since the advent of prenatal detection of fetal anomalies in the 1980s, we have tabulated separately information on those infants whose mothers had planned to deliver at another hospital but delivered at Brigham and Women's Hospital because a fetal malformation had been detected. We have designated these as maternal transfers. Some transfers were for an elective termination of pregnancy before 24 weeks of gestation and some were to facilitate surgical treatment immediately after birth at the adjacent Boston Children's Hospital. Each mother was interviewed, if possible, to determine: her transfer status, the results of all prenatal diagnosis studies, the family's medical history, any chronic diseases which the mother had, including diabetes mellitus and epilepsy, demographic characteristics and parental occupations. If the mother could not be interviewed in person, she was interviewed by telephone or the information was obtained from her medical record. The infant's medical record was used to determine birth status, length, head circumference, the major malformations that were present, the results of all diagnostic studies and, if the infant had died, the finding in the autopsy. Infants with unusual or atypical malformations were often examined personally or if they had died or had an elective termination the findings were reviewed with a pathologist.

Two additional sources of case finding were reviewed for this report. Details of infants of diabetic mothers with major malformations have been tabulated by Dr Michael F. Greene since 1983; some of his findings have been published (Greene et al 1989). We compared his list with that of the Malformations Surveillance Program and added any additional cases he had described. A cohort study of all infants born to women with epilepsy has been carried out since 1986 at Brigham and Women's Hospital and four other Boston area hospitals (Holmes et al 1990). We compared the list of malformed, drug-exposed infants identified by this cohort study with those identified by the Malformations Surveillance Program.

Information on other maternal conditions alleged to be associated with an increased risk of neural tube defects, including hyperthermia, the use of drugs to stimulate ovulation and other maternal conditions, has been obtained from a review of the medical literature.

Results

Infants of diabetic mothers

During the 15 years the Malformations Surveillance Program was conducted from 1972–1974 and 1979–1990, there were 123 489 live-born or still-born infants of at least 20 weeks gestational age born at Brigham and Women's Hospital in Boston. In addition, malformed infants delivered by elective termination in the second trimester of pregnancy were identified after 1979.

Among the mothers who had planned to deliver at this hospital there were 54 who had infants with spina bifida, 74 who had infants with anencephaly and 19 whose infants had encephalocele. None of the affected infants was a twin. The total prevalence of this group of neural tube defects was 146/123 489 or 1.2 per 1000. Excluded from this list are infants with lipomeningocele (3), cloacal exstrophy (4), chromosome abnormalities (6) and specific syndromes, such as the Meckel syndrome.

The frequency of insulin-dependent diabetes mellitus among the mothers of these infants with neural tube defects was: 12.2% for the infants with anencephaly; 3.7% for those with spina bifida and zero for the 19 infants with encephalocele. These malformations occurred also among infants whose diabetic mothers had planned to deliver at another hospital before the prenatal detection of the fetal abnormality, i.e. maternal transfers (Table 1).

TABLE 1 Association of two maternal conditions (insulin-dependent diabetes mellitus and maternal epilepsy treated with anticonvulsants) with the occurrence of infants with neural tube defects at Brigham and Women's Hospital, Boston among 123 489 births in the years 1972–1974 and 1979–1990

		Anencephaly		*Spina bifida*		*Encephalocele*	
Year	*Number of births*	*Non-transfer*	*Transfer*	*Non-transfer*	*Transfer*	*Non-transfer*	*Transfer*
1972	6052[a]	4	0	0	0	0	0
1973	6052[a]	9 (1)[b]	0	4	0	0	0
1974	6051[a]	4	0	3	0	3	0
1979	6263	6 (1)[b]	4	2	0	0	0
1980	6533	2 (1)[b]	4	2	0	0	0
1981	7363	2	0	4 (1)[b]	0	2	1
1982	7599	3	3 (1)[b]	3	1	0	0
1983	8207	2 (1)[b]	7	3	4 (1)[b]	1	0
1984	8803	6	12 (2)[b]	7	3	1	1
1985	9534	5 (1)[b]	5 (1)[b]	3	6	2	3
1986	9900	3	12	1	12 (1)[c]	2	2
1987	10 183	8 (2)[b]	10	4	13	1	0
1988	10 109	8	9	7 (1)[b]	22 (1)[c], (1)[b]	3	2
1989	10 349	2	18 (1)[c]	4	17 (1)[c]	1	0
1990	10 491	10 (2)[b]	16	7	12	1	3 (1)[b]
Total	123 489	74 (9)[b]	100 (4)[b] (1)[c]	54 (2)[b]	90 (2)[b] (3)[c]	19	12 (1)[b]

[a]The average annual total of the total births for the years 1972–1974.
[b]Mother had insulin-dependent diabetes mellitus.
[c]Infant exposed to anticonvulsant.

Maternal epilepsy

Among the infants ascertained by the Malformations Surveillance Program as having either spina bifida, anencephaly or encephalocele, there were four infants whose mothers had taken anticonvulsants during pregnancy. All of the infants were born to mothers who had been transferred, that is had intended to deliver elsewhere (Table 1). There were three valproate-exposed infants with spina bifida and one infant with anencephaly whose mother had taken carbamazepine and lithium. Since their mothers had planned to deliver at other hospitals, these affected infants cannot be used to estimate the frequency of occurrence of neural tube defects among the infants of women with epilepsy at this hospital.

The systematic study of maternal epilepsy (Holmes et al 1990) provides theoretically a better assessment of the association of anticonvulsant therapy with the occurrence of anencephaly, spina bifida and encephalocele. Between 1 August 1986 and 31 December 1992 we surveyed the postpartum medical records of 113 062 live-born and still-born infants at five Boston area hospitals, the largest of which was Brigham and Women's Hospital, the site of the Malformations Surveillance Program. 453 of these infants had been exposed to anticonvulsants during pregnancy. The specific exposures were: a) monotherapy with phenytoin (136), phenobarbital (101), carbamazepine (89), valproic acid (8), diazepam (6), clonazepam (5) and switching between drugs (29); b) polytherapy (79), which included eight infants exposed to valproic acid. 31 major malformations were identified among these 113 062 infants. Only one of the 31 infants with a major malformation had a neural tube defect, an infant with spina bifida whose mother had taken valproic acid as part of anticonvulsant polytherapy. While this cohort study provides a tabulation of the occurrence of major malformations among all the live-born and still-born infants surveyed, it does not identify those women with epilepsy who had planned to deliver at one of these hospitals but had an elective termination elsewhere after fetal anomalies had been detected by prenatal screening.

Discussion

In assessing the infants with anencephaly, spina bifida and encephalocele in association with diseases in their mothers, we have considered whether there is anything notable about the level of the lesion, the presence of other malformations or the relative frequency of specific phenotypes.

Insulin-dependent diabetes mellitus in the mother has been shown to be associated with a two- to threefold increase in the frequency of a wide variety of malformations (Greene et al 1989, Mills et al 1979). Mills and his associates (1979) noted that the period of development of these malformations appears to be quite early in embryogenesis, possibly 3–5 weeks post-fertilization. A higher frequency of anencephaly than spina bifida among infants of diabetic mothers was also noted and was confirmed in our tabulation at a Boston hospital (Table 1). One might expect to see more affected infants with multiple non-neural malformations and/or vertebral anomalies because these are more common among infants of diabetic mothers. We could not look for this correlation because most of the affected fetuses in the 1980s were detected prenatally and delivered electively by a destructive procedure before 24 weeks of gestation. Having her diabetes mellitus well controlled decreases significantly the risk of all types of fetal abnormalities for the pregnant diabetic (Mills et al 1988).

The association between exposure to the anticonvulsant valproic acid and the occurrence of spina bifida was observed in 1982 in a malformations surveillance programme in Lyon, France (Robert & Guibaud 1982). Retrospective studies of other malformations surveillance programmes were used to estimate that the absolute risk of occurrence was 1–2% (Lindhout & Meinardi 1984). In a recent prospective cohort study of 297 pregnancies in 261 women who had taken anticonvulsants in Holland, it was observed that spina bifida occurred in 5.4% of the pregnancies. Spina bifida occurred only among the 92 infants whose mothers had taken valproate, 60 as monotherapy and 32 as polytherapy. A correlation with the dose of valproic acid taken during the pregnancy was noted. None of the 92 fetuses exposed to valproic acid had anencephaly, which is consistent with the very low rate of occurrence of anencephaly in infants of women exposed to this drug. The ratio of spina bifida:anencephaly is 33:1 (Lindhout et al 1992). This experience confirms that valproic acid, as a human teratogen, seems most likely to be associated with spina bifida rather than other types of neural tube defects. Infants exposed to valproic acid also have an increased frequency of craniofacial abnormalities, heart defects and longitudinal limb deficiencies (Omtzigt et al 1992, Verloes et al 1990, Winter et al 1987), but the overall risk of all malformations associated with valproic acid has not been established.

Pregnant women who take the anticonvulsant carbamazepine also have an increased risk that their infant will have a neural tube defect. The absolute risk was estimated to be 1.0% in a review of 21 cohort studies in which 10 of 1132 drug-exposed infants were identified as having a neural tube defect (Rosa 1991). The frequency of anencephaly among fetuses exposed to carbamazepine appears to be much less than expected, but the rate has not been determined.

The folic acid antagonist aminopterin and the fertility drugs or drugs used to stimulate ovulation, such as clomiphene, have also been postulated to cause neural tube defects. Aminopterin was used many years ago as an abortifacient. Several infants have been reported who were born from pregnancies which continued after the drug was used. Two infants with anencephaly and encephalocele had been exposed at 4–6 and 7 weeks of gestation, respectively; they had no other malformations (Milunsky et al 1968). Aminopterin-exposed infants often had other non-neural malformations.

There is controversy as to whether exposure to drugs used to stimulate ovulation is associated with an increased risk for neural tube defects (Cornel et al 1990, Mills et al 1990, Mills 1990, Milunsky et al 1992, Van Loon et al 1992). Six follow-up studies of 1642 children born after stimulation of ovulation with clomiphene identified three children with a neural tube defect. While this would mean a birth prevalence rate of 0.18%, that is higher than expected, the 95% confidence interval was large and the findings were considered equivocal (Cornel et al 1990). In a case-control study of 571 women who had had a fetus with a neural tube defect, 546 women with a fetus with other abnormalities and 573 women whose infants were normal, the use of fertility drugs was not significantly higher in the first group (Mills et al 1990). An important potential confounder in these studies is the possibility that the causes of the infertility in some women could increase their risk of having an infant with a neural tube defect. This question will continue to confront families who utilize *in vitro* fertilization, as some studies have shown an increased rate of affected infants (Cornel et al 1990). However, a much larger number of exposed infants would be needed to determine the risk with adequate certainty.

Hyperthermia is another postulated cause of neural tube defects in humans. Studies on animals show clearly that short periods of hyperthermia are associated with a significant increase in the frequency of exencephaly and spina bifida. Some studies of human pregnancies have suggested that this association is significant, but others have not (Milunsky et al 1992, Smith et al 1978, Saxen et al 1982). Unfortunately, it has been very difficult to obtain accurate information on a sufficient number of exposed and unexposed human pregnancies.

Women with morbid obesity and who have had gastric bypass surgery have been observed to have more infants with neural tube defects than would be expected by chance (Haddow et al 1986, Martin et al 1988). This correlation could reflect a vitamin deficiency in these women, as those with affected infants had not taken a recommended vitamin supplement.

Two general observations suggest that there may be other unidentified genetic or environmental factors which could affect pregnant women and their risk of having children with neural tube defects. Seasonal fluctuations have been observed in the birth of infants with neural tube defects (Fraser et al 1986). These birth defects are also more common among infants from lower socioeconomic classes in England and in the United States (Blackburn et al 1987).

References

Blackburn BL, Fineman RM, Ward RH 1987 Sociodemographic factors influence incidence of neural tube defects (NTD) and congenital hydrocephalus (CH) in Utah, 1940–1979. Am J Hum Genet (suppl) 41:A49 (abstr)

Copp AJ, Brook FA, Estibeiro P, Shum ASW, Cockroft DL 1990 The embryonic development of mammalian neural tube defects. Prog Neurobiol 35:363–403

Cornel MC, Ten Tate LP, Te Meerman GJ 1990 Association between ovulation stimulation, in vitro fertilization, and neural tube defects. Teratology 42: 201–203

Fraser FC, Frecker M, Allderdice P 1986 Seasonal variation in neural tube defects in Newfoundland and elsewhere. Teratology 33:299–303

Greene MF, Hare JW, Cloherty JP, Benacerraf BR, Soeldner JS 1989 First-trimester hemoglobin A_1 and risk for major malformation and spontaneous abortion in diabetic pregnancy. Teratology 39:225–231

Haddow JE, Hill LE, Kloza EM, Thanhauser D 1986 Neural tube defects after gastric bypass. Lancet 1:1330

Holmes LB, Harvey EA, Hayes AM, Brown KS, Schoenfeld DA, Khoshbin S 1990 The teratogenic effects of anticonvulsant monotherapy: phenobarbital (Pb), carbamazepine (CBZ) and phenytoin (PHT). Teratology 41:565

Lindhout D, Meinardi H 1984 Spina bifida and in utero exposure to valproate. Lancet 2:396

Lindhout D, Omtzigt JGC, Cornel MC 1992 Spectrum of neural tube defects in 34 infants prenatally exposed to antiepileptic drugs. Neurology 42(suppl 5):111–118

Martin L, Chavez CF, Adams MJ Jr et al 1988 Gastric bypass surgery as maternal risk factor for neural tube defects. Lancet 1:640–641

Mills JL 1990 Fertility drugs and neural tube defects: why can't we make up our minds? Teratology 42:595–596

Mills JL, Baker L, Goldman AS 1979 Malformations in infants of diabetic mothers occur before the seventh gestational week. Diabetes 28:292–293

Mills JL, Knopp RH, Simpson JL et al 1988 Lack of relation of increased malformations rates in infants of diabetic mothers to glycemic control during organogenesis. N Engl J Med 318:671–676

Mills JL, Simpson JL, Rhoads GG et al 1990 Risk of neural tube defects in relation to maternal fertility and fertility drug use. Lancet 336:103–104

Milunsky A, Graef JW, Gaynor MF Jr 1968 Methotrexate-induced congenital malformations. J Pediatr 72:790–795

Milunsky A, Ulcickas M, Rothman KJ, Willett W, Jick SS, Jick H 1992 Maternal heat exposure and neural tube defects. JAMA (J Am Med Assoc) 268:882–885

Nelson K, Holmes LB 1989 Malformations due to presumed spontaneous mutations in newborn infants. N Engl J Med 320:19–23

Omtzigt JGC, Los FJ, Grobbee DE et al 1992 The risk of spina bifida aperta after first-trimester exposure to valproate in a prenatal cohort. Neurology 42(suppl 5): 119–125

Robert E, Guibaud P 1982 Maternal valproic acid and congenital neural tube defects. Lancet 2:937

Rosa FW 1991 Spina bifida in infants of women treated with carbamazepine during pregnancy. N Engl J Med 324:674–677

Saxen L, Holmberg PC, Nurminen M, Koosma E 1982 Sauna and congenital defects. Teratology 25:309–313

Smith DW, Clarren SK, Harvey MAS 1978 Hyperthermia as a possible teratogenic agent. J Pediatr 92:878–883

Van Loon K, Besseghir K, Eshkol A 1992 Neural tube defects after infertility treatment: a review. Fertil Steril 58:875–884

Verloes A, Frikiche A, Gremillet C et al 1990 Proximal phocomelia and radial ray aplasia in fetal valproic syndrome. Eur J Pediatr 149:266–267

Winter RM, Donnai D, Burn J, Tucker SM 1987 Fetal valproate syndrome: is there a recognizable phenotype? J Med Genet 24:692–695

DISCUSSION

Stanley: The rates of birth defects in the infants of women with epilepsy that you showed are not particularly higher than the rates that we find in our Birth Defects register in Western Australia. Apart from neural tube defects, what were the rates of malformations such as heart defects?

Holmes: There were 10 infants with heart defects out of 453 infants.

Stanley: This rate is not greater than that we see for cardiac defects in the general population. Your work agrees with the results of our studies in Western Australia of diabetes being an important maternal teratogen, but epilepsy contributing far less to the incidence of neural tube defects than we had expected.

Opitz: Do mothers with well-controlled insulin-dependent diabetes still have a risk of bearing an infant with anencephaly or spina bifida? If the levels of haemoglobin A_1c before conception and during early pregnancy are normal, is there still a risk?

Holmes: The Diabetes in Early Pregnancy Study was carried out in five centres in the US between 1980 and 1985 (Mills et al 1988a). Almost 400 women with insulin-dependent diabetes were recruited soon after they conceived. These women, as a group, had their diabetes under better control than does the average diabetic woman. Yet, they still had a frequency of major malformations roughly twice the rate for the non-diabetic group. The rates were 2.1% among non-diabetic controls, 4.9% among the women who enrolled very early and tended to have pretty good control, and 9.0% among the women who enrolled late. The women who enrolled late often told us that they had not wanted to enrol early in the study, when they were considered as part of a recruitment effort.

At the beginning of the study, we were concerned that the examiner of these infants would not be blind as to the status of the mother, because an experienced paediatrician would recognize the physical features of an infant of a diabetic mother. However, we found this was not the case, as the mothers who enrolled in the study were more likely to have their disease under good control and their infants did not show plethora or macrosomy, the recognizable features of this phenotype. So, the 'blind' experimenter remained unaware of the exposure status of these infants. Yet, as a group they still had twice the frequency of major malformations.

This study did not show a correlation between haemoglobin A_1 levels and the occurrence of major malformations (Mills et al 1988a). Others feel that there is one.

Opitz: Fuhrmann et al (1983) in former East Germany studied a large cohort and found that if the haemoglobin A_1c levels are reduced to as near normal as possible, one can reduce the incidence of anomalies in infants of those mothers.

Mills: Our diabetic group wasn't perfectly controlled, so there is the possibility that if you could normalize metabolic control absolutely in the organogenesis period, you could reduce that doubled risk down to no increased risk. However, no study to my knowledge has ever been done showing excellent control and no excess risk of malformations that had any kind of control group. The Fuhrmann et al (1983) study found an astonishingly low rate of malformations—one malformation in 128 pregnancies, which is even better than you would expect in a non-diabetic population. So I'm reluctant to say that if you can control diabetes perfectly, you can eliminate the risk, but there is some reason to think that that could be the case.

Stanley: The results from our studies in Western Australia confirm that infants of diabetic mothers have a much higher rate of all birth defects, not just neural tube defects. The group that are at extremely high risk in Western Australia (and I presume in some parts of America as well) are the diabetic Aboriginal women. Aboriginal people in Australia have extraordinarily high rates of diabetes. We seem to get an association of birth defects even with gestational diabetes in these women. Carol Bower's work suggests that perhaps up to 7% of birth defects in children of Aboriginal women could be prevented, if there was better diagnosis and control of diabetes.

Opitz: The paper of Mills et al (1979) was to me almost revolutionary. The point made was that the malformations all arose during blastogenesis—during the first seven gestational weeks. The specific combination of anomalies comprises quintessentially those seen in associations. Since nobody at this moment can say which factors in diabetic women are responsible for these anomalies, it is probably useful to think of the constellation of anomalies of infants of diabetic mothers as a true association.

In native American populations, there is good evidence that insulin-resistant diabetes (not insulin-dependent diabetes) is associated with an increased incidence of spina bifida. In this population there is a confounding factor of alcohol intake. Frequently, when I see a native American child with spina bifida, the mother was diabetic and exposed to alcohol.

Holmes: Are the Australian aborigines afflicted with a high rate of alcoholism?

Stanley: Yes, but the lack of diabetic control is unbelievable—and often in pregnancy. There is an extraordinarily high rate of alcoholism as well.

Mills: The NICHD Walnut Creek birth defects study looked at a primarily middle class population in California. The data were collected on first trimester drinking at the time of registration for prenatal care (Mills & Graubard 1987). There was a self-administered questionnaire, which made people a little less likely to lie about how much they drink. There were about 34 000 pregnancies. There was one child with a neural tube defect in the heaviest drinking groups. These contained about 950 women who drank at least one drink per day on the average every day during pregnancy. In California the general population rate of neural tube defects is about 0.9 per 1000. So this rate in the people who were consistent drinkers is about what's expected for the Californian population.

Another way to look at these results is that in about 34 000 pregnancies, you might expect at least 35 cases of fetal alcohol syndrome (assuming a rate of one or greater per 1000 pregnancies). We found only one baby with a neural tube defect. So there is not a close association between fetal alcohol syndrome and neural tube defects in this population.

This population is not native American; it was 13% Hispanic. It was probably very well nourished. So there may very well be an interaction between the diet and other risk factors.

Shurtleff: Reports from San Francisco (Strassburg et al 1983) and covering the Los Angeles areas of California (Sever et al 1982) show that Hispanic women of Mexican origin have a higher rate of pregnancies affected by neural tube defects than women of Caucasian origin. Clarren & Smith (1978) reported only one in 35 infants with fetal alcohol syndrome will have meningomyelocele.

Oakley: This discussion about haemoglobin A_1c levels and the control of diabetes was going on at the the same time as the discussion on Smithells' studies and other observational studies with regard to folate. Almost all the studies are observational; the data are not randomized. Yet, the medical community has behaved as if the diabetes study were a randomized controlled trial—they have assumed that it is known that control of diabetes prevents birth defects. The medical community did not conclude that folic acid prevents neural tube defects until after the randomized controlled trial.

Copp: I'm interested in the residual high risk of neural tube defects, even when the diabetic condition is controlled. Have there been any studies of infants of first degree relatives of diabetics in which there is no actual diabetes? Diabetes is a polygenic condition; some of the genes could be those we mentioned earlier in the meeting as modifying susceptibility to neural tube defects.

Holmes: We could not address this question because we did not enrol enough families. As there is a 5% frequency of major malformations among well-controlled diabetics and a 2% frequency in the general population, it is very hard to find enough diabetic mothers who had a child with a major malformation and whose non-diabetic sister also had a child with a major malformation.

Hall: In British Columbia we have good data on 700 families with a child with a neural tube defect. We have looked for several suggested related familial factors, such as Downs' syndrome, twinning, other anomalies and diabetes. We find no increased risk.

Mills: Peter Bennett and his colleagues (1979) looked at families in which the father had diabetes but not the mother. They found no excess risk of a child with birth defects. Other studies looked at families where the mother developed diabetes after already having children. There was not an increased risk in the children born after the diabetes started (Mills 1982). These studies suggest that the association between diabetes and birth defects is not primarily genetic.

HLA type is an important risk factor for type I diabetes. We wondered whether genetic factors could play a secondary role in teratogenesis. We have found somewhat higher malformation rates in infants whose mothers had both of the HLA types, DR3 and DR4, that are associated with diabetes (Simpson et al 1993). However, the numbers are very modest; we've never been able to convince ourselves that there was a definite association.

In the Diabetes in Early Pregnancy study, there are a few other things of interest. One is the risk for spontaneous abortion (Mills et al 1988b). All the subjects enrolled within 21 days of fertilization, so we were able to look at loss rates in both the diabetic and the control groups, starting within 21 days. Overall, there was no difference. However, among the small number of diabetic women who were not well controlled there was a remarkable increase in losses with increasing level of glycosylated haemoglobins. The poorer the mother's metabolic control, the higher the loss rate. There was a very strong correlation there. As shown in our paper (Fig. 1), within the normal range of glycosylated haemoglobin (−3 to +3 standard deviations from the control mean), the loss rate is low in both the diabetic and control groups. It is slightly lower in the diabetic group, but the difference is not statistically significant. As the glycosylated haemoglobin goes above 3 SD from the control mean, indicating worsening diabetic control, the loss rate increases dramatically. Thus, diabetic women are at greatly increased risk for miscarriage if they are in poor control, but no increased risk if they are in excellent control.

I looked at several of the basal body temperature charts, which included information on which days intercourse occurred, and I looked at the HCG (human chorionic gonadotropin) tests indicating when the women had a positive pregnancy test. Despite having all this information, there were a number of cases where I could not tell when the woman had conceived. So reports that state categorically a woman was 13 or 27 days postconception, I take with a grain of salt. I question whether anyone can date conception *in vivo* that accurately except in special cases. We were also surprised that the loss rate starting within 21 days of fertilization was only 16%.

Oakley: Lew, could you speculate on what the phenotype of the folic acid-preventable neural tube defects is going to be?

Holmes: We don't see many live-born, term infants with neural tube defects any more. This reflects the fact that most affected infants are diagnosed prenatally and their parents opt for an elective termination of pregnancy. Because of this trend, we will not be able to look clinically for a change in the phenotype once all pregnancies have been supplemented with folic acid. One could look for an increase in, for example, the frequency of non-neural anomalies. Hopefully, investigators in countries where prenatal diagnosis and elective termination of affected infants is less common will be able to look for such a change.

Oakley: Presumably, the other causes would not be part of the folic acid-preventable phenotype. Dr Ron Lemire thought that skin-covered defects were not likely to be prevented by folic acid.

Shurtleff: I agree. The sacral-level lesions that occur in the area of secondary neurulation, both skin-covered and non-skin-covered, might be the phenotype that escapes prevention by folate supplementation. I don't know whether sacral-type lesions in the area of secondary neurulation are due to the same mechanisms as lesions that occur higher up the spinal cord in the region of primary neurulation.

Holmes: One of the mild phenotypes which I excluded was the lipomeningocele. We don't have a good way of knowing whether that's part of the anencephaly–meningomyelocele spectrum.

Seller: In the early days of the vitamin supplementation project, I had the opportunity to examine the fetuses of women who had recurrences despite full supplementation. There were not many and they had just ordinary neural tube defects. There was one with an ordinary anencephaly, one with complete craniorachischisis. We had speculated that the ones that weren't prevented by vitamin and folate supplementation might be a minor form of neural tube defect that would otherwise have been more severe. We thought that if supplementation were not completely successful, we might have a lot of very small lesions, but that's not what we found.

Copp: So it is not that folate is preventing the least severe defects and the ones getting through are the most severe ones that you really can't do anything about.

Holmes: Andrew, when you decrease the frequency of spina bifida in *curly tail* mice by various means, do you see a shift in the severity?

Copp: Yes, towards the less severe phenotype. A higher proportion have only curly tails and fewer have open spina bifida. But there you are dealing with a single aetiological group. In humans, we are probably dealing with multiple aetiological groups. It could be that folate is acting selectively on one aetiology or a group of aetiologies and others are unresponsive.

Holmes: Godfrey, you will have an excellent opportunity to answer your own question through the study in northern China.

References

Bennett PH, Webner C, Miller M 1979 Pregnancy, metabolism, diabetes and the fetus. Excerpta Medica, New York

Clarren SK, Smith DW 1978 The fetal alcohol syndrome. N Engl J Med 98:1063–1067

Fuhrmann K, Reiher H, Semmler K, Fischer F, Fischer M, Glöckner E 1983 Prevention of congenital malformations in infants of diabetic mothers. Diabetes Care 6:219–223

Mills JL 1982 Malformations in infants of diabetic mothers. Teratology 25:385–394

Mills JL, Graubard BI 1987 Is moderate drinking during pregnancy associated with an increased risk for malformations? Pediatrics 80:309–314

Mills JL, Baker L, Goldman AS 1979 Malformations in infants of diabetic mothers occur before the seventh gestational week. Diabetes 28:292–293

Mills JL, Knopp RH, Simpson JL et al 1988a Lack of relation of increased malformations rates in infants of diabetic mothers to glycemic control during organogenesis. N Engl J Med 318:671–676

Mills JL, Simpson JL, Driscoll SG et al 1988b Incidence of spontaneous abortion among normal women and insulin-dependent diabetic women whose pregnancies were identified within 21 days of conception. N Engl J Med 319:1617–1623
Sever LE, Sanders M, Monsen R 1982 An epidemiological study of neural tube defects in Los Angeles County. Teratology 25:315–321
Simpson JL, Mills JL, Ober C et al 1993 Is there a genetic basis for diabetic teratogenicity? In: Teoh E-S, Ratnam SS, MacNaughton M (eds) Fetal physiology and pathology. Parthenon Publishing, New York, p 119–125
Strassburg MA, Greenland S, Portigal LD 1983 A population-based case-control study of anencephalus and spina bifida in a low-risk area. Dev Med Child Neurol 25:632–641

General discussion

Environmental factors affecting neural tube defects

Oakley: We have heard at this symposium that retinoic acid (vitamin A) affects normal neurulation (Morriss-Kay et al, this volume). It might therefore be important epidemiologically in the incidence of neural tube defects. My understanding is that not only is there a developmental stage-specificity about whether retinoic acid causes a given anomaly, but that the dose required changes dramatically with time. Early embryos are exquisitely sensitive to retinoic acid; for older embryos you need larger doses to have an effect. Would low levels of retinoic acid in humans cause something like holoprosencephaly? The question arises because of the development of a skin cream that contains retinoic acid, which could be distributed worldwide to almost any woman who didn't want to have wrinkles. What implication would that have for the incidence of neural tube defects?

Nau: We studied quite a few patients who had been treated with this cream. We did not see a significant change in endogenous retinoic acid levels. In normal human plasma, there is about 1 ng/ml all-*trans*-retinoic acid, about 1 ng/ml 13-*cis*-retinoic acid and about 2 ng/ml 4-oxo-retinoic acid. Other possible sources of retinoic acid, such as vitamin pills or eating vitamin A-rich food (e.g. liver), will have a much greater influence on endogenous retinoic acid levels than topical application of a cream.

Hall: What scares me is the prospect of a spectrum of effects from prosencephaly to just a few neurons being knocked out. If the spectrum was that broad, the effect might not be recognized in surveys looking for more specific deleterious effects, such as neural tube defects.

Stanley: We have at least one Australian case report of a woman who reported using this cream excessively and who had a baby with a major birth defect involving neural crest-derived structures.

Opitz: With respect to public health hazards in neural tube defects, in our part of the world (Montana, USA) alcohol consumption appears to be the worst offender. Among native American populations, I consider all neural tube defects to be alcohol related until proven otherwise. It's not just the open meningomyeloceles that we see or spina bifida occulta, there have even been a few cases of holoprosencephaly due to alcohol. In the native American population that may amount to many individuals.

Holmes: John, do you see the other morphological signs of the fetal effects of alcohol in addition to the neural tube defects?

Opitz: Yes. The literature is now suggesting more and more that you can see the central nervous system behavioural phenotype without seeing any gross or minor anomalies. I concur 100% with that. Sterling Clarren's experimental work in monkeys has shown that also. But usually, where there is a severe lesion, such as a neural tube defect, you see the minor anomalies as well.

Hall: There are two aspects to consider. One is that the phenotypic effects are probably ethnospecific; for instance, North American Indians seem to show the features of fetal alcohol syndrome to a greater degree than other ethnic groups. Secondly, the phenotypic features seem to become less obvious with age; children are more likely to show the 'fetal alcohol' phenotype than teenagers or adults. So we could be in danger of missing it.

Shurtleff: The cases of fetal alcohol syndrome that we called to the attention of Dave Smith and Sterling Clarren were seen amongst native Americans or Athabascan nations in the central plains of North America. Most of these families were of the Crow nation of Montana: they are of the same genetic stock that John Opitz sees.

The neural tube defects that we see in the west coast native Americans, Salish, and going north to the Kuakiautle, Haida and Tshmanians, the other west coast natives, are much more like those in the Polynesians who have frontonasal encephaloceles. Documentation goes back into antiquity from the ancient designs that B. Holm has described amongst totem poles erected to commemorate their families (personal communication 1985). We have observed the fetal alcohol syndrome only in infants of very severely alcoholic mothers. The defect almost always involves the caudal, not the cephalic, neuropore. The offspring are microcephalic and have marked evidence of the fetal alcohol syndrome. I have not seen a child who has a mild form of fetal alcohol syndrome and a neural tube defect.

Opitz: We have several such mild cases in Montana. One is roller skating champion with her lame legs.

Mills: Where does this linkage between fetal alcohol syndrome and neural tube defects come from? Is it just something that is seen clinically or have there been large-scale studies that link fetal alcohol syndrome and neural tube defects?

Opitz: These are just clinical observations.

Mills: We looked at the birth defects data from the Kaiser Permanente Walnut Creek Birth Defects Study. This study was commissioned by the NICHD for 1974–1977. The analysis that I did looked at major birth defects in relation to reported alcohol consumption. We found no association between neural tube defects and heavy levels of drinking in this population. Genitourinary defects were the only ones that seemed to be related among the major gross defects. These data were collected before fetal alcohol syndrome was widely recognized, so we didn't get any specific information on that. This covered over 30 000 pregnancies, so there should have been enough cases to have shown an effect, if it was pronounced.

Stanley: If one in 1000 pregnancies might have fetal alcohol syndrome (depending on the level of drinking in the population), and if neural tube defects are associated with 5–10% of cases of fetal alcohol syndrome, one might not detect an association amongst 30 000 pregnancies. In most of the studies that are done on Caucasian pregnant women, there are very low levels of alcohol consumption, but I don't know about Walnut Creek. The only groups in which you could do those sorts of studies are those that report that they drink excessively or are known to do so. The studies on Australian Aborigines find that there is a significantly higher rate. I think we probably have a diagnostic bias towards over-diagnosis of fetal alcohol syndrome in Aborigines, because we look for it more in them. But there is definitely a higher proportion of neural tube defects in Aboriginal births than we would expect in the total population. Of course, some of that might relate to nutritional problems as well.

Mills: We did power calculations when we did our study; we wouldn't have found a subtle difference, but we would have found a major increase. There may be genetic susceptibility factors, but I'm not aware of any case-control study or cohort study that reported an excess of neural tube defects associated with fetal alcohol syndrome. It might be useful to look at this in your population, where you have a good denominator.

Stanley: The numbers are only small, as there are only 1400 Aboriginal births per year in our Western Australian population, but there is certainly an association there. We do collect total malformation data up to age six years, so we could look at that in more depth.

Scott: There is a correlation between alcohol and folate deficiency and it's difficult to distinguish which is having an effect. In the American population, folate deficiency in men is almost diagnostic of alcoholics.

Stanley: I agree that alcoholism is very important, but retinoic acid is a known teratogen. Before there is widespread introduction of a teratogen amongst the female population, it is important to know about its absorption.

Hall: I would have thought the mouse mutants would be particularly important for this sort of study. If a mouse mutant could be identified that is susceptible to retinoic acid, then there might be a way of screening which people would be most susceptible to folic acid deficiency.

Juriloff: We have created, by selection and inbreeding, a mouse strain, SELH, in which the frequency of spontaneous exencephaly is high (Juriloff et al 1989). This exencephaly is not a single gene trait; it is a multifactorial threshold trait that has been fixed in the strain. We gave these mice a low dose (5 mg/kg) of retinoic acid to parallel Mary Seller's study on *curly tail* mice (Seller et al 1979). After treatment on Day 8.5 of gestation, the frequency of exencephaly was increased from 10% to about 50% by a relatively low dose of retinoic acid that we hoped would prevent neural tube defects. This teratogenic response was much higher than that of two normal strains in the study.

Hall: How many genes are involved?

Juriloff: Two or three additive loci. Basically, the risk of exencephaly goes up as the mice accumulate more of the 'liability' alleles at these loci.

Hall: The message I get from the early part of this meeting is that in animal models, there are not many neural tube defects attributable to single genes. We find single gene defects in the mouse and in humans, but in the other animal models not many single genes responsible for defects have been recognized yet. The studies in other animals are much more of the type which correlate factors that are interacting at a particular stage in embryogenesis. This discussion is about specific environmental factors and how, potentially, they interfere with neural tube development. We haven't been hearing about specific genes for susceptibility to particular environmental factors, such as alcohol or vitamin A.

Shum: In the *curly tail* mouse embryos, mild hyperthermia prevents spinal neural tube defects (Copp et al 1988).

Copp: It can also cause them: there is a craniocaudal difference. The incidence of neural tube defects is increased in the cranial region (Seller & Perkins-Cole 1987) but decreased in the caudal region by hyperthermia (Copp et al 1988). There are many studies showing that hyperthermia predisposes to cranial neural tube defects in mice. There are also strain differences (Finnell et al 1986), therefore multiple genes influence susceptibility to hyperthemia.

Trasler: Our *splotch* mouse is very susceptible to retinoic acid-induced neural tube defects. We now know that the human Waardenburg syndrome is homologous with *splotch*; they are both due to mutations in *Pax-3*. One could speculate that someone with Waardenburg syndrome might be more susceptible to retinoic acid. Once we start getting the mouse homologues of the human genes and develop tests in the mouse for major, dominant genes as well as for the multifactorial genes studied by Diana Juriloff, we may be able to predict which individuals are predisposed to produce children with neural tube defects.

Oakley: Unlike cystic fibrosis, most human neural tube defects are highly unlikely to be single gene lesions. The message from the folic acid trials is that folic acid prevents 50–66% of neural tube defects (Czeizel & Dudás 1992, MRC Vitamin Study Research Group 1991). So the issue is folic acid. No matter what the other models are, the question is: what can we learn that will help us understand more about the mechanism related to folic acid? It may be like the anaemias: there may be 500 different genetic causes for anencephaly and we will find them as we go along, but overall the non-folic acid-preventable spina bifida and anencephaly are going to contribute a small proportion of the total human burden of neural tube defects.

Juriloff: Are you asking for an animal model that responds to folic acid?

Oakley: Yes; to help us determine how folate works.

Juriloff: I imagine most of us are now trying folic acid on our various models. We have to wait for the results of those experiments.

Oakley: There is some relevant epidemiological evidence. The rate of neural tube defects among black infants, who in the US are usually disproportionately poor, is one-third that in the white population. There is some important biology underlying that, which nobody has ever studied. It would be nice to work out how the protective effect works. It would be nice to have a mouse model to give us a clue. One of the negative things about mouse models is that it isn't yet clear which mouse models are actually important for what's observed in people.

Another interesting observation relates to valproic acid. Only about 1% of fetuses exposed to valproic acid have neural tube defects (Lindhout & Meinardi 1984). This suggests there is some human phenotype which is susceptible.

Hall: Those are also lower defects aren't they? Doesn't valproic acid cause almost entirely spina bifida?

Lindhout: As far as I know, only one case of anencephaly has been reported, against 50–80 cases of spina bifida.

I agree there must be a reason only such a small proportion (1–2%) of fetuses exposed to valproic acid react with a neural tube defect. In a prospective prenatal cohort, we evaluated during five years more than 80 pregnancies exposed to valproate. These were taken into the cohort without any prior knowledge of fetal conditions. We were surprised that five pregnancies were affected, including a monozygotic twin pair concordantly. Within this small cohort, we found a clear dose relationship: fetuses with spina bifida were exposed to 1640 mg valproate/day; the normal cases were exposed to 960 mg/day. We are convinced that the risk is higher than 1%, especially with high daily doses.

Oakley: In humans, even where there are epidemics of neural tube defects, the maximum absolute risk is around 1%. This is another 1% kind of number. Is it the same susceptibility or not? Presumably, a lot of the epidemic form is somehow related to folic acid. This would be my bet. Again, is there a sensitive strain of humans? Finally, we need to find out where and how folic acid acts.

Lindhout: The maternal metabolism of methionine/homocysteine and the role of folic acid is intriguing because it might help explain genetic as well as environmental effects.

A group in Nijmegen in The Netherlands examined the homocysteine levels in the blood of mothers who have had at least one child with a neural tube defect (Steegers-Theunissen et al 1991). They discovered maternal cystinaemia after oral methionine loading in 20–30% of the cases. The levels are in the same range as those in heterozygotes for homocystinuria. However, the cystathionine synthase levels are normal, indicating another kind of defect. This study needs to be replicated by other groups. If the finding is true, it would explain at least 20% of cases of neural tube defects in the population, the sensitivity to folic acid supplementation and the relative predominance of positive maternal family histories.

Holmes: Is it only after oral loading with homocysteine that this maternal cystinaemia is observed?

Lindhout: Yes.

Hall: There is incredible heterogeneity in a given family with Waardenburg syndrome. Is it possible that the difference in expression of Waardenburg syndrome without neural tube defects is also somehow related to the level of vitamin A? This variation in expression seems to be separate from neural tube defects, but it would be really important for those families. If one could modulate the mother's environment during the pregnancy, it may make a big difference in how the gene is expressed in the fetus with Waardenburg syndrome.

Trasler: Since the *Pax-3* mutation was published as the cause of Waardenburg syndrome, clinicians are noting neural tube defects associated with Waardenburg syndrome families. This was reported only very seldom before, now it is being reported quite frequently in the literature.

Oakley: What kind of neural tube defects?

Opitz: Lumbosacral. Chatkupt et al (1993) reported a family with at least two, possibly three, such cases.

Lindhout: This is an important issue. Of course there is a reporting bias, but the speed with which these reports appeared and the number of them suggest that there may be an increased risk. If this can be confirmed, we may compare those cases with neural tube defects with other patients with Waardenburg syndrome who have normal neural tube development. Then we can look for differences in exposure to, for example, retinoic acid during pregnancy. This can be a very powerful tool to sort out whether these kinds of mechanisms play a role. Approaching the problem in the monogenic model may help you to solve the multifactorial cases. Finding a solution for the multifactorial ones directly will be very difficult.

Copp: It is not surprising that there is a very low frequency of neural tube defects in families with Waardenburg syndrome, because the families consist predominantly of heterozygotes. *splotch* heterozygotes do not have neural tube defects unless they are treated with retinoic acid.

Oakley: Does anybody know what the *splotch* gene does?

Copp: It encodes a transcription factor.

Lindhout: And it's expressed especially in those tissues that are affected in three types of Waardenburg syndrome.

Opitz: What are the racial differences in neural tube defects that are observed throughout the world? As I understand, in Thailand there are mostly frontonasal encephaloceles, is that correct? What other such populations exist?

Hall: That's true for many Asians—from the North-West Indians of North America, across the Aleutians and all the way down the Asian coast, also for some of the Japanese. Several Asian groups have an increased incidence of frontonasal encephaloceles. Interestingly, the incidence of lumbosacral defects appears to be low in some of those populations.

Oakley: They might be different disorders even though we call them all neural tube defects. They could have different causal mechanisms. They might not be related to folate at all.

Opitz: Has folic acid been tested yet on those populations?

Oakley: Not that I'm aware of.

Opitz: Last week, I evaluated a new patient with a large frontonasal/ethmoidal encephalocele, non-syndromal, non-native American. The mother is 16 years old. Is this the kind of mother we give folic acid to prevent recurrence?

Hall: Of course, you would use folic acid preconceptionally and hope it was helpful. In our experience in British Columbia, we have seen no familial recurrences of non-syndromic encephaloceles and no familial recurrences of craniorachischisis. I can't say there are no recurrences, but we haven't seen any. It seems to me that encephaloceles are a different type of neural tube disorder. But I would still put a mother on folic acid preconceptionally in any future pregnancies.

Lindhout: What's your denominator?

Hall: About 35 cases of encephalocele and about 20 cases of craniospinal-rachischisis.

Lindhout: So a recurrence risk of about 1–2% is still possible?

Hall: Yes. I was suspicious enough to look in the literature for recurrence in a family with encephalocele of a regular neural tube defect. I can find only one case, which I think might represent a chance occurrence and therefore I believe non-syndromic encephaloceles may well have a different mechanism.

Shurtleff: We have a similar population as Judith Hall and we have never seen or obtained a history of a recurrence of an encephalocele in a family amongst a maximum of 150 cases.

Lindhout: These numbers are small, but they add up.

References

Chatkupt S, Chatkupt S, Johnson WG 1993 Waardenburg syndrome and myelomeningocele in a family. J Med Genet 30:83–84

Copp AJ, Crolla JA, Brook FA 1988 Prevention of spinal neural tube defects in the mouse embryo by growth retardation during neurulation. Development 104: 297–303

Czeizel AE, Dudás I 1992 Prevention of the first occurrence of neural-tube defects by periconceptional vitamin supplementation. N Engl J Med 327:1832–1835

Finnell RH, Moon SP, Abbott LC, Golden JA, Chernoff GF 1986 Strain differences in heat-induced neural tube defects in mice. Teratology 33:247–252

Juriloff DM, Macdonald KB, Harris MJ 1989 Genetic analysis of the cause of exencephaly in the SELH/Bc mouse stock. Teratology 40:395–405

Lindhout D, Meinardi H 1984 Spina bifida and in utero exposure to valproate. Lancet 2:396

Morriss-Kay GM, Wood H, Chen W-H 1994 Normal neurulation in mammals. In: Neural tube defects. Wiley, Chichester (Ciba Found Symp 181) p 51–69

MRC Vitamin Study Research Group 1991 Prevention of neural tube defects: results of the MRC vitamin study. Lancet 338:132–137

Seller MJ, Perkins-Cole KJ 1987 Hyperthermia and neural tube defects of the curly-tail mouse. J Craniofacial Genet Dev Biol 7:321–330

Seller MJ, Embury S, Polani PE, Adinolfi M 1979 Neural tube defects in curly-tail mice. II. Effects of maternal administration of vitamin A. Proc R Soc Lond B Biol Sci 206:95–107

Steegers-Theunissen RPM, Boers GHJ, Trijbels FJM, Eskes TKAB 1991 Neural-tube defects and derangement of homocysteine metabolism. N Engl J Med 324:199–200

Screening for neural tube defects

H. S. Cuckle

Institute of Epidemiology and Health Services Research, Department of Clinical Medicine, University of Leeds, 34 Hyde Terrace, Leeds LS2 9LN, UK

Abstract. The birth prevalence of neural tube defects fell by 95% in England and Wales between 1970 and 1990 largely as a result of antenatal screening, subsequent diagnosis and selective abortion of affected pregnancies. Anencephaly can be diagnosed using ultrasound, whereas both amniotic fluid biochemistry and ultrasound are required for the diagnosis of spina bifida. Both methods have a false-positive rate of about two per 1000 when carried out in high-risk pregnancies. The diagnostic results are likely to be better if the two methods are used in parallel. Maternal serum α-fetoprotein screening at 16–18 weeks gestation can detect about 75% of pregnancies affected by spina bifida. Routine ultrasound anomaly screening, usually performed at 18–20 weeks gestation, can detect a similar proportion. New developments indicate that earlier, more accurate detection may be possible but more research will be needed before this can be established.

1994 Neural tube defects. Wiley, Chichester (Ciba Foundation Symposium 181) p 253–269

In England and Wales the birth prevalence of neural tube defects fell by 95% between 1970 and 1990. Although other factors will have contributed, most of this decline was due to antenatal screening and the selective abortion of affected pregnancies. This paper compares the different methods of screening and diagnosis currently available and considers possible future developments.

Available methods

Screening differs from the other procedures carried out in medicine in that it is performed routinely and not in response to specific symptoms. Usually, the aim is not to make a diagnosis *per se* but to select individuals for diagnostic procedures, too expensive or hazardous to offer to everyone. The discriminatory powers of both screening and diagnostic tests are assessed in the same way, by determining the detection rate (proportion of affected individuals with positive results) and the false-positive rate (proportion of unaffected individuals with positive results).

Diagnosis of neural tube defects is performed by analysis of two markers in amniotic fluid, α-fetoprotein (AFP) and acetylcholinesterase (AChE), or by a detailed ultrasound 'anomaly' scan; screening uses either determination of maternal serum AFP or routine ultrasound. The best estimate of discriminatory power for the biochemical methods comes from large multicentre collaborative studies, whereas sufficient cases to assess adequately the ultrasound methods can come only from meta-analysis of separate small studies.

Biochemical diagnosis

Anencephaly is readily diagnosed using ultrasound: the principal purpose of biochemical methods is the diagnosis of 'open' spina bifida. An open spinal lesion is one in which the neural tissue is exposed or covered with a thin, transparent membrane. About one in six spina bifida lesions is 'closed'; these cases are associated with a relatively good prognosis.

α-fetoprotein in the amniotic fluid

AFP is an α-globulin of largely fetal hepatic origin. In unaffected pregnancies a small amount of AFP is voided into the amniotic fluid and subsequently broken down by the fetal kidneys. On average, the concentration of AFP in the amniotic fluid decreases by 10–15% per week at mid-trimester: to allow for this and to take account of differences between laboratories, levels are expressed in multiples of the normal median (MoM) level for the relevant gestation and laboratory.

The concentration of AFP in fetal cerebrospinal fluid is more than 100 times greater than in amniotic fluid, so leakage through an open spinal lesion leads to an increase in the concentration in the latter. The UK Collaborative Study on Alpha-fetoprotein (1979) measured concentrations of AFP in the amniotic fluid surrounding 123 fetuses with open spina bifida and 12 804 unaffected fetuses. 97% of the women carrying a fetus with spina bifida had AFP levels above 3.0 MoM, but there was no cut-off level that could completely separate affected and unaffected pregnancies.

Acetylcholinesterase

AChE was developed as a second marker because of the shortcomings of AFP measurement. The best established analytical technique is polyacrilamide gel electrophoresis. When amniotic fluid from a normal pregnancy is analysed, detection of AChE on a gel reveals a single band; this is not AChE but a pseudo-cholinesterase. If the fetus has an open spina bifida, the amniotic fluid contains a second protein that migrates on gels to the same position as AChE from cerebrospinal fluid.

The discriminatory power of AChE analysis is greater than that of AFP measurement. However, the gel method is labour intensive and particular care is required when the AChE band is faint. Therefore it is not practical for AChE to replace AFP as a marker. A sequential policy is recommended, with AFP measurement used initially and detection of AChE only if the AFP level is raised. The Collaborative AChE Study (Wald et al 1989) collected data from samples that had been routinely tested for both markers: 428 pregnancies in which the fetus had open spina bifida and 31 801 pregnancies in which the fetus had no neural tube defect were included.

Detection and false-positive rates

For a similar false-positive rate, the detection rate of open spina bifida was 14% higher for routine analysis of AChE than for routine measurement of AFP (Table 1). A sequential policy would have led to a small loss in detection compared with routine AChE determination but the false-positive rate would have been almost halved. This policy would require only 5% of samples assayed for AFP to be tested for AChE.

Table 1 also shows that the discriminatory power of biochemical analysis of amniotic fluid is poorer for women having the tests because of a positive maternal serum AFP screening result than for those having the tests for other reasons. When the amniocentesis was performed because of the serum AFP level the false-positive rate was 2–4 times higher than for other indications, although the detection rate was not materially altered.

Causes of false-positive results

There are three principal reasons for false-positives: blood contamination, non-viable pregnancy and serious abnormality other than a neural tube defect.

Blood contamination. This occurs if the amniocentesis needle passes through the placenta, but there may be other causes. As the fetus-to-amniotic fluid AFP gradient is so great, a small amount of fetal blood can substantially increase the concentration of AFP in the amniotic fluid. The AChE in fetal blood is also likely to yield a false-positive result. Maternal blood contamination will not raise AFP levels but it too can be a source of AChE false-positives.

In practice, most of these problems can be avoided by further analysis of visibly bloodstained amniotic fluids in the small number of positive cases. Total red cell count and the Kleihaur haemological test will indicate if fetal blood is the problem. Samples that were positive in the AChE test can be reanalysed using an inhibitor (BW284C51) that is specific for the pseudo-cholinesterase of maternal blood. Fetal blood contamination as a cause of AChE false-positives cannot be dealt with biochemically and a repeat amniocentesis may be indicated.

Fetal death or serious malformation. A common cause of false-positives is impending fetal death, presumably because of autolysis; serious fetal malformations other than a neural tube defect occur less frequently. There could be leakage

TABLE 1 Determination of α-fetoprotein levels and acetylcholinesterase in amniotic fluid: positive rate (%) according to pregnancy outcome, reason for amniocentesis and quality of sample

Outcome[a]	*No. tested*	*Routine AFP*[b]	*Routine AChE*	*Sequential AChE if AFP raised*[c]
Spina bifida	428	85	99	96
No neural tube defect	1801	0.52	0.49	0.27
Amniocentesis for raised serum AFP				
Spina bifida	344	84	99	97
No neural tube defect	5688	1.6	0.88	0.69
Amniocentesis for other reasons				
Spina bifida	84	86	96	95
No neural tube defect	26113	0.35	0.40	0.18
Amniocentesis for raised serum AFP: clear samples				
Spina bifida	298	85	99	98
Viable, unaffected	4785	1.0	0.23	0.19
Amniocentesis for other reasons: clear samples				
Spina bifida	76	87	96	95
Viable, unaffected	24223	0.09	0.15	0.03

Based on the results of the Collaborative AChE Study (Wald et al 1989).
[a]Closed spina bifida lesions are excluded; unaffected pregnancies exclude all serious abnormalities.
[b]Gestation-specific cut-off levels were used: 3.0 MoM (13–15 weeks); 3.5 (16–18); 4.0 (19–21); 4.5 (22–24).
[c]An overall 2.0 MoM cut-off level was used.

through an open lesion (e.g. abdominal wall defects) or skin; reduced amniotic fluid production leading to increased concentration of AFP (e.g. urethral obstruction or renal agenesis); abnormal fetal swallowing and protein breakdown (e.g. atresia or Finnish-style nephrosis). Most of the abdominal wall defects that give false-positive AFP results will also be positive in the AChE test, as will about half the other defects. Abdominal wall defects can be distinguished from neural tube defects by scanning densitometry of the gel bands, because the AChE band is relatively dense compared with the pseudo-cholinesterase band.

Since the erroneous termination of pregnancies that are non-viable or in which the fetus has a serious abnormality may be considered a desirable outcome of the diagnostic process, these cases are sometimes excluded when calculating false-positive rates. The Collaborative AChE Study detection rate and the false-positive rate for viable pregnancies without serious abnormality when the amniotic fluid was clear (i.e. not bloodstained) are also shown in Table 1. With sequential AChE testing the false-positive rate was about two per 1000 for those having amniocentesis following a positive AFP screening test and less than one half per 1000 for other indications.

TABLE 2 Diagnostic ultrasound: spina bifida detection and false-positive rates

Study	*Main reason for scan*	*% Detection*	*% False-positive*
Pre-1985			
Robinson et al (1980)	AFP, FH	79 (15/19)	0.0 (0/302)
Hobbins et al (1982)	AFP	100 (6/6)	0.0 (0/22)
Roberts et al (1983)	AFP, FH	58 (22/38)	2.2 (53/2414)
Campbell et al (1984)	AFP, FH, US	94 (112/119)	0.3 (3/1306)
Gough (1984)	AFP	100 (19/19)	3.0 (6/201)
Sabbagha (1985)	AFP, US	75 (6/8)	2.2 (2/92)
Hashimoto et al (1985)	AFP, US, FH	100 (32/32)	0.0 (0/94)
Combined	—	88 (212/241)	1.4 (64/4431)
Post-1985			
Lindfors et al (1987)	AFP	75 (9/12)	0.3 (2/649)
Richards et al (1988)	AFP	80 (8/10)	0.0 (0/587)
Hogge et al (1989)	AFP, FH	100 (10/10)	0.0 (0/167)
Nadel et al (1990)	NS	100 (31/31)	NS
Morrow et al (1991)	AFP	97 (38/39)	0.0 (0/875)
Platt et al (1992)	AFP	92 (148/161)	NS
Combined	—	93 (244/263)	0.1 (3/2278)

AFP, raised concentration of α-fetoprotein in maternal serum or amniotic fluid; FH, family history; NS, not specified; US, suspicious ultrasound scan in other institution.

Ultrasound diagnosis

Detailed examination of the fetal spine has become routine in many centres but that is more like screening than diagnosis and is dealt with below. This section relates to such scans performed selectively on women at high risk of having a child with a neural tube defect. Table 2 summarizes the reported discriminatory power for detection of spina bifida in 13 studies; all of the anencephalic fetuses in these studies were detected. The rates were more favourable in the post-1985 studies, presumably because of improved ultrasound equipment with better resolution and because of greater experience. When the results of the post-1985 studies are combined, the spina bifida detection rate is 92% and the false-positive rate is fewer than two per 1000; this is 10-fold less than for the earlier period.

Comparison with biochemical diagnosis

The rates of detection and of false-positives for women with positive maternal serum AFP results using the biochemical assays (Table 1) are comparable to those achieved using ultrasound diagnosis after 1985, when most examinations were done for this reason. Both methods appear to have a similar discriminatory power. Because of its subjective nature, the rates for ultrasound are not as applicable as for the biochemical studies. However, the value of ultrasound is unlikely to have been substantially overestimated and for a fair comparison with the biochemical tests the hazards of amniocentesis must also be considered.

The principal hazard is miscarriage, but the excess risk is difficult to quantify precisely. 3–4% of mid-trimester pregnancies will miscarry without amniocentesis: it is possible only rarely to attribute the adverse outcome directly to the procedure (e.g. when there is amnionitis or chronic amniotic fluid leakage). There has been only one randomized trial of amniocentesis: the fetal loss rate in the 2302 women randomly assigned to undergo the procedure was 0.8% higher than in the control group (Tabor et al 1986). Whilst this is necessarily limited to the skills and experience of a single obstetric unit, the results provide the only unbiased estimate of hazard.

Even if individual practitioners were confident of a 0.5% miscarriage rate in their hands, there would be more iatrogenic fetal losses with biochemical diagnosis than with ultrasound.

Combining biochemical and ultrasound results

Some centres rely entirely on ultrasound for diagnosis of neural tube defects, whilst others act on a biochemical result regardless of the ultrasound findings. The best results could potentially be obtained if both methods were used in parallel. The problem is knowing how to interpret cases in which there are conflicting results. Various policies are possible. For example, whether a borderline biochemical result is positive could be determined by the strength of the ultrasound suspicion and vice versa. It is not possible to estimate precisely the consequences of such a policy, because there are few data on the correlation between the two methods.

Maternal serum AFP screening

In the serum of non-pregnant women there is a negligible amount of AFP; as pregnancy progresses the level begins to rise, reaching a rate of increase of 15–20% a week by mid-trimester. As with estimation of the amount of AFP in amniotic fluid, MoMs are used to allow for this. When the fetus has an open neural tube defect, the AFP level is increased on average, mainly owing to perfusion across the amniotic membranes. The UK Collaborative Study on Alpha-fetoprotein (1977) includes data on 146 women carrying an anencephalic fetus, 142 carrying one with spina bifida and 18 684 with unaffected pregnancies. The results strongly suggest that the discriminatory power of maternal serum AFP screening is higher at 16–18 weeks than at other times in gestation, with an average level of 6.4 MoM for anencephalic fetuses and 3.8 MoM for those with open spina bifida. Table 3 shows the discriminatory power estimated from a Gaussian model fitted to the data in the study. Even when performed in the optimal gestational week, there is still considerable overlap between the distribution of levels in affected and unaffected pregnancies.

Refinements

The AFP results can be improved by allowing for a series of factors.

Gestational age error. This is by far the most important factor. The rapid increase in AFP levels means that a woman who has underestimated her gestational age by two weeks will have a 30–40% higher MoM value than if the correct one was used. It follows that women with raised AFP levels include a disproportionate number who have substantially underestimated their gestational age. When such women have their gestational age reassessed by ultrasound, the recalculated MoM will be in expectation lower than the original MoM, leading to as many as one half of the positive results becoming negative. Not all the pregnancies for which the results are changed from positive to negative will be unaffected; however, provided the biparietal diameter rather than another biometric measure is used to estimate gestational age, a loss of detection will not be frequent. The reason is that the biparietal diameter is reduced on average in spina bifida by the equivalent of about two weeks gestation. Therefore, in affected pregnancies, gestational age will be systematically underestimated and the MoM value will be increased further.

Errors in the estimation of gestational age apply equally to women with low levels of AFP in their serum. The full benefits of ultrasound dating would be achieved if all women had a routine scan prior to the AFP test. Table 3 shows the estimated effect on detection if this were to be done. Although most women have a dating scan, increasingly it is delayed until after the AFP test so that it can be combined with a more detailed anomaly scan. The cost of an early scan may be regarded as prohibitive, especially in centres that are confident in the ability of the anomaly scan to detect neural tube defects. A compromise would be to ensure that the 10–20% of women with uncertain dates of conception are scanned early.

Maternal weight. A small but readily remediable source of AFP variability is differences in maternal weight. The concentration of AFP in maternal blood is negatively correlated with maternal weight, presumably due to a fixed mass of AFP produced by the fetus being concentrated in different volumes of body tissues. Published formulae are available to calculate the expected MoM value for a given weight. Dividing the observed MoM by this expected value is a simple method of adjustment for differences in weight.

Ethnic origin. Women of Afro-Caribbean origin have AFP levels on average about 15% higher than other women. Therefore, unless special allowance is

TABLE 3 Measurement of α-fetoprotein concentration in maternal serum: positive rate (%) according to pregnancy outcome, gestation and basis for the normal median level (MoM)

	Gestational age (weeks)		
Outcome	*15*	*17*	*19*
MoM based on gestational age estimated by menses			
Anencephaly	74	89	89
Open spina bifida	57	72	66
Closed spina bifida	2.5	2.5	2.5
All spina bifida[a]	48	60	55
Unaffected	2.5	2.5	2.5
MoM based on gestational age estimated by measurement of the biparietal diameter			
Open spina bifida	73	85	81
Closed spina bifida	25	25	25
All spina bifida[a]	65	75	72
Unaffected	1.9	1.9	1.9

A 2.5 MoM cut-off was used; the rates are estimated from Gaussian frequency distributions with the parameters in Wald et al (1992) derived from the UK Collaborative Study on Alpha-fetoprotein (1977).
[a]Assuming that one spina bifida in six has a closed lesion.

made a disproportionate number will be assumed to have raised AFP levels. This is particularly unfortunate because offspring of such women have a low prevalence of neural tube defects. Simply reducing the MoM by 15% or increasing the cut-off is sufficient to allow for this; centres serving a population with a sizeable ethnic minority can use separate medians for the MoM calculation.

Repeat testing. Another source of AFP variability is fluctuations within an individual. However, repeat testing on a fresh sample for women with positive results is not recommended because of an unacceptable loss of detection. Routine repeating would be too costly.

Multiple pregnancy. On average, the maternal serum AFP level in women carrying twins is twice normal. The discovery of twins in a woman with a high AFP level is often taken to be an explanation for the positive result. However, if the level is high enough, say above 5.0 MoM, there may be a high risk that one of the fetuses is affected. When it is known that there is a multiple pregnancy prior to screening a high cut-off can be chosen in the first place. For twins, detailed ultrasound may be a more appropriate diagnostic procedure than amniocentesis, especially if there is a single sac.

Insulin-dependent diabetes. Women with insulin-dependent diabetes are at increased risk of having children with a neural tube defect or other structural

TABLE 4 Screening ultrasound: spina bifida detection rate

Study	*No. women screened*	*% Detection*
Persson et al (1983)	10 147	25 (2/8)
Campbell & Smith (1984)	11 664	100 (8/8)
Hill et al (1985)	5 420	100 (1/1)
Rosendahl & Kivinen (1989)	9 012	75 (3/4)
Levi et al (1991)	16 072	77 (10/13)
Neven et al (1991)	23 369	86 (36/42)
Chitty et al (1991)	8 342	100 (7/7)
Constantine & McCormack (1991)	4 984	100 (5/5)
Luck (1992)	8 523	100 (3/3)
Shirley et al (1992)	6 366	100 (4/4)
Combined	103 899	83 (79/95)

abnormalities. Whilst it is usual practice to refer them directly for detailed ultrasound, most will also have an AFP test. This presents a problem of interpretation, since maternal serum AFP levels are reduced on average in diabetics by about one-quarter. As with Afro-Caribbeans, this can be dealt with by increasing the MoM or reducing the cut-off accordingly.

Ultrasound screening

Nearly all women will have an ultrasound scan at some stage of their pregnancy, for dating, to check for twins, confirm viability and locate the placenta. Occasionally, a neural tube defect will be identified at this time but it is an incidental finding and not the result of screening.

In many centres women will also be offered an anomaly scan at about 18–20 weeks gestation to detect structural abnormalities, including neural tube defects. The discriminatory power achievable by ultrasound when performed routinely is unlikely to be as great as when it is used selectively. However, in published studies it is not substantially less and is greater than for AFP screening.

Table 4 shows the results of eight studies in which routine anomaly scanning was performed in a total of more than 90 000 pregnancies. The combined spina bifida detection rate, albeit on fewer than 100 cases, is encouragingly high, but these studies need to be interpreted with caution. There is a possibility that the ultrasound observations were influenced by the degree of previously established risk, including AFP results. Also, it is remarkable that in no study were neural tube defect false-positives reported. This could reflect a low prior suspicion of neural tube defects, on average, but it may also indicate bias.

New developments

AChE radioimmunoassay

Monoclonal antibodies have been raised against the cerebrospinal isoform of AChE and an immunoassay has been developed. This has not been tested in enough women to be certain that it could replace gel electrophoresis but the results so far are encouraging (Rasmussen-Loft et al 1990).

Early amniocentesis

Until recently, biochemical diagnosis has been of amniotic fluid obtained at 13–24 weeks gestation. Now, amniocentesis can be performed in the first trimester for the diagnosis of chromosomal defects. Too few data are available on affected pregnancies to judge the value of AFP measurement in these samples, but it is already clear they are unsuitable for AChE diagnosis because of an extremely high false-positive rate.

Cranial and cerebellar signs of spina bifida

The anomaly scan is time consuming, expensive in operator skills and normally undertaken rather late in pregnancy. A simpler ultrasound screening method has now been developed, based on fetal cranial and cerebellar signs (the 'fruit' signs) detectable at the dating scan.

Table 5 summarizes the results of retrospective studies, based mainly on examination of photographs or scans carried out when the presence of abnormality had been established and on prospective studies of high-risk pregnancies. The overall detection rate of the cerebellar sign is much higher than for AFP screening and the false-positive rate appears to be negligible. It is not yet known whether the same results can be achieved in routine practice.

Transvaginal scanning

Using a transvaginal probe, one can now obtain a detailed picture of fetuses in the first trimester and detect structural abnormalities such as neural tube defects. However, routine use cannot be recommended until more is known about early normal development. A lesion present very early in pregnancy may subsequently regress.

Down's syndrome screening

AFP is one of the analytes measured in multi-marker maternal serum screening for aneuploidy. Some centres that would otherwise have preferred ultrasound

TABLE 5 Ultrasound screening using cranial and cerebellar signs: spina bifida detection and false-positive rates

Study	*% Detection* *Cranial*	*Cerebellar*	*% False-positive*[a]
Retrospective			
Nicolaides et al (1986)	100 (54/54)	95 (20/21)	0.0 (0/100)
Furness et al (1987)	100 (13/13)	—	NS
Penso et al (1987)	67 (16/24)	—	0.0 (0/12)
Pilu et al (1988)	—	100 (19/19)	0.0 (0/17)
Nyberg et al (1988)	55 (17/31)	—	0.0 (0/30)
Goldstein et al (1989)	70 (14/20)	100 (19/19)	3.3 (1/33)
Combined	80 (114/142)	98 (58/59)	0.52 (1/192)
Prospective			
Campbell et al (1987)	100 (26/26)	96 (25/26)	1.2 (5/410)
Chambers et al (1988)	75 (9/12)	—	0.0 (0/169)
Nyberg et al (1988)	79 (15/19)	—	1.3 (3/230)
Petrikovsky (1990)	67 (10/15)	53 (8/15)	1.6 (5/313)
Thiagarajah et al (1990)	75 (18/24)	100 (22/22)	NS
Van den Hof et al (1990)	83 (108/130)	95 (124/130)	0.7 (9/1367)
Combined	82 (186/226)	93 (179/193)	0.88 (22/2489)

NS, not specified.
[a]All the false-positives except that in Goldstein et al (1989) were due to cranial signs.

screening for neural tube defects have retained AFP measurement so that both kinds of screening can be done together. With the advent of first trimester multi-marker screening for Down's syndrome such centres may reassess their policy.

Conclusions

With the use of established methods, antenatal screening can be expected to detect all anencephalic fetuses and at least three-quarters of those with spina bifida, with few diagnostic false-positive results. New developments are likely to lead to an earlier and more accurate diagnosis than hitherto but more research is required before this can be quantified.

References

Campbell S, Smith P 1984 The routine screening for congenital abnormalities by ultrasound. In: Rodeck CH, Nicolaides KH (eds) Prenatal diagnosis. Royal College of Obstetrics and Gynaecology, London (Proc 11th Study Group R Coll Obstet Gynaecol) p 325–330

Campbell S, Smith P, Pearce JM 1984 The ultrasound diagnosis of neural tube defects and other cranio-spinal abnormalities. In: Rodeck CH, Nicolaides KH (eds) Prenatal diagnosis. Royal College of Obstetrics and Gynaecology, London (Proc 11th Study Group R Coll Obstet Gynaecol) p 245–257

Campbell J, Gilbert WM, Nicolaides KH, Campbell S 1987 Ultrasound screening for spina bifida: cranial and cerebellar signs in a high risk population. Obstet Gynecol 70:247–250

Chambers SE, Muir BB, Bell JE 1988 'Bullet'-shaped head in fetuses with spina bifida: a pointer to the spinal lesion. JCU (J Clin Ultrasound) 16:25–28

Chitty LS, Hunt GH, Moore J, Lobb MO 1991 Effectiveness of routine ultrasonography in detecting fetal structural abnormalities in a low risk population. Br Med J 303:1165–1169

Constantine G, McCormack J 1991 Comparative audit of booking and mid-trimester ultrasound scans in the prenatal diagnosis of congenital anomalies. Prenatal Diagn 11:905–914

Furness ME, Barbary JE, Verco PW 1987 Fetal head shape in spina bifida in the second trimester. JCU (J Clin Ultrasound) 15:451–455

Goldstein RB, Podrasky AE, Filly RA, Callen PW 1989 Effacement of the fetal cisterna magna in association with myelomeningocele. Radiology 172:409–413

Gough JD 1984 Ultrasound. In: Wald NJ (ed) Antenatal and neonatal screening. Oxford University Press, Oxford, p 423–444

Hashimoto BE, Mahoney BS, Filly RA, Golbus MS, Anderson RL, Callen PW 1985 Sonography, a complementary examination to alpha-fetoprotein testing for fetal neural tube defects. J Ultrasound Med 4:307–310

Hill LM, Breckle R, Gehrking RT 1985 Prenatal detection of congenital malformations by ultrasonography. Am J Obstet Gynecol 151:44–50

Hobbins JD, Venus I, Tortara M, Mayden K, Mahoney MJ 1982 Stage II ultrasound examination for the diagnosis of fetal abnormalities with an elevated amniotic fluid alpha-fetoprotein concentration. Am J Obstet Gynecol 142:1026–1029

Hogge WA, Thiagarajah S, Ferguson JE, Schnatterly PT, Harbert GM 1989 The role of ultrasonography and amniocentesis in the evaluation of pregnancies at risk for neural tube defects. Am J Obstet Gynecol 161:520–524

Levi S, Hyjazi Y, Schapps J-P, Defoort P, Coulon R, Buekens P 1991 Sensitivity and specificity of routine antenatal screening for congenital anomalies by ultrasound: the Belgian Multicentre Study. Ultrasound Obstet Gynecol 1:102–110

Lindfors KK, Gorczyca DP, Hanson FW, Tennant FR, McGahan JP, Peterson AG 1987 Midtrimester screening for open neural tube defects: correlation of sonography with amniocentesis results. AJR (Am J Roentgenol) 149:141–145

Luck CA 1992 Value of routine ultrasound scanning at 19 weeks: a four year study of 8849 deliveries. Br Med J 304:1474–1478

Morrow RJ, McNay MB, Whittle MJ 1991 Ultrasound detection of neural tube defects in patients with elevated maternal serum alpha-fetoprotein. Obstet Gynecol 78:1055–1057

Nadel AS, Green JK, Holmes LB, Frigoletto FD, Benacerraf BR 1990 Absence of need for amniocentesis in patients with elevated levels of maternal serum alpha-fetoprotein and normal ultrasonic examinations. N Engl J Med 323:557–561

Neven P, Ricketts NEM, Geirsson RT, Smith R, Crawford JW 1991 Screening for neural tube defect with maternal serum α-fetoprotein and ultrasound without the use of amniocentesis. J Obstet Gynaecol 11:5–8

Nicolaides KH, Campbell S, Gabbe SG, Guidetti R 1986 Ultrasound screening for spina bifida: cranial and cerebellar signs. Lancet 2:72–74

Nyberg DA, Mack LA, Hirsch J, Mahoney BS 1988 Abnormalities of fetal cranial contour in sonographic detection of spina bifida: evaluation of the 'lemon' sign. Radiology 167:387–392
Penso D, Redline RW, Benacerraf BR 1987 A sonographic sign which predicts which fetuses with hydrocephalus have an associated neural tube defect. J Ultrasound Med 6:307–311
Persson PH, Kullander S, Gennser G, Grennert L, Laurell CB 1983 Screening for fetal malformations using ultrasound and measurements of α-fetoprotein in maternal serum. Br Med J 286:747–749
Petrikovsky BM 1990 'Fruit' signs and neural tube defects. Prenatal Diagn 10:134
Pilu G, Romero R, Reece EA, Goldstein I, Hobbins JC, Bovicelli L 1988 Subnormal cerebellum in fetuses with spina bifida. Am J Obstet Gynecol 58:1052–1056
Platt LD, Feuchtbaum L, Filly R, Lustig L, Simon M, Cunningham MD 1992 The California maternal serum α-fetoprotein screening program: the role of ultrasonography in the detection of spina bifida. Am J Obstet Gynecol 166:1328–1329
Rasmussen-Loft AG, Nanchahal K, Cuckle HS, Wald NJ, Hulten M, Norgaard-Pedersen B 1990 Amniotic fluid acetylcholinesterase in the antenatal diagnosis of open neural tube defects and abdominal wall defects: a comparison of gel electrophoresis and a monoclonal antibody immunoassay. Prenatal Diagn 10:449–459
Richards DS, Seeds JW, Katz VL, Lingley LH, Albright SG, Cefalo RC 1988 Elevated maternal serum alpha-fetoprotein with normal ultrasound: is amniocentesis always appropriate? A review of 26,069 screened patients. Obstet Gynecol 71: 203–207
Roberts CJ, Evans KT, Hibbard BM, Lawrence KM, Roberts EE, Robertson IB 1983 Diagnostic effectiveness of ultrasound in detection of neural tube defect: the South Wales experience of 2509 scans (1977–1982) in high risk mothers. Lancet 2:1068–1069
Robinson HP, Hood VD, Adam AH, Gibson AAM, Ferguson-Smith MA 1980 Diagnostic ultrasound: early detection of fetal neural tube defects. Obstet Gynecol 56:705–710
Rosendahl H, Kivinen S 1989 Antenatal detection of congenital malformations by routine ultrasonography. Obstet Gynecol 73:947–951
Sabbagha RE, Sheikh Z, Tamura RK et al 1985 Predictive value, sensitivity and specificity of ultrasonic targeted imaging for fetal anomalies in gravid women at high risk for birth defects. Am J Obstet Gynecol 152:822–827
Shirley MI, Bottomley F, Robinson VP 1992 Routine radiographer screening for fetal abnormalities by ultrasound in an unselected low risk population. Br J Radiogr 65:654–659
Tabor A, Philip J, Madsen M, Bang J, Obel EB, Norgaard-Pedersen B 1986 Randomised controlled trial of genetic amniocentesis in 4606 low-risk women. Lancet 2:1287–1293
Thiagarajah S, Henke J, Hogge WA, Abbitt PL, Breeden N, Ferguson JE 1990 Early diagnosis of spina bifida: the value of cranial ultrasound markers. Obstet Gynecol 76:54–57
UK Collaborative Study on Alpha-fetoprotein 1977 Maternal serum-alpha-fetoprotein measurement in antenatal screening for anencephaly and spina bifida in early pregnancy. Lancet 1:1323–1332
UK Collaborative Study on Alpha-fetoprotein 1979 Amniotic-fluid alpha-fetoprotein measurement in antenatal diagnosis of anencephaly and open spina bifida in early pregnancy. Lancet 2:651–662
Van den Hof MD, Nicolaides KH, Campbell J, Campbell S 1990 Evaluation of the lemon and banana signs in one hundred thirty fetuses with open spina bifida. Am J Obstet Gynecol 162:322–327

Wald N, Cuckle H, Nanchahal 1989 Amniotic fluid acetylcholinesterase measurement in the prenatal diagnosis of open neural tube defects. Second report of the Collaborative Acetylcholinesterase Study. Prenatal Diagn 9:813–829
Wald NJ, Cuckle HS, Densem JW, Kennard A, Smith D 1992 Maternal serum screening for Down syndrome: the effect of routine ultrasound scan determination of gestational age and adjustment for maternal weight. Br J Obstet Gynaecol 99:144–149

DISCUSSION

Dolk: You said most of the fall in the prevalence of neural tube defects in the UK is due to antenatal screening and termination. In some centres in the British Isles there are data which would give you the exact figures (Stone et al 1988, EUROCAT Working Group 1991).

Cuckle: There are individual studies in well defined areas (e.g. south Wales and west Scotland) that have tried to find all the terminations as well as births. But it's not easy to be sure that all terminations are accounted for. The notification forms for termination of pregnancy can't be studied because of confidentiality. Hospital ward records can be used but that is difficult.

Dolk: I don't think the records are so very incomplete. The data fit well with the fact that in Dublin, for example, where there isn't any screening, a very similar natural fall in prevalence has been shown (Radic et al 1987).

Cuckle: Ultrasound is used in Ireland and, although abortion is illegal, cases detected can be terminated in England and Wales. Several thousand terminations are done here each year to Irish residents.

Dolk: In Dublin in the early 1980s, the incidence of neural tube defects was halved; not half the women were being screened. Moreover, one would expect this to affect the ratio of spina bifida:anencephaly, because anencephaly is much more likely to be detected by screening.

Oakley: Professor Cuckle, when we have eliminated folic acid-preventable spina bifida, how is that going to affect the screening process?

Cuckle: I should imagine that there will be less invasive diagnosis. The justification for carrying out amniocentesis in women with positive results from screening is that they are a high-risk group. As the prior prevalence falls, so will the risk in the high-risk group.

Hall: Have you looked at the cost:benefit ratio? Right now, the ratio probably is positive, but once there aren't so many neural tube defects because of folic acid treatment it may not be.

Cuckle: The financial benefits of screening relate to the one third to one half of spina bifida cases surviving early childhood. The cost of screening is not great: the AFP test costs about £10 and ultrasound is being done anyway for a variety of reasons. As ultrasound becomes the modality of first choice, the marginal costs will be negligible compared to the benefits, even if prevalence falls by three quarters.

Brenner: Do you have figures for predictive values of positive and negative tests depending on the prevalence of neural tube defects?

Cuckle: Yes, but only for biochemical screening alone or ultrasound alone. When the modalities are combined, as is increasingly the case, the predictive value cannot be calculated, because the extent of correlation between the results of ultrasound screening and biochemical tests is not known.

Nau: Which of the closed forms of spina bifida can you pick up with ultrasound?

Cuckle: The published ultrasound studies are not broken down in that way. I think they will have detected a fair spread of different types of spina bifida.

Stanley: How good is the follow-up in these particular studies in terms of the false negatives? Did you ascertain all of the outcomes of pregnancies to see which ones were missed?

Cuckle: In deriving the tabulated detection rates, I assumed 100% follow-up. I cannot be sure that this was achieved, but equally the biochemical results may also be subject to incomplete follow-up.

Hall: If the limb defects associated with chorionic villus sampling (reviewed by Rodeck 1993) are an indication, you may well have missed a lot of abnormalities.

Wald: There's a credibility gap concerning the ultrasound results. There are cases of spina bifida that were missed by ultrasound screening. We also know anecdotally that some fetuses were thought to have spina bifida by ultrasound screening, but didn't.

Cuckle: That was the case, but things have changed since the late 1980s. The equipment is better and operators are much more experienced.

Seller: The surveys on which you based the results on ultrasound diagnosis were probably from large regional centres. These are centres of excellence and they have queries referred to them and so on.

Cuckle: That's true for the diagnostic ultrasound. The more recent screening ultrasound is, however, based on a spectrum of practice, including District General Hospitals.

Seller: For any woman in the United Kingdom having a neural tube defect in her fetus diagnosed by scanning, the position has changed radically over the last few years. This is due particularly to these banana and lemon signs in the ultrasound scan, which anyone can see. When the original published surveys were done in specialized centres, those particular centres were getting a 100% diagnosis, but this wasn't the case everywhere. Now, any woman anywhere is likely to have spina bifida in her fetus detected.

Cuckle: The studies of routine anomaly ultrasound were carried out before the discovery of the 'fruit' signs.

Opitz: Please could you explain these signs?

Cuckle: The lemon sign is pinching of the frontal bones so that the cranium is lemon shaped. The banana sign is an area of echogeneity in the cerebellum in the shape of a banana.

Opitz: Indicating an Arnold Chiari malformation?

Cuckle: Possibly. It may be to do with the general tortions experienced by the brain as a result of changes in pressure.

Czeizel: We have a unique opportunity to evaluate the false-positive and false-negative cases of prenatal diagnosis in our family planning system. We follow the women from before conception to after the pregnancy outcome. In the first 5000 pregnancies we had 18 selective abortions (i.e. terminations after prenatal diagnosis of defects), two of which were false positive. One diagnosis was gastroschisis and the fetal pathologist didn't find anything. The other diagnosis was hydrocephalus: the fetal pathologist found obstructive abnormality of kidney. We had six cases with a neural tube defect, including two with spina bifida aperta that were not diagnosed prenatally. This indicates that the Hungarian standard of maternal serum AFP and ultrasound screening is probably low.

Wald: There are really only two reasons for screening, the financial cost of the diagnostic procedure and the harm due to the diagnostic procedure. If the diagnostic procedure were cheap and safe, there would be no point in pre-selecting people through screening. An amniotic fluid sample is the most valuable specimen for diagnostic purposes; it is much better than the serum tests and better than ultrasound if information on Down's syndrome is also required. If one were to perform an amniocentesis on a greater proportion of women, but remove only, say, 1 ml of amniotic fluid, could amniocentesis be comparably safe? The current risk of amniocentesis is just under 1%. Professional groups quote figures of 0.5%. With the improvements in ultrasound guidance of the needle, use of a smaller needle (there is an association between needle size and hazard) and by removing a very small quantity of amniotic fluid, one may be able to have a diagnostic test for neural tube defects and for Down's syndrome (possibly also for trisomy 18), if *in situ* hybridization was used on the fetal amniotic fluid cells.

Shurtleff: I think the risk of amniocentesis as a screening procedure far outweighs the value. In our centres, it is strictly a follow-up procedure at the present time.

Cuckle: You may not even need to deplete the amniotic fluid greatly to obtain material for karyotyping or FISH (fluorescence *in situ* hybridization), if the new amniofiltration technique is used. New methods may make amniocentesis safer but there has been only one randomized trial of amniocentesis (Tabor et al 1986) and I don't believe there will ever be another one just to test new methods.

Seller: Amniocentesis samples are expensive, because an invasive procedure requires an obstetrician. Scanning is done by radiographers and people who are less expensive.

References

EUROCAT Working Group 1991 Prevalence of neural tube defects in 20 regions of Europe and the impact of prenatal diagnosis, 1980–1986. J Epidemiol Community Health 45:52–58

Radic A, Dolk H, De Wals P 1987 Declining rate of neural tube defects in three eastern counties of Ireland: 1979–1984. Ir Med J 80:226–228
Rodeck CH 1993 Prenatal diagnosis. Fetal development after chorionic villus sampling. Lancet 341:468–469
Stone DH, Smalls MJ, Rosenberg K, Womersley J 1988 Screening for congenital neural tube defects in a high-risk area: an epidemiological perspective. J Epidemiol Community Health 42:271–273
Tabor A, Philip J, Madsen M, Bang J, Obel EB, Norgaard-Pedersen B 1986 Randomised controlled trial of genetic amniocentesis in 4606 low-risk women. Lancet 2:1287–1293

Meningomyelocele: management *in utero* and post natum

David B. Shurtleff*, David A. Luthy+, David A. Nyberg++, Thomas J. Benedetti° and Laurence A. Mack**

**Birth Defects Clinic, Division of Congenital Defects, Department of Pediatrics, University of Washington, Seattle, WA 98105, +Division of Perinatal Medicine, ++Department of Ultrasound, Swedish Medical Center, 747 Summit Avenue, Seattle, WA 98104, °Department of Obstetrics & Gynecology, **Departments of Diagnostic Ultrasound and Radiology, University of Washington, Seattle, WA 98195, USA*

Abstract. We report a four year follow-up of 39 of 47 infants born after pre-labour Caesarean section and 68 of 79 born vaginally. Loss of motor function due to late complications was more frequent in the Caesarean section group (Fisher's Exact; $P = 0.004$). However, the means of the differences between the X-ray levels (measured as the last intact vertebral arch seen on standard anteroposterior roentgenograms of the spinal column) subtracted from the motor levels still favour Caesarean section (mean = 3.24; SD = 2.7) over vaginal delivery (mean = 1.2; SD = 2.7) (Student's t-test; $P = 0.0003$). The frequencies of other complications, death and neonatal meningitis, were not significantly different. Another 38 infants born by Caesarean section after labour were more paralysed (mean of X-ray and motor difference = 1.8, SD = 2.2) following rupture of amniotic membranes than those with intact amniotic membranes with or without labour (mean = 3.4; SD = 2.2) (Student's t-test; $P = 0.0067$). The differences between X-ray and motor levels for patients born by Caesarean section with intact amniotic membranes and without labour (mean = 3.6; SD = 2.4) were not significantly different from those with labour and intact amniotic membranes (mean = 2.89; SD = 1.5). The number of new cases of meningomyelocele presenting to our clinic has decreased from an average of 30 per year between 1970 and 1987 to 14 between 1988 and 1992.

1994 Neural tube defects. Wiley, Chichester (Ciba Foundation Symposium 181) p 270–286

This paper presents recent advances in the management of fetuses and infants with meningomyelocele (spina bifida aperta). The pathoembryology presented in this symposium clearly describes the variation in types of lesions and the heterogeneity of the aetiology of neural tube defects along the lower spine (Wald, this volume, Holmes, this volume, Oakley et al, this volume). These myelodysplastic lesions include failures of neurulation such as meningomyelocele and meningocele, as well as errors of canalization such as myelocystoceles, lipomeningomyeloceles and other types of lipomas of the canda equina and filum terminale. Skin-covered lesions are not detectable by maternal serum α-fetoprotein screening and some spina bifida occulta lesions neither are detectable by ultrasound nor do they place the patient at added risk for nerve damage at the time of delivery. Stark & Drummond (1970) and Ralis (1975), however, propose that trauma at the time of vaginal delivery adds to the congenital paralysis associated with meningomyelocele. Until the prenatal diagnostic methods described previously in this symposium (Cuckle, this volume), there were no opportunities to explore their hypothesis, the subject of this paper.

Background

We shall first briefly address the history of the treatment of meningomyelocele. Scientific advances in understanding the cause of infection and the introduction of aseptic surgery were followed shortly by the first reported series of treated meningomyelocele by Bayer (1892). His early enthusiasm was tempered by a high death rate from infection and hydrocephalus, as reported five years later (Bayer 1897). These two complications are the topics of many series of studies in the late 19th and early 20th centuries that reported relatively similar results (Shurtleff 1986a). After the introduction of the first effective silicone cerebrospinal fluid shunting systems to treat hydrocephalus, Sharrard et al (1963) advocated urgent treatment of all, just as Bayer had done 70 years earlier. Eight years later, one of the Sheffield group reported the need to establish selection criteria and not urgently treat all neonates born with meningomyelocele (Lorber 1971). Lorber's hypothesis emphasized the limited value of the lives of these severely disabled children and their cost to family and society. Others in the United States published series suggesting treatment did not lead to such poor results as described by Lorber (McLone 1983, O'Brien & McLanahan 1981, Reigel 1982, Shurtleff et al 1974). During the next decade, cranial ultrasound and computer-assisted tomography allowed serial observations of intracranial fluid accumulations. These techniques avoided complications previously associated with pneumography, to perform which air is exchanged for cerebrospinal fluid. A broad armamentarium of antibiotics and improved cerebrospinal fluid shunt systems developed during the same period have greatly reduced the morbidity and mortality from the hydrocephalus and infection associated with meningomyelocele (Leonard & Freeman 1981, McLaughlin et al 1985, Shurtleff et al 1985, 1986) (Fig. 1).

Despite remarkable improvement in the results of treatment of meningomyelocele in the 1970s and 1980s, the complex interacting influences of cultural beliefs, significance of birth defects, residual congenital impairment, availability of resources and evolving public policy led to much debate in the United States about which infants with meningomyelocele should be treated (Shurtleff & Shurtleff 1986). In 1983, Gross et al recommended use of Lorber's selection criteria, which led to severe criticism (Freeman 1984). Many agreed

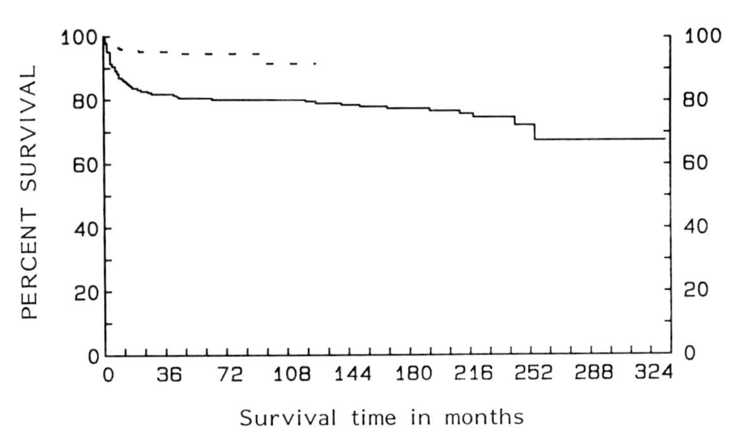

FIG. 1. Cumulative survival of infants with meningomyelocele. Early deaths were due to infection of the central nervous system or failure to shunt cerebrospinal fluid. All infants were seen at or before one week of age and before treatment. —, 222 infants seen 1955–1965; - -, 360 infants seen 1975–1985. Later deaths in the 1955–1965 group included renal failure, now a rare complication. Deaths in both groups continue to occur as a result of midbrain failure associated with the Arnold Chiari II malformation (Hays et al 1989). $P<0.0001$; Log rank statistic = 23.003, calculated from a Mantel-Cox Chi-square analysis for unequal censorship.

with Veatch (1977) that the Oklahoma Group (Gross et al 1983) had become 'so infatuated with their technical abilities to accumulate data and tally scores that they misunderstood the nature of the difficult decisions they were making'.

Prenatal diagnosis does not eliminate the difficult moral issues surrounding the treatment of congenitally defective infants, it only places the decision making in a different legal setting that allows termination of the fetus' life (Robertson 1982). Some prospective parents, however, refuse both prenatal diagnostic tests and abortion of an affected fetus because of personal beliefs (Lippman-Hand & Piper 1981, Marion et al 1980). Both our moral codes (Fletcher 1954) and the legal system in the United States (Robertson 1982) require that we offer parents of a defective fetus a full range of options as to what might be done for their child.

The Seattle experience with prenatal diagnosis of meningomyelocele

Our experience in the Pacific Northwest began in 1980, when we first had the opportunity to present the hypothesis that less paralysis may be associated with delivery of a fetus with meningomyelocele diagnosed *in utero* by pre-labour Caesarean section. Our subsequent experience suggested that less paralysis followed pre-labour Caesarean section than might be expected after vaginal delivery or Caesarean section with labour (Luthy et al 1991).

TABLE 1 Methods of pregnancy management with a fetal diagnosis of meningomyelocele

Intake:

Questionable or high level of maternal serum α-fetoprotein or questionable prenatal ultrasound.

Initial evaluation:

1. Repeat measurement of maternal serum α-fetoprotein, if other examinations normal.
2. Detailed ultrasound examination.
3. Determination of amniotic fluid α-fetoprotein and acetylcholinesterase.
4. Genetic amniocentesis.
5. Management options:
   a) termination;
   b) pre-labour Caesarean section;
   c) cephalocentesis;
   d) vaginal delivery;
   e) consult with Birth Defects team concerning treatment of affected fetus if desired by parents.
6. If pregnancy is to be continued with pre-labour Caesarean section as a possible choice:
   a) serial ultrasound examinations to assess progress of hydrocephalus and leg movement;
   b) amniocentesis beginning at 36 weeks to evaluate fetal lung maturity;
   c) consultations with neonatologist, neurosurgeon, nurse coordinator and paediatrician in Birth Defects Clinic to explain and plan post-delivery care.

Method

In this follow-up two years later, we have included all women pregnant with a fetus with meningomyelocele presenting to our two high-risk prenatal clinics and all infants and children with meningomyelocele seen in our University of Washington Birth Defects Clinics. Pregnancies were managed according to the outline in Table 1.

Data concerning type of delivery, duration of labour, presentation by breech or head and the indication for Caesarean section were obtained for infants not diagnosed prenatally. These data, together with the clinical status of the patients, were stored in our patient Data Management System for later retrieval and analysis (Shurtleff 1991).

Results

As previously reported, of 35 fetuses diagnosed prior to 24 weeks gestational age, 13 (37%) were brought to term and one was delivered vaginally after cephalocentesis (Luthy et al 1991). The remaining 12 were born by Caesarean section without labour (9) or with minimal labour (3). Of 22 pregnancies that were terminated, five (23%) fetuses had trisomy 13 or 18, nine (26%) fetuses

TABLE 2 Number of new patients with meningomyelocele seen per year

Year	Number of new cases	Born by Caesarean section because of fetal meningomyelocele	
		No.	*%*
1970	32	0	0
1971	30	0	0
1972	36	0	0
1973	29	0	0
1974	28	0	0
1975	21	0	0
1976	28	0	0
1977	34	0	0
1978	30	0	0
1979	21	0	0
1980	38	2	5
1981	28	1	4
1982	37	3	8
1983	42	3	7
1984	25	4	16
1985	31	7	23
1986	21	7	33
1987	30	8	27
1988	15	5	33
1989	16	10	63
1990	16	8	50
1991	13	6	46
1992	10	5	50

1970–1987 more new cases per year than 1988–1992, $t = 5.8$; df = 21; $P = 0.000008$.

had other fatal fetal anomalies and eight (23%) fetuses had isolated meningomyelocele. Three pregnancies from 46 diagnoses after 24 weeks gestational age were terminated because of trisomy (2) or other fatal fetal anomalies (1). Severe hydrocephalus led to delivery of 12 (26%) by Caesarean section or vaginally after cephalocentesis for cephalopelvic disproportion. These twelve patients were not considered when evaluating motor function results, because the far advanced hydrocephalus was associated with completely paralysed legs and because cephalopelvic disproportion was the indication for intervention with delivery rather than the preservation of nerve function. 31 (68%) infants were delivered by Caesarean section with only three (7%) experiencing minimal labour.

Termination of affected pregnancies has significantly reduced the numbers of new cases of meningomyelocele presenting to our Birth Defects Clinic from a mean of 30 per year between 1970 and 1987 to 14 between 1988 and 1992 (Table 2).

Since Cochrane et al (1991) have reported failure of improved muscle strength to persist long term after pre-labour Caesarean section, we report the same cohorts we described previously (Luthy et al 1991), but after two years additional follow-up. Our four-year follow-up includes 39 of the original 47 infants born by pre-labour Caesarean section and 68 of the 78 born vaginally (Table 3). The anatomical level was measured as the last intact vertebral arch on a standard anteroposterior roentgenogram of the spine. The eight cervical, 12 thoracic, four lumbar and five sacral dorsal laminae are all normally closed, adding up to 29 levels. The degree of paralysis (motor level) was measured by trained physiotherapists unaware of the status of the patient's method of delivery, according to a standardized, internationally used protocol (McDonald et al 1986). Neuromotor levels were converted to the same numerical designation as were the vertebral levels. Completely normal strength was recorded as 30. Patients with levels of paralysis cephalad (toward the head) or higher (smaller numbers) than the vertebral levels were assigned negative values. Patients with lower levels of paralysis (higher numbers) than the vertebral levels were assigned positive values. The difference between the anatomical and motor level for the pre-labour Caesarean group ($n = 34$; mean = 3.24; SD = 2.7) remains better than for the vaginally delivered group ($n = 68$; mean = 1.2; SD = 2.7) (Student's t-test; $t = 3.7$; df = 105; $P = 0.0003$). Of the babies delivered by pre-labour Caesarean section, the proportion in the least severely paralysed group (sacral or no loss) was 45% after two years. This has decreased slightly to 38% after four years. The proportion in the more severely paralysed group (L3 or above) has risen from 21 to 26%. Of the babies delivered vaginally, the proportion in the least severely paralysed group has increased from 14 to 25% and that in the most severely paralysed group has increased from 31 to 35%. These changes have obliterated any significance in the differences between the groups delivered by Caesarean section and vaginally. Two observations explain these shifts in proportions.

Firstly, patients remaining in the follow-up group born after pre-labour Caesarean section have a mean anatomical level by roentgenogram that is significantly higher (mean = 21.3; SD = 2.7, range 14–25) than those born vaginally (mean = 22.4; SD = 2.07; range 16–26) ($t = 2.3$; df = 105; $P = 0.023$). These data suggest that fetuses with minimal, low-level lesions are not more likely to be selected for pre-labour Caesarean section than fetuses with higher-level lesions who, theoretically, would have more paralysis, more associated abnormalities and be selected more often for termination. On the other hand, small, low-sacral-level lesions may be more readily overlooked by maternal serum α-fetoprotein and ultrasound screening. Low-sacral lesions arising from abnormalities in the canalization process distal to the mid-lumbar area are more likely to be skin covered, not to be associated with hydrocephalus and not to be associated with cranial anomalies such as the Arnold Chiari II malformation and the 'lemon' and 'banana' ultrasonographic signs (Nicolaides et al 1986, Nyberg et al 1988).

TABLE 3 Level of paralysis (motor level) at two[a] and four years of age according to type of delivery

	Total No.	*Sacral or no loss No. (%)*	*L4 or L5 No. (%)*	*L3 or above No. (%)*	*Relative risk*	*95% Confidence interval*
Pre-labour Caesarean section						
2 year	47	21 (45)	16 (34)	10 (21)	1.0	—
4 year	39	15 (38)	14 (36)	10 (26)		
Vaginal delivery						
2 year	78	11 (14)	43 (55)	24 (31)	2.1	1.8–2.5
4 year	68	17 (25)	27 (40)	24 (35)		

[a]Two year data were fully reported previously (Luthy et al 1991).

Secondly, loss of motor function due to complications of hydromyelia (syrinx of the cord) (1), cord tethering (5) or spinal cord hypoplasia (3) (Just et al 1990, McEnery et al 1992, McDonald et al 1992) was more frequent in the pre-labour Caesarean section group: nine of 39 (23%) patients compared to two of 68 (3%) patients in the vaginally delivered group (Fischer's Exact: $P=0.004$). Reflexive muscle function in fetuses and infants with these complications may have been initially confused with useful function. We have reported the relatively poor predictability of muscle strength in later life from examinations conducted before 60 months of age (McDonald et al 1986). All the patients with a large gibbus and an apparent low level of muscle function shortly after birth were finally classified in the most severely paralysed group (L3 or above). No child in our experience has maintained a low-lumbar-sacral or no-loss-level of paralysis in the presence of congenital gibbus. We therefore recommend considering a gibbus detected by ultrasound *in utero* as a contraindication for pre-labour Caesarean section to preserve motor function. Determination of whether spinal cord hypoplasia, identified in four of our patients postnatally, can be recognized *in utero* will require further study.

Our final analysis for this report included additional cases with short-term follow-up for a total of 65 infants born by Caesarean section with or without labour and ruptured amniotic membranes. For 28 infants born by Caesarean section without labour, the mean difference of the anatomical level subtracted from the motor level was 3.63 (SD = 2.44). For the 14 born by Caesarean section after a period of labour, the mean was 2.89 (SD = 1.5). The difference in values between the two groups was insignificant ($t=1.0$, $df=40$, $P=0.31$). None in these two groups experienced ruptured amniotic membranes. The mean for the 28 infants in the Caesarean section group with intact membranes and no labour did significantly exceed the mean of the 23 infants in the group born by Caesarean section that experienced ruptured amniotic membranes and labour (mean = 1.78; SD = 2.16; $t=2.78$; $df=50$; $P=0.0076$). We interpret these data to mean labour with intact amniotic membranes can be tolerated by nerves protruding in a sac dorsal to the infants back. When rupture of the amniotic membranes allows amniotic fluid to escape from around the meningomyelocele sac, however, the force changes from pressure on all sides of the exposed placode and nerves to a plane of force that may crush the nervous elements.

Summary

We conclude that, as predicted in 1986 (Shurtleff 1986b), patients with meningomyelocele are now being born significantly less frequently than prior to the introduction of prenatal diagnosis. The myelodysplasia lesions presenting now tend to be smaller, lower on the spine, associated with less paralysis and more frequently covered with skin. Personal beliefs influence some parents to bring their fetus to term, regardless of how severely deformed. Pre-labour Caesarean section is recommended for fetuses with a protruding meningomyelocele sac, knee or ankle motion, normal chromosomes, absence of fatal fetal anomalies and absence of a gibbus.

Further study is necessary to determine: whether a flat or depressed meningomyelocele sac obviates the need for Caesarean section; if labour, as measured by cervical dilation, adversely affects protruding meningomyelocele lesions; if spinal cord atrophy or hypoplasia can be detected *in utero*; if spinal cord atrophy or hypoplasia, hydromyelia and cord tethering are more common following Caesarean section birth; and, finally, whether breech presentation before onset of labour injures protruding nerves in a meningomyelocele sac.

Acknowledgements

Funds for this study were provided by a gift from Steve Largent and gifts to the Birth Defects Research Fund. We thank Holly Kaopuiki for preparation of the manuscript.

References

Bayer C 1892 Zur Technite der Operation des Spina Bifida und Encephalocele. Prag Med Wochenschr 17:317–320

Bayer C 1897 Ztchr f Heille Berl Vor der Operation der Spina Bifida and Encephalocele. Z Herlkronen Berlin 18:405–416

Cochrane D, Aaronyk K, Sawatsky B, Wilson D, Steinbok P 1991 The effects of labour and delivery on spinal cord function and ambulation in patients with meningomyelocele. Child's Nerv Syst 7:312–315

Cuckle HS 1994 Screening for neural tube defects. In: Neural tube defects. Wiley, Chichester (Ciba Found Symp 181) p 253–269

Fletcher J 1954 Morals and medicine. Princeton University Press, Princeton

Freeman JM 1984 Early management and decision making for the treatment of myelomeningocele: a critique. Pediatrics 73:564–566

Gross RH, Cox A, Tatyrek R, Pollay M, Barnes WA 1983 Early management and decision making for treatment of myelomeningocele. Pediatrics 72:450–458

Hays RM, Jordan RA, McLaughlin JF, Nickel RE, Fisher LD 1989 Central ventilatory dysfunction in myelodysplasia: an independent determinant of survival. Dev Med Child Neurol 3:366–370

Holmes LB 1994 Spina bifida: anticonvulsants and other maternal influences. In: Neural tube defects. Wiley, Chichester (Ciba Found Symp 181) p 232–244

Just M, Schwartz M, Ludwig B, Ermert J, Thelen M 1990 Cerebral and spinal MRI findings in patients with post repair myelomeningocele. Pediatr Radiol 20:262–266

Leonard CO, Freeman JM 1981 Spina bifida, a new disease. Pediatrics 68:136–137

Lippman-Hand A, Piper M 1981 Prenatal diagnosis for the detection of Down syndrome: why are so few eligible women tested? Prenatal Diagn 1:249–257

Lorber J 1971 Results of treatment of myelomeningocele: an analysis of 524 unselected cases with special reference to possible selection for treatment. Dev Med Child Neurol 13:279–303

Luthy DA, Wardinsky T, Shurtleff DB et al 1991 Cesarean section before onset of labour and subsequent motor function in infants with meningomyelocele diagnosed antenatally. N Engl J Med 324:662–666

Marion JP, Kassman G, Farnhoft PM 1980 Acceptance of amniocentesis by low-income patients in an urban hospital. Am J Obstet Gynecol 138:11–15
McEnery G, Borzyskowski M, Cox TC, Neville BG 1992 The spinal cord in neurologically stable spina bifida: a clinical and MRI study. Dev Med Child Neurol 34:342–347
McDonald CM, Jaffe KM, Shurtleff DB 1986 Assessment of muscle strength in children with meningomyelocele: accuracy and stability of measurements over time. Arch Phys Med Rehabil 67:855–861
McDonald CM, Shurtleff DB, Weinberger E 1992 Diffuse spinal cord atrophy/hypoplasia. Dev Med Child Neurol 34(suppl 66):4(abstr)
McLaughlin JF, Shurtleff DB, Lamers JY, Stuntz TJ, Hayden PW, Kropp RJ 1985 Influence of prognosis on decisions regarding the care of newborns with myelodysplasia. N Engl J Med 312:1589–1594
McLone DG 1983 Results of treatment of children with myelomeningocele. Clin Neurosurg 30:407–412
Nicolaides KH, Campbell S, Gabbe SG, Guidetti R 1986 Ultrasound screening for spina bifida: cranial and cerebellar signs. Lancet 2:71–74
Nyberg D, Mack L, Hersh J, Mahoney B 1988 Abnormalities of fetal cranial contour in sonographic detection of spina bifida: evaluation of the 'lemon' sign. Radiology 167:387–392
Oakley GP Jr, Erickson JD, James LM, Mulinare J, Cordero JF 1994 Eradication of folic acid-preventable spina bifida and anencephaly. In: Neural tube defects. Wiley, Chichester (Ciba Found Symp 181) p 212–231
O'Brien MS, McLanahan C 1981 Review of the neurosurgical management of myelomeningocele at a regional pediatric medical center. In: Epstein F, Hoffman HJ, Raimondi AJ (eds) Concepts in pediatric neurosurgery. Karger, Basel, p 202–215
Ralis ZA 1975 Traumatizing effect of breech delivery on infants with spina bifida. J Pediatr 87:613
Reigel D 1982 Spina bifida. In: Section of Pediatric Neurosurgery of the American Association of Neurological Surgeons (eds) Pediatric neurosurgery: surgery of the developing nervous system. Grune & Stratton, New York, p 23–47
Robertson JA 1982 The right to procreate and in utero fetal therapy. Int J Leg Med 3:333–366
Sharrard WJW, Zachary RB, Lorber J, Bruce AM 1963 A controlled trial of immediate and delayed closure of spina bifida cystica. Arch Dis Child 38:18–22
Shurtleff DB 1986a Meningomyelocele: a new or vanishing disease? Z Kinderchir (suppl) 1:5–9
Shurtleff DB 1986b Selection process for the care of congenitally malformed infants. In: Shurtleff DB (Ed) Myelodysplasias and exstrophies. Grune & Stratton, Orlando, FL, p 89–116
Shurtleff DB 1991 Computer data bases for pediatric disability. Phys Med Rehabil Clin NA 2:665–687
Shurtleff H, Shurtleff DB 1986 Decision making for treatment or nontreatment of congenitally malformed individuals. In: Shurtleff DB (Ed) Myelodysplasias and exstrophies. Grune & Stratton, Orlando, FL, p 3–24
Shurtleff DB, Stuntz JT, Hayden PW 1985 Experience with 1201 cerebrospinal fluid shunt procedures. Pediatr Neurosci 12:49–57
Shurtleff DB, Hayden PW, Loeser JD, Kronmal RA 1974 Myelodysplasia: decision for death or disability. N Engl J Med 291:1005–1011
Shurtleff DB, Stuntz JT, Hayden PW 1986 Hydrocephalus. In: Shurtleff DB (ed) Myelodysplasias and exstrophies. Grune & Stratton, Orlando, FL, p 139–179

Stark G, Drummond M 1970 Spina bifida as an obstetric problem. Dev Med Child Neurol (suppl) 22:157–160
Veatch RM 1977 The technical criteria fallacy. Hastings Cent Rep 7:15–16
Wald NJ 1994 Folic acid and neural tube defects: the current evidence and implications for prevention. In: Neural tube defects. Wiley, Chichester (Ciba Found Symp 181) p 192–211

DISCUSSION

Wald: The principal concern in this comparison of infants born by Caesarean section or vaginally is that you are not comparing like with like. In clinical medicine, doctors try to choose the most suitable patients for a particular procedure. There are all kinds of informal and formal selective processes involved in helping to determine whether a person is the right candidate for the procedure or operation. Various factors are used to determine whether a woman would be suitable for a Caesarean section, for example whether the fetus has signs that would indicate that its prognosis might be better than average. Even if the selection processes were less formal in earlier years, when you developed the technique, I suspect they were still there. How confident can one be that you would have got the same results had you done a randomized trial?

Shurtleff: Unfortunately, our human studies review committee refused a randomized trial. I agree that's what should be done.

The other answer is that we proved our null hypothesis. We expected those cases that were detected *in utero* and born by Caesarean section to have better motor function and lower level anatomical lesions than those who were born vaginally. Even though that hypothesis is in the 1991 paper, I can tell you now that we were wrong. We've had women with infants with high-level lesions, including two with anencephaly, detected *in utero* carry to term because of personal beliefs. Hence, amongst our prenatally diagnosed group, more babies with severe and higher level lesions were brought to term than amongst those not diagnosed. There is also a lower likelihood of identifying small, low lesions, hence these were not detected prenatally more frequently than expected and were, therefore, born vaginally.

Trasler: Rick Finnell, at Texas A & M University, claimed that anencephalic babies can live for several months. He showed us pictures of these children, 3–4 months old or more.

Shurtleff: That's extremely unusual. The majority of the anencephalics will die fairly promptly.

Seller: Pat Baird in Vancouver did a study on the survival of babies with anencephaly. 4–5 days is not unusual; several months is not unknown but it is rare (Baird & Sadovnik 1984).

O'Rahilly: There is an account in the French literature (André-Thomas & de Ajuriaguerra 1959) where several neurological tests were done on babies with anencephaly days and weeks after birth.

Stanley: We have just done a study in Western Australia looking at survival (Kalucy 1992) and we were surprised at the length of survival of children with anencephaly, but it certainly wasn't anything like months—the longest was five days.

Lindhout: In the UK, the number of children born each year with neural tube defects has fallen from 2000–3000 to fewer than 200; it may drop even further. What kind of measure should we take to keep the experience available for the remaining children to be treated?

Shurtleff: We are now collaborating with about 35 centres in Europe, Australia, Canada, Argentina and the United States. We have completed several multicentre studies using a standardized, computer-based protocol (Liptak et al 1992, McDonald et al 1991, Shurtleff 1991, Shurtleff et al 1986, Wright et al 1991). The progress that has been made with screening and with other techniques means that there will be fewer centres seeing fewer of these increasingly rare disorders. In addition, these children are extremely variable in presentation of their medical problems. Even with hundreds of cases, we still have to combine data from several centres to be able to sort out different variables that may contribute to outcome, to determine what is the result of intervention and what is the result of natural history. This is a very important point. I hope to leave this field not only with almost complete absence of the disease which I started to take care of 35 years ago, but also to leave to the young people treating these patients a concept of collaboration, not competition.

Hall : The need for designated centres caring for individuals with neural tube defects will become even more important as the incidence of the defects decreases. The families that have affected children are very committed to research, particularly if they have had more than one. My guess is, five years from now when we have identified 10 relevant human genes involved in neural tube closure and we really start to sort out what's going on, the families followed in special neural tube defect clinics will be the families that will really want to cooperate. We should encourage the development within countries of 2–3 centres where kids are cared for and where the proper expertise can be concentrated and maintained.

Copp: I discussed this with my paediatric neurosurgeon colleagues at Great Ormond Street Hospital in London. Previously, they very rarely received referrals of children with straightforward myelomenigocele for operation, because most of those cases were dealt with by general paediatric surgeons throughout the UK. Recently, as the numbers have fallen, they have begun to receive referrals. They are now experiencing an increased workload of this condition, presumably because the general paediatric surgeons are seeing so few they don't feel able to deal with them. So, experience of these defects is being concentrated in certain specialist centres.

Lindhout: In The Netherlands, there are a number of self-supporting academic and non-academic so-called myelodysplasia working groups. Our centre in

Rotterdam has 100–200 patients in care. Until recently, there was no really collaborative approach from the scientific point of view, although there were informal exchanges.

Hall: Certainly, the quality of life is markedly improved for these kids by centralized clinics.

Shurtleff: If you look at children with the same lesion entering the treatment care programme over the past 30 years, those with minimal involvement do much better today. On the other hand, those who in the past died because of their complex multiple problems are now being kept alive. As a result, the total population still includes a considerable number who are significantly impaired. There are several things in the United States that have helped these children; these are not related to their medical care but are general changes in society. There is now better access to school and to jobs. This has been promoted by people who sustained spinal injuries in the Vietnam War. These veterans have won access to society and have swept the disabled children along.

Unfortunately, with primary neurulation defects of the spine, faulty enzymes are malfunctioning in the central nervous system as well as at the site of the spina bifida. Almost all children with primary neurulation defects of the spine will have central nervous system impairments that adversely affect their intellect and ability to cope in a highly technical society.

Hall: Andrew Copp said we could divide neural tube defects into problems with initiation of the anterior or of the posterior areas. Are you saying that we can't do that, because in children with lumbar spina bifida, the anterior part of the neural plate also didn't close properly when the neural tube was closing?. In children with posterior closure defects, is the brain also functionally abnormal?

Shurtleff: Malformation of the central nervous system is both anatomical and functional. Even when the brain looks fairly normal by magnetic resonance imaging or by computerized axial tomography, the children can have functional abnormalities in their brain tissue. In the presence of polymicrogyri and the Arnold Chiari II malformation, all have central nervous system impairments.

Holmes: What percent of individuals with spina bifida would you say have structural cortical abnormalities?

Shurtleff: Magnetic resonance imaging can demonstrate absence or marked thinning of the corpus callosum, beaking of the tectum and a small cerebellum pressed down into the cervical area as described in the Arnold Chiari type II malformation in approximately 98% of patients with meningomyelocele. Functional abnormalities with primary neurulation defects include impaired hand–eye coordination, fine motor deficits and inability to code short memory into permanent memory easily (Shaffer et al 1985). The severity of functional impairment does not necessarily correlate with the extent of anatomical variation seen by imaging techniques.

Holmes: If a woman had a prenatal ultrasonographic scan at 19 weeks of gestation which showed the fetus had a lumbosacral meningomyelocele, what could we tell her and the father of the baby about the difference in IQ between the affected child and a healthy sibling?

Shurtleff: Adina Sella tested the siblings of our patients (Sella et al 1966). She found that 35% of patients were at least 15 IQ points below their siblings. Only 15% of children scored the same difference from normal standards. In addition, children with neuromotor levels at lumbar 2 or proximally and those with central nervous system infection have lower IQ test scores. These results have since been confirmed by others. We warn all the parents that they should expect to have a child with some degree of mental impairment, although 50% of patients with a high-level lesion and 85% of patients with a low-lumbar and sacral level lesion will have 'normal intellect'.

Secondly, the standard IQ tests for these children depend to a significant degree on fine motor control. Nowadays, children can compensate for their deficit by using a computer. In middle school, they can write using a word processor. The first year that one patient had a computer available, he caught up three grades at school. So we don't know how much of this deficit in IQ tests is based on fine motor inability and how much relates to intellectual deficit.

Hall: If you look at neurulation in animals and interfere with it, does it make a difference where you do this?

Jacobson: I have worked on amphibia for 40 years now and I have yet to see a normal unclosed neural plate. They don't have these problems. Chickens, sometimes 30–50%, will have open plates in the middle of a Texas summer. I have never seen it in amphibia.

Copp: In mice, we do see some of the secondary effects of low spina bifida that are seen in humans. We see hydrocephalus and perhaps something that corresponds to Arnold Chiari malformation.

Hall: The ataxia in Diana Juriloff's mice would be the same (see p 136).

Lindhout: I can add some experience from cases with valproate-related neural tube defects. The current cases cannot be examined in detail because of termination of pregnancy, which usually makes brain morphology very difficult to study. I remember three earlier cases: one had agenesis of the corpus callosum in addition to the neural tube defect. One had lissencephalic changes at the medial sides of the occipital lobes. There were also one or two cases of agenesis of septum pellucidum.

Holmes: The cohort study of pregnancies exposed to valproic acid (VPA) that is needed would identify and enrol women early in the pregnancies. Those whose fetuses were found to have spina bifida through prenatal screening could opt for an elective termination of the pregnancy. Nevertheless, these outcomes of pregnancy could be tabulated in addition to the pre- and postnatal findings in the other VPA-exposed children. With sufficient numbers and follow-up, we would determine the portion of VPA-exposed children with spina bifida, other major malformations and developmental abnormalities.

Lindhout: From the various prospective Dutch cohort studies, I can trace more than 300 children who are grown up now who were exposed to valproate. So far, we have not studied them for specific intellectual or motor deficits because we don't have the money to do that. The patient population is available, with all the required prenatal parameters already acquired before outcome was known.

Hall: Howard, do most people in the UK terminate the pregnancy if an abnormality is diagnosed early?

Cuckle: There are various ethnic and religious minority groups in the UK that won't tolerate termination of pregnancy. Nonetheless, births of anencephalic infants have fallen to under 50 a year and many of those are twins, who are left for the sake of the normal co-twin. Therefore, at least for this defect, termination must be widely accepted.

O'Rahilly: I would like to make a plea that whenever weeks are mentioned in prenatal life, it should be specified which kind of weeks. 'Menstrual' weeks would be a perfectly appropriate term, but weeks of gestation is not. The latter term is used for both weeks since menstruation and estimated weeks since fertilization. This is very confusing.

Cuckle: I can understand your concern but there is a general usage in obstetrics in the UK that gestational means post-menstrual days or weeks. If there is a scan, the scan is calibrated to post-menstrual days.

Shurtleff: The gestational ages in weeks I referred to are post-conception.

Cuckle: I used 'gestational' for menstrual weeks, because in epidemiological studies the days of conception are not known in general.

O'Rahilly: I have no complaint with menstrual weeks, but why not specify it?

Wald: There is added confusion, which is between completed weeks and the nearest week. If you say someone is at 17 weeks, are they in expectation 17 weeks and 0 days or 17 weeks and 3 days?

O'Rahilly: 17th week is another confusing term.

Wald: Do we know whether there is an association in the general population between spina bifida occulta and spina bifida cystica? If there were, some of the things we have been talking about, for example, looking at the risk of women on anticonvulsants or of diabetics, could be examined more economically by looking at the prevalence of spina bifida occulta as a surrogate for spina bifida cystica.

Czeizel: There was a very good study by Carter et al (1975). Their message was that it is necessary to separate the spinal dysraphism from the spina bifida occulta. Spinal dysraphism has three criteria: (1) more than two vertebrae are affected; (2) there are other morphological manifestations, for example sacral skin hair; and (3) there are some functional symptoms, e.g. difficulties when learning to walk or weakness of the lower limbs.

In Hungary the prevalence of spina bifida occulta in adults is about 21% (Vajda et al 1977). From the Warkany (1971) book it appears that at birth all new-borns have a spina bifida occulta, because they have incomplete ossification of the vertebral laminae. The term spina bifida occulta is wrong, because it suggests some similarity between spina bifida occulta and spina bifida aperta cystica, but these two entities are completely different.

O'Rahilly: We have to distinguish more carefully what we mean by spina bifida occulta. At the end of the embryonic period proper, at eight weeks, there is a complete spina bifida occulta, there is not even cartilage behind the spinal cord (O'Rahilly et al 1980). At term, there is cartilage between the two halves of the neural arch. So the X-ray appearance at term shows spina bifida occulta, but there is cartilaginous completion.

Shurtleff: Laurence et al (1971) compared roentgenograms of first-degree relatives of patients with meningomyelocele to those of age-matched controls. Sutow & Pryde (1956) reported that ossification of posterior laminae cartilage occurs until age 18 in females and about 21 years in males in the US.

O'Rahilly: Particularly in the sacral region. About 20–25% of the adult population has something missing in the sacral region from an X-ray point of view.

Shurtleff: The studies I referred to above suggest the common, benign form of spina bifida occulta occurs at approximately L5 or S1. When L3 or higher and S2 or lower is bifid, you have an uncommon abnormality more likely to be associated with functional abnormalities. It is the location on the spine as well as the type of abnormality that is important. This correlates very well with your concept of where primary neurulation occurs.

Copp: Ronan, do you feel that in older children and adults who have defects of the bony neural arches, there was originally a defect of cartilaginous formation, which should be present at birth? Or does the defect arise after birth, during the transformation from cartilage to bone?

O'Rahilly: We don't know. I've asked orthopaedic surgeons and they don't seem to know either.

Copp: This makes a big difference in terms of how we view the postnatal conditions of spina bifida occulta.

Stanley: Judith, what do clinical geneticists do in terms of counselling people they find have spina bifida occulta?

Hall: They totally ignore it. Or they may say to a patient that spina bifida occulta had been observed but it is thought to mean nothing, because as far as we can tell it has nothing to do with recurrence or occurrence risks for neural tube defects.

From the study we did in the Sikhs in British Columbia, it looks as if the background incidence of spina bifida occulta is increased in that population. We found it in all of the parents of Sikh children with neural tube defects, but there was also a very high frequency in the general population. So spina bifida occulta may involve a different gene in that Sikh population, which could be useful. I think that once we have eliminated folic acid-preventable neural tube defects, we will find some other very interesting disorders in humans.

References

André-Thomas, de Ajuriaguerra J 1959 Étude anatomo-clinique de l'anencéphalie. In: Heuyer G, Feld M, Gruner J Malformations congénitales du cerveau. Masson, Paris, p 207–267

Baird PA, Sadovnick AD 1984 Survival in infants with anencephaly. Clin Pediatr 23:268–271

Carter CO, Evans KA, Till K 1975 Spinal dysraphism: genetic relation to neural tube malformations. J Med Genet 13:343–350

Kalucy MJ 1992 A retrospective study of the survival and long term outcome of infants born with neural tube defects in Western Australia 1966–1990. Dissertation for B Med Sci (Hons) degree. University of South Australia, Adelaide

Laurence KM, Bligh AS, Evans KT, Shurtleff DB 1971 Vertebral abnormalities in parents and sibs of cases of spina bifida cystica, encephalocele and anencephaly. In: Proceedings of the International Congress for Paediatrics, Vienna, Austria (29 August–4 September 1971). p 415–421

Liptak G, Shurtleff DB, Bloss JW, Baltus-Hebert E, Manitta P 1992 Mobility aids for children with high-level myelomeningocele: parapodium versus wheelchair. Dev Med Child Neurol 34:787–796

McDonald CM, Jaffe KM, Shurtleff DB, Menelaus MB 1991 Modifications to the traditional description of neurosegmental innervation in myelomeningocele. Dev Med Child Neurol 33:473–481

O'Rahilly R, Müller F, Meyer DB 1980 The human vertebral column at the end of the embryonic period proper. J Anat 131:565–575

Sella A, Foltz EL, Shurtleff DB 1966 A three-year developmental study of treated and untreated hydrocephalic children. J Pediatr 69:887–891

Shaffer J, Friedrich WN, Shurtleff DB, Wolf L 1985 Cognitive and achievement status of children with meningomyelocele. J Pediatr Psychol 10:325–330

Shurtleff DB 1991 Computer data bases for pediatric disability: clinical and research applications. Phys Med Rehabil Clinics N Am 2:665–687

Shurtleff DB, Menelaus MB, Staheli LT 1986 The natural history of flexion deformity of the hip in myelodysplasia. J Pediatr Orthop 6:666–673

Sutow WW, Pryde AW 1956 Incidence of spina bifida occulta in relation to age. Am J Dis Child 91:211–217

Vajda R, Göblyös P, László J et al 1977 The radiographic findings of parents of patients with neural tube defects. (In Hungarian) Orv Hetil 118:2341–2343

Warkany J 1971 Congenital malformations. Notes and comments. Year Book Medical Publishers, Chicago, IL

Wright JG, Menelaus MB, Broughton NS, Shurtleff DB 1991 Natural history of knee contractures in myelomeningocele. J Pediatr Orthop 11:725–730

Final discussion

Prevention of neural tube defects by folic acid supplementation

Hall: I would like to propose the following recommendations, with which all those present at this symposium seem to agree. I suspect that each of us would like to see them applied in our own country.

1) There should be a programme of public education concerning the importance of folic acid in the prevention of neural tube defects.

2) The population should be encouraged to eat a healthy, appropriate diet that is rich in folic acid, for example green vegetables, fruits and fruit juice.

3) There should be fortification with folic acid of a staple food (e.g. the grain chain, including bread, flour, rice, etc) to a level that will be beneficial in preventing neural tube defects but will not be likely to cause problems in the elderly who might have vitamin B_{12} deficiency.

4) Folic acid capsules (containing 0.4 to 0.8 mg) should be generally available, without prescription, to women of reproductive age.

5) A committee or agency should be established to assess the effectiveness of this prevention programme: its impact on folic acid levels in the general population, the effect on the incidence of neural tube defects and other congenital abnormalities, any possible side effects, and safety.

6) Further research should be supported to investigate the way in which folic acid prevents neural tube defects and to study the defects that do not respond to folic acid, such as those identified in the mouse.

Mills: Fortification of food with folic acid should be at twice the replacement level (i.e. adding an amount equal to twice that which is lost during milling). As a trial, this could be done in a defined area, with a control population also being monitored.

Oakley: Fortification should be sufficient for the vast majority of women of reproductive age to consume 0.4 mg per day of folic acid more than that derived from the naturally occurring folates in food, because folic acid added to a regular diet has been shown to prevent spina bifida and anencephaly. Fortification should proceed as rapidly as possible in all countries where this is practical. There should then be a serious evaluation of the impact in all those countries.

Stanley: One could monitor antenatal blood samples, which would assess the target population.

Wald: We also need investigations that would give us guidance as to the most appropriate dose.

Czeizel: In Europe the present recommended daily dose of folic acid for pregnant women is 0.8 mg; in the US it was 0.4 mg. It might be better to recommend folic acid as part of a multivitamin package. The Smithells et al (1983) study and our study (Czeizel & Dudás 1992) used multivitamins. Folic acid is a key component, but I am not sure about its effectiveness alone, at a low dose.

Scott: The European Commission's new uniform recommendations for daily doses of vitamins in all EC countries recommend 0.2 mg of folic acid for non-pregnant women and 0.3 mg for pregnant women.

Mills: The FDA educational material makes it clear that not all women will benefit and not all neural tube defects will be prevented; other birth defects will also not be prevented. Local recommendations should state what the risk of having a child with a neural tube defect is in that particular country, so that women can evaluate the risk for themselves.

Copp: It is important to stress the need for further research because, although we know that folic acid has this preventive effect, we have no idea how it works.

Stanley: We should also stress that birth defects are relatively common. 5% of all pregnancies have a major structural abnormality.

Scott: Pernicious anaemia causing vitamin B_{12} deficiency through malabsorption is usually a disease of the elderly, but some populations are prone to getting pernicious anaemia at a younger age. Some concern has been expressed about combining food fortification with the availability of folic acid capsules: this would clearly only occur in women of childbearing age who have a very low risk of vitamin B_{12} deficiency.

Mills: That is probably safe. There is some debate in the US about whether black and Hispanic women are more at risk of developing vitamin B_{12} deficiency.

Hall: The Sikh population in British Columbia also appears to have a higher risk of vitamin B_{12} deficiency. This may be true of most Sikh populations, because of their diet, and possibly of vegetarians in general.

Oakley: I don't think we should get involved in the details. Different circumstances will prevail in different countries. The main point is that research has shown folic acid prevents a large proportion of neural tube defects. Where possible, fortification of a food staple with sufficient folic acid should start immediately.. There should be a surveillance programme to evaluate the effectiveness and safety of fortification and there should be regular reviews of the data.

References

Czeizel AE, Dudás I 1992 Prevention of the first occurrence of neural-tube defects by periconceptional vitamin supplementation. N Engl J Med 327:1832–1835

Smithells RW, Seller MJ, Harris R et al 1983 Further experience of vitamin supplementation for prevention of neural tube defect recurrences. Lancet 1:1027–1031

Index of contributors

Non-participating co-authors are indicated by asterisks. Entries in bold type indicate papers; other entries refer to discussion contributions.

Indexes compiled by Liza Weinkove

Subject index